Trustworthy Machine Learning under Imperfect Data

Bo Han • Tongliang Liu

Trustworthy Machine Learning under Imperfect Data

Springer

Bo Han
Department of Computer Science
Hong Kong Baptist University
Hong Kong, Kowloon, Hong Kong

Tongliang Liu
School of Computer Science
The University of Sydney
Darlington, NSW, Australia

ISBN 978-981-96-9395-5 ISBN 978-981-96-9396-2 (eBook)
https://doi.org/10.1007/978-981-96-9396-2

This Springer imprint is published by the registered company Springer Nature Singapore Pte Ltd.
The registered company address is: 152 Beach Road, #21-01/04 Gateway East, Singapore 189721, Singapore

Preface

Machine learning has become increasingly prevalent in decision-critical applications. However, real-world deployment scenarios often deviate from idealized assumptions where data are perfectly labeled, identically distributed, and free from malicious interference. Therefore, ensuring the trustworthiness of machine learning has emerged as one of the most pressing challenges in machine learning research.

In practice, machine learning systems frequently encounter imperfect data in three main forms: noisy labels due to annotation errors, adversarial examples crafted to mislead predictions, and out-of-distribution samples that differ from training data. These challenges represent critical vulnerabilities that can lead to catastrophic failures, such as misdiagnosis in medical imaging, fraudulent transactions bypassing financial safety systems, and safety hazards in autonomous vehicles.

This book provides a comprehensive exploration of trustworthy machine learning in imperfect data scenarios. Through six structured chapters, we present a unified framework encompassing the major categories of data imperfections and their mitigation strategies.

Chapter 1 establishes the fundamental concepts of trustworthy machine learning and provides a comprehensive taxonomy of imperfect data scenarios. We introduce the core principles of robustness, safety, and reliability that define trustworthiness, while categorizing the three primary types of imperfect data: noisy labels, adversarial examples, and out-of-distribution samples.

Chapter 2 focuses on trustworthy machine learning with noisy labels, exploring statistical learning approaches, deep learning methodologies, applications to over-parameterized models, and the natural transition from noisy label learning to weakly supervised learning paradigms. We examine how different noise assumptions—symmetric, class-dependent, and instance-dependent noise—lead to distinct theoretical properties and practical solutions, providing both theoretical guarantees and empirical validation.

Chapter 3 addresses adversarial examples from three complementary perspectives: statistical learning methods that model the transition relationships between adversarial and natural labels, deep learning approaches that enhance robustness through adversarial training, and diffusion-based methods that purify or detect adversarial perturbations in the preprocessing stage. This multifaceted approach provides comprehensive coverage of defense strategies throughout the entire machine learning pipeline.

Chapter 4 examines out-of-distribution data challenges through two perspectives: detection and generalization. In out-of-distribution detection, we explore methods for identifying when test samples fall outside the training label space, enabling models to abstain from making unreliable predictions. In out-of-distribution generalization, we investigate techniques to improve generalization across different domains and distributions, focusing on invariant risk minimization and contrastive-based approaches.

Chapter 5 extends our exploration to advanced application domains that represent the cutting edge of trustworthy machine learning research: federated learning, graph learning, and foundation models. These domains present unique challenges that amplify the importance of trustworthiness considerations. Federated learning introduces privacy and safety concerns in distributed settings, graph learning requires handling complex relational data structures with novel vulnerability patterns, and foundation models demand trustworthiness at unprecedented scale.

Chapter 6 synthesizes the key insights from previous chapters and provides a forward-looking perspective on the future directions of trustworthy machine learning research. We discuss emerging challenges, identify promising research opportunities, and offer practical guidance for implementing trustworthy machine learning systems in real-world applications.

We hope this book provides an introductory background to the concept and coverage of trustworthy machine learning and summarizes some of the cutting-edge progress in trustworthy machine learning research. This can prepare the readers to embark on a fast track to keep up with the current dilemmas and challenges in this research area, conducting new research upon existing methods, and adapting relevant heuristics to scenarios that require trustworthy machine learning systems.

Finally, we would like to thank the members of the Sydney AI Centre, TML Lab, and TMLR Group who have contributed to this book. They are: Yingbin Bai, Zhuo Huang, Muyang Li, Xiuchuan Li, Runqi Lin, Yuhao Wu, Yu Yao, Chaojian Yu, Suqin Yuan, and Dawei Zhou from the Sydney AI Centre and TML Lab; and Chentao Cao, Yongqiang Chen, Haoang Chi, Rong Dai, Zhen Fang, Xiao Feng, Xue Jiang, Xuan Li, Chenxi Liu, Feng Liu, Xiangyu Lu, Zhenheng Tang, Qizhou Wang, Binghui Xie, Zhiqin Yang, Jiangchao Yao, Shuhai Zhang, Yonggang Zhang, Zhun Zhong, Zhanke Zhou, and Jianing Zhu from the TMLR Group (listed alphabetically within each group).

Hong Kong, China — Bo Han
Sydney, NSW, Australia — Tongliang Liu

Contents

Chapter 1
Introduction

Artificial intelligence (AI) has profoundly transformed our lives, with pervasive applications ranging from content creation, medical diagnostics, and so on. In the wave of AI, machine learning has laid the foundation for the rapid development of AI and aroused great interest. However, in real environments, the assumptions of machine learning are often not met. There are various imperfect scenarios that can cause the performance of machine learning models to drop significantly or unpredictable behaviors. Developing trustworthy machine learning systems in imperfect scenarios is important. In this chapter, we will briefly introduce the basic concepts of machine learning and trustworthy machine learning, laying the foundation for the content of subsequent chapters.

1.1 Trustworthy Machine Learning

This section will briefly introduce the foundation of machine learning and trustworthy machine learning.

1.1.1 Machine Learning

The core goal of machine learning [186] is to enable the model to learn the features in the data, so as to achieve classification, regression, clustering, and generation. The implementation of machine learning depends on two core components: data and models. Data is the foundation of machine learning. In the real world, there is data of different modalities, such as images, texts, tables, etc. Datasets usually consist of two parts, namely, input data and labels. The input data contains features that we expect the model to learn, while the labels provide a reference for the output.

B. Han, T. Liu, *Trustworthy Machine Learning under Imperfect Data*,
https://doi.org/10.1007/978-981-96-9396-2_1

For some datasets, labels may be difficult to obtain. In this case, we can usually only train without labels. High-quality data collection requires a lot of effort and is usually difficult to obtain. Models are the core tools of machine learning, through which we can learn implicit features from data. In modern machine learning, models mainly use complex neural network architectures, such as ResNet [144] and Transformer [377]. Machine learning finds the best parameters of the model by optimizing a specific objective function, and the specific parameter adjustment process is mainly achieved by the backpropagation algorithm.

Machine learning is often divided into three categories: supervised learning, unsupervised learning, and reinforcement learning [354]. Supervised learning is performed using labeled data, where labels serve as the output supervisory signal. While supervised learning excels, it necessitates a substantial quantity of high-quality labeled data, hence significantly constraining model performance in domains where labels are challenging to acquire. Unsupervised learning substantially mitigates this constraint by acquiring the inherent information and characteristics of the data without requiring labeled data. Nonetheless, unsupervised learning lacks explicit direction, rendering its outcomes comparatively challenging to validate. Reinforcement learning depends on the agent's interaction with the environment to acquire knowledge, and via multiple explorations and rewards, the model determines the optimal strategy. However, reinforcement learning training is expensive, and the reward mechanism needs to be carefully designed to avoid reward hacking.

Although machine learning has achieved great success, practical scenarios often fail to meet the assumptions of machine learning. For example, the training data may contain noise or even adversarial samples, which may cause the performance of the trained model to degrade or even fail completely. In addition, there may be a large distribution gap between the training and test data, and the performance of the model on out-of-distribution data is likely to be limited. In practical applications, there are also problems such as user data privacy and decision fairness, which highlight the vulnerability of traditional machine learning models in imperfect scenarios. This raises an important question, namely, *how to ensure that machine learning systems can perform trustworthily in imperfect scenarios?* Based on the above challenges, trustworthy machine learning came into being. Trustworthy machine learning hopes to build a machine learning system that can still operate reliably in imperfect scenarios. While maintaining the performance of the model, it focuses on the robustness, security, reliability, fairness, and interpretability of the model, providing a more solid guarantee for the practical application of machine learning.

1.1.2 Trustworthy Machine Learning

Trustworthy machine learning represents an emerging direction in the field of machine learning, with the objective of developing systems that are reliable, secure, and ethically sound in real-world environments. Traditional machine learning usually focuses on improving model performance when designing algorithms and assumes high data quality, often ignoring the inherent flaws in actual situations. Unlike traditional machine learning, trustworthy machine learning studies how to ensure the effectiveness of models under imperfect scenarios. We summarize imperfect scenarios into the following three categories:

- **Noisy labels**: Noisy label refers to the incorrect annotations that may be contained in the training data, such as some incorrectly labeled samples in ImageNet.
- **Adversarial samples**: Adversarial samples refer to samples that may contain malicious designs in the data, which may cause the model to output dangerous or noncompliant content.
- **Out-of-distribution samples**: Out-of-distribution samples refer to samples that may contain samples outside the distribution of the training dataset in the test data, including data outside the label space or training domain.

Trustworthy machine learning requires the model to have the following properties:

- **Robustness**: Robustness requires that the model maintains stable performance in the face of changes, interference, or uncertainty in input data.
- **Security**: Security requires models to possess the capability to withstand malicious attacks.
- **Reliability**: Reliability entails the ability to detect samples outside the training label space, enabling the model to explicitly indicate "unknown" rather than forcing a prediction when encountering inputs beyond its known scope.
- **Interpretability**: Interpretability emphasizes transparency of the decision-making process, helping us understand the reasons behind the model's predictions.

As machine learning becomes more common in all aspects of life, the model's inaccurate or unpredictable output may have major effects, particularly in high-risk sectors such as autonomous driving, medical diagnostics, and financial decision-making. In these high-risk scenarios, the accuracy and reliability of the model are directly related to human life safety or economic stability. Ensuring the trustworthiness of machine learning systems in imperfect scenarios is an urgent problem to be solved. In subsequent chapters of this book, we will delve into how to achieve trustworthy machine learning in imperfect scenarios and provide insights for future theoretical research and practical applications based on our exploration.

1.2 Imperfect Data

This section will introduce the concept of imperfect data since it matters in trustworthy machine learning. We will first explore why imperfect data is real and common and then outline three common types of imperfect data in the real world: data with noisy labels, data with adversarial samples, and data with OOD samples.

1.2.1 *Why Imperfect Data*

In the real world, imperfect data originates from numerous sources. First, during the data collection and annotation phase, it is challenging to entirely eliminate dataset annotation errors. For example, studies [280] show that ImageNet [91] comprises a significant quantity of inaccurately labeled items, such as mislabeling images of one animal category as another animal category, and rectifying such mislabeling in extensive datasets is prohibitively costly. Second, with the advancement of machine learning, the demand for large-scale datasets has grown rapidly, but it is very difficult to ensure the data quality of large-scale datasets. For example, annotations on crowd-sourcing platforms are often difficult to maintain consistent quality.

Furthermore, during the model deployment phase, imperfect data can manifest in even more complex ways. On one hand, models are vulnerable to malicious interference—malicious users might input adversarial examples to elicit illicit outputs, such as adding imperceptible perturbations to images to trick the model into producing incorrect results or crafting carefully designed prompts to jailbreak large language models and generate prohibited content. In addition, during the model deployment phase, the model is likely to encounter samples that are beyond the distribution of the training data; that is, different domains will also lead to a significant decline in model performance. For example, even if the categories of the test samples are the same, an autonomous driving model trained only on sunny data may be difficult to generalize to wet or snowy weather. In addition, the model may also encounter test samples outside the label space of the training set. If the model directly makes judgments, unacceptable consequences may occur.

1.2.2 *Different Types of Imperfect Data*

1.2.2.1 Data with Noisy Labels

One primary reason for the success of the machine learning models in the past decades has been the availability of large-scale, carefully annotated data, so that one can perform supervised training of machine learning models at an enormous scale to learn complex input-output patterns. However, as the scale of the data escalates, it becomes apparent that we often cannot guarantee the quality of all the annotations, hence creating the problem of label noise. Under the presence of label noise, machine learning models will treat those incorrect labels as the true supervision signal and eventually overfit to them, hence adversely impacts the generalization of machine learning models to the underlying true distribution. This creates significant challenges for the trustworthiness and reliability of the machine learning models developed in real-world applications.

Label noise [137] manifests in various forms, each presenting unique challenges to learning algorithms. In the simplest case, random noise occurs when labels are assigned incorrectly with no particular pattern, essentially introducing statistical uncertainty into the learning process. Despite its simplicity, random label noise has been beneficial for the research community to understand the basic qualities of label noise. However, in real-world scenarios, the distribution of label noise could be more complex, often following specific patterns—perhaps consistently mislabeling particular classes or features, hence creating more realistic types of label noise such as class-dependent label noise and data-dependent label noise. For class-dependent noise, where certain classes are more prone to mislabeling than others, this creates imbalanced learning difficulties. Data-dependent noise, where labeling errors correlate with specific input features, can lead models to learn spurious correlations rather than meaningful patterns.

As the development of machine learning progresses and more learning tasks emerge, the concept of label noise not only describes the incorrect annotations in the classification task but can also be set to include a wide spectrum of input-annotation pairs, to name a few examples, in cross-modal learning, images pairs with incorrect or low-quality text description can often be viewed as a variant of label noise; in the supervised fine-tuning process of large language models, it is also possible that for a given input prompt, the corresponding desired output is incorrect.

For decision-critical applications such as healthcare, autonomous driving, finance, and security, the stakes of label noise are particularly high. In those domains, where AI models are currently playing vital roles, the consequences of incorrect labels in the training process could be dire. For example, in the domain of medical imaging, an image with an incorrect label could potentially mislead the model and give an incorrect diagnosis for patients meanwhile keeping such harm almost at stealth. These decision-critical contexts underscore why addressing label noise is not merely an academic concern but an essential component of responsible machine learning practice.

We aim to equip researchers and practitioners with a comprehensive understanding of label noise—its origins, implications, and remedies—enabling more resilient machine learning systems that can thrive despite the inevitable imperfections in real-world data. As the scale and importance of machine learning applications continue to grow, so too does the imperative to address the fundamental challenge of learning from noisy supervision.

1.2.2.2 Data with Adversarial Examples

Modern machine learning models have achieved unprecedented success and have been widely deployed in decision-critical areas, such as autonomous driving [129, 238], facial recognition [22, 332], and medical imaging diagnosis [101]. With our society's increasing reliance on these algorithms, ensuring their trustworthy in worst-case scenarios is crucial to proactively mitigate the risks associated with their potential failures. However, most machine learning algorithms exhibit significant vulnerabilities to malicious attacks, failing to meet the broad societal security expectations. Numerous red-teaming studies have demonstrated that introducing human-imperceptible but meticulously crafted adversarial perturbations to the original input can completely compromise the model's prediction ability, enabling adversaries to manipulate the output toward any specific target. In a real-world example, by applying adversarial perturbations to a stop sign, an autonomous driving system may mistakenly recognize it as a speed limit sign indicating 120 km/h, thus posing substantial public safety hazards.

In response to the urgent need for trustworthy machine learning systems that can remain dependable in adversarial scenarios, various defense techniques have been implemented throughout their entire life cycle, including during preprocessing, development, and deployment stages. In this book, we introduce adversarial defense methods within three principal frameworks: statistical learning, deep learning, and diffusion-based approaches, each specifically tailored to a specific phase of the machine learning life cycle.

Statistical learning-based defenses focus on discovering and exploiting the relationships between adversarial and clean labels. By establishing the transition relationship between adversarial and natural labels, statistical defenses can effectively correct mispredictions caused by adversarial examples. In the post-processing stage, this allows for the rectification of errors before the machine learning models actually deliver incorrect results to users.

Deep learning-based defenses concentrate on enhancing robustness by on-the-fly identifying and remediating adversarial vulnerabilities during the development phase. These approaches proactively expose the machine learning model to adversarial examples to identify weaknesses and, correspondingly, apply remedies to mitigate these vulnerabilities, thereby continually strengthening the model's robustness.

Diffusion-based defenses are designed to purify and detect adversarial examples before they can reach the decision-making model. In the preprocessing stage, these methods can effectively restore the distribution shift induced by adversarial examples to their natural data manifold, making the models impervious to attacks. Furthermore, these methods enable nondestructive implementation as they do not require modifications to the backbone model.

By systematically incorporating these three types of strategies into the machine learning life cycle, we establish a comprehensive framework for building adversarially robust systems. Finally, we aim to provide practical guidance for a future blueprint toward adversarial robustness in machine learning.

1.2.2.3 Data with Out-of-Distribution Samples

Machine learning models are typically evaluated using data presumed to be taken from the same distribution as the training data, thereby quantifying the model effectiveness accordingly. Nevertheless, in practical scenarios, this assumption is often overly restrictive and frequently proves incorrect. Specifically, when deploying machine learning models, we often encounter out-of-distribution (OOD) data, which refers to those drawn from distributions that are very different from the training cases. Such discrepancies can arise due to various factors, including environmental changes, bias in data collection procedures, and the emergence of new, previously unseen scenarios. According to classical statistical learning theory, such a difference between training and deployment distributions results in a distribution gap, which impairs the model's overall performance.

For example, in healthcare applications, machine learning models developed using patient data from the past might not be able to identify symptoms of recently discovered illnesses, which could result in incorrect diagnoses or postponed treatments. Similarly, when applied to patient groups that were not represented during training, demographic changes in patient populations—such as shifting age distributions or ethnic compositions—can have a substantial impact on model dependability and generalization, degrading performance. OOD scenarios also commonly appear in autonomous driving applications. On the road, vehicles may run against unforeseen barriers like garbage, animals, or road dangers, which are very different from the things they would encounter when training a model. Furthermore, the resilience and safety of autonomous driving systems are put to the test by environmental changes, such as unfavorable weather conditions like fog, intense rain, or snow, which can drastically modify sensor inputs and impair model performance.

In the literature, semantic and covariate OOD data are the two common types of OOD data that are extensively studied. Semantic OOD data [431] refer to data points for which ground truth labels are not in the considered label space. For example, in a cat-and-dog classification model, an image of a bird constitutes a semantic OOD case. No matter how strong our classification model is, it will always wrongly predict the bird as either a cat or a dog. In such cases, it is critical for the model to have the ability to detect these anomalies and avoid making subsequent predictions, leading to the field of OOD detection.

On the other side, covariate OOD data refers to data whose styles or patterns were not seen during training. Also, taking our cat-and-dog classification model as an example, if we train the model using real images of cats and dogs but then encounter a cartoon image of a dog, this cartoon image is considered a covariate OOD. For a robust and general model, we typically hope it can still make the correct prediction in such cases, leading to the literature on OOD generalization. OOD detection and OOD generalization focus on different perspectives of OOD in practice, yet both are critical for the trustworthiness of real-world machine learning systems. These topics will be discussed in detail in Sects. 4.1 and 4.2.

1.2.3 The Focus of the Book and Why

From the above introduction, we know that imperfect data is ubiquitous. Therefore, in order to make machine learning better serve society, an urgent problem to be solved is how to build a trustworthy machine learning system in imperfect scenarios. These three types of imperfect data introduced above are common in practical applications and may cause model performance to be far below expectations. Therefore, this book aims to systematically explore how to achieve trustworthy machine learning systems in the presence of imperfect data.

In addition, the book extends the scope to advanced application areas of trustworthy machine learning, including three research areas: federated learning, graph learning, and foundation models. These advanced topics reflect emerging needs in modern machine learning scenarios but also face the challenges of imperfect scenarios. Machine learning systems face challenges such as security challenges in distributed environments, robustness issues in complex data structures, and trustworthy assurance in large language model applications. This book further explores how to achieve trustworthy machine learning in these areas.

1.3 Organization

The book is organized as shown in the Fig. 1.1. The content structure of each chapter is described below.

Chapter 2 focuses on trustworthy machine learning with noisy labels. In this section, we analyze the impact of label noise on model performance and introduce some of our explorations in this direction, such as algorithms to mitigate label effects. The main content includes statistical learning methods, deep learning approaches, and the application of over-parameterized models while also exploring the transition from learning with noisy labels to weakly supervised learning.

In Chap. 3, we study trustworthy machine learning with adversarial examples. In this section, we discuss three categories of methods to enhance the security and robustness of models under malicious interference, namely, statistical learning methods, deep learning techniques, and diffusion models.

In Chap. 4, we discuss trustworthy machine learning with out-of-distribution (OOD) data, which is divided into two key areas: detection and generalization. In the OOD detection subsection, we introduce three categories of detection methods: post hoc detection, outlier exposure, and representation-based detection methods to improve the reliability of models when encountering samples outside the training set label space. In the OOD generalization subsection, we introduce invariant risk minimization and contrastive learning-based methods to improve the adaptability of models to unknown distributions.

Chapter 5 focuses on three research directions: federated learning, graph learning, and foundation models. By exploring these advanced topics, we will show how to deal with security challenges in distributed environments, robustness issues in complex data structures, and trustworthiness assurance in large-scale model applications, providing more comprehensive experimental support and theoretical basis for the practical application of trustworthy machine learning.

In summary, this book will start with three imperfect scenarios and gradually go deeper into advanced applications, aiming to provide comprehensive theoretical guidance and practical insights for researchers and practitioners. In each subse-

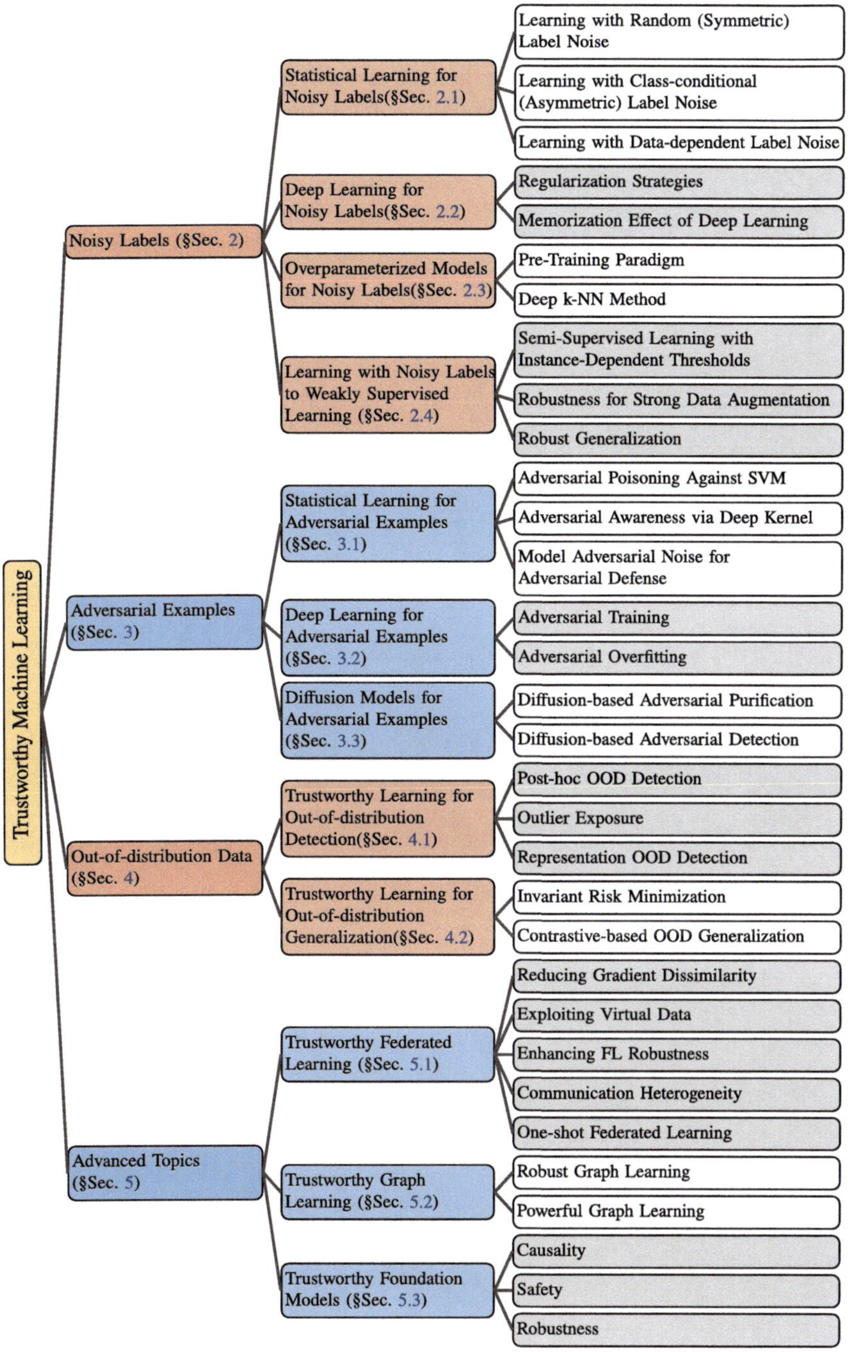

Fig. 1.1 Organization of this book

quent chapter, we will analyze specific problems independently while maintaining interrelationships to form a complete framework for trustworthy machine learning under imperfect data conditions. In the end, we will also give our conclusions and prospects for future research. The Notation table is provided in Table 1.1.

Table 1.1 Notation table

Symbol	Description
$\mathbf{x}$	Input data, single instance, vector form
$\mathbf{X}$	Input data, batched instances, matrix form
y	Label, single instance, scalar form
$\mathbf{y}$	Label, batched instance, vector form
X	Random variable corresponding to the input data
Y	Random variable corresponding to the label
n	Number of samples
$\mathcal{X}$	Instance domain
$\mathcal{Y}$	Label domain
$\lvert \cdot \rvert$	The cardinality of a given set
$\mathcal{D}$	Underlying distribution where data are sampled from
D	Input dataset, sampled from $\mathcal{D}$, $\lvert D \rvert = n$
$\mathcal{H}$	Hypothesis class
f_θ	Hypothesis function, parameterized by θ
f^*	Bayes optimal hypothesis function
$\hat{y}$	Predicted label, $\hat{y}_i = f_\theta(\mathbf{x}_i)$
ℓ	Loss function
$\mathcal{L}_D(\theta)$	Empirical risk (supervised learning), $\mathcal{L}_S = \frac{1}{n}\sum_{i=1}^{n} \ell(f_\theta(\mathbf{x}_i), y_i)$
$\mathcal{L}_{\mathcal{D}}(\theta)$	Expected risk (supervised learning), $\mathcal{L}_{\mathcal{D}} = \mathbb{E}_{(\mathbf{x},y)\sim\mathcal{D}}\left[\ell(f_\theta(\mathbf{x}), y)\right]$
$P(\cdot)$	Probability distribution
$p(\cdot)$	Probability value
$\mathbb{R}^d$	d-dimensional real vector space
Chap. 2 notations	
$\tilde{y}$	Noisy label
$\tilde{Y}$	Random variable corresponding to the noisy label
Chap. 3 notations	
$\tilde{\mathbf{x}}$	Adversarial example
δ	Adversarial example perturbation
v	Adversarial weight perturbation
$\mathcal{V}$	Adversarial weight perturbation budget, $v \in \mathcal{V}$
Chap. 4 notations	
S	Scoring function used in the OOD detection process
D_{id}	In-distribution dataset
D_{ood}	Out-of-distribution dataset
G	OOD detector
$\mathcal{Z}$	Latent domain
φ	Feature extractor

(continued)

Table 1.1 (continued)

Symbol	Description
Chap. 5 notations	
$\mathcal{M}$	Set of clients
$\frac{n_m}{N}$	Training data ratio of client m
$\mathcal{L}_m$	Empirical risk or client m
r	Training round in federated learning
$\mathcal{V} = \{v_i\}_{i=1}^{N}$	The collection of nodes, where N denotes number of nodes
$\mathcal{E} = \{e_{ij}\}_{ij=1}^{M}$	The collection of edges, where M denotes number of edges
$\mathbf{A}$	The adjacency matrix where $A_{ij} = 1$ indicating an edge between v_i and v_j
$I(\cdot;\cdot)$	Mutual information between two variables
$\mathcal{R}$	Set of relations
$\mathcal{G}$	Graph
e_a	Target entity
e_q	Query entity
r_q	Query relation
$\widehat{\mathcal{G}}^L$	Propagation path, consisting of entity sets $\{\mathcal{V}^0, \mathcal{V}^1, \ldots, \mathcal{V}^L\}$ at each propagation step

Chapter 2
Trustworthy Machine Learning with Noisy Labels

In the development of machine learning models, we often presume that our training data represents ground truth—a flawless reflection of reality against which our models learn to make predictions. However, this assumption rarely holds in real-world applications. Label noise is virtually unavoidable in large-scale datasets due to various factors, such as the lack of expertise from annotators, ambiguous labeling guidelines, inherent uncertainty in the labeling process itself, etc.

As the machine learning community begins to appreciate the importance of this problem, researchers have developed a diverse range of approaches to address label noise under different generative assumptions. These assumptions are crucial, as they formalize an otherwise arbitrary phenomenon and enable systematic investigation of the problem. Specifically, during the course of development, three primary assumptions are proposed: random (symmetric) label noise, class-conditional (asymmetric) label noise, and instance-dependent label noise. Under each framework, researchers have studied the theoretical properties of label noise and developed risk-consistent solutions with demonstrable generalization guarantees.

Under the influence of label noise, machine learning models tend to adversely overfit to corrupted labels, resulting in poor generalization to the underlying true distribution. Consequently, mitigation strategies typically fall into two categories: regularization and sample selection. Regularization techniques aim to minimize the negative impact of label noise by preventing overfitting, while sample selection methods attempt to distinguish between clean and noisy samples, thereby avoiding the learning of incorrect patterns altogether.

With the emergence of deep learning, how deep neural networks (DNNs) cope with label noise has revealed some unique and intriguing properties, for instance, as complex hypothesis classes, DNNs can achieve complete memorization of random label while being able to generalize better than other hypothesis classes, making the classical interpretations of the generalization of DNN doubtable [461]. Also, the implicit preferences of DNNs when learning with a distribution that contains "hard example" (such as noisy samples), known as the memorization effect, suggest

B. Han, T. Liu, *Trustworthy Machine Learning under Imperfect Data*,
https://doi.org/10.1007/978-981-96-9396-2_2

that DNNs tend to learn common patterns from data first and then move on to memorizing the specific mappings on an instance level [10].

In this chapter, we will explore these topics in depth, highlighting how advances in this field have enhanced the reliability and trustworthiness of machine learning models when facing label noise.

2.1 Statistical Learning for Noisy Labels

Learning with noisy labels has been one of the long-concerning questions in the field of statistical learning theories, while numerous analyses have been established upon the assumption that if the supervision signal is accurate, the learned models can produce a bounded rate of errors with certain probabilities [24], or the produced models can be asymptotically consistent [25]. However, making those results hold under the presence of label noise is a challenging problem, specifically, without further assumption on the noise generation process or learning algorithm, the obtained model can easily overfit to the noisy labels and generate predictions that deviate significantly from the underlying true distribution.

For the better characterization of the problem, the notion of **label noise transition matrix** has been proposed to instantiate the generation process of label noise. Under some context, the label noise transition matrix can also be described as the **flip rate** equivalently. Specifically, the generation process of label noise can be defined as

$$P(\tilde{Y}|X) = T^{\top} P(Y|X), \tag{2.1}$$

where $P(Y|X)$ represents the clean class posterior, i.e., the underlying true conditional probability of the clean class given input data. This is characterized by a transition process we term as T, which models the transition probability from clean class to noisy class; consequently, the $P(\tilde{Y}|X)$ represents the noisy class posterior, where our training data are sampled from.

Concretely, suppose the cardinality of the label space is C, and the noise transition process T is a mapping, we have $|Y| = |\tilde{Y}| = C$, which means the transition matrix will be a C-dimensional square matrix:

$$T = \begin{bmatrix} e_{1,1} & e_{1,2} & \cdots & e_{1,C} \\ e_{2,1} & e_{2,2} & \cdots & e_{2,C} \\ \vdots & \vdots & \ddots & \vdots \\ e_{C,1} & e_{C,2} & \cdots & e_{C,C} \end{bmatrix} \in \mathbb{R}^{C \times C}$$

By incorporating T into the learning framework, one can develop methods that aim to recover the true class posterior from noisy observations, thereby restoring many of the statistical guarantees (such as bounded error rates and asymptotic consistency) associated with noise-free learning scenarios.

For the following subsections, we will focus on different assumptions of the T, and what sorts of theoretical guarantees can be derived under different assumptions.

2.1.1 *Learning with Random (Symmetric) Label Noise*

Random label noise, also known as symmetric label noise, uniform label noise, or class-independent label noise assumes that, for a given class, the probabilities of it being corrupted to any other class are uniform. This is arguably the strongest assumption on the transition process of label noise, also regarded as the most benign case. In symmetric label noise, the transition matrix can be denoted as

$$T = \begin{bmatrix} 1-e & \frac{e}{C-1} & \cdots & \frac{e}{C-1} \\ \frac{e}{C-1} & 1-e & \cdots & \frac{e}{C-1} \\ \vdots & \vdots & \ddots & \vdots \\ \frac{e}{C-1} & \frac{e}{C-1} & \cdots & 1-e \end{bmatrix} \in \mathbb{R}^{C \times C}.$$

Earlier line of works mostly focuses on answering the question "what kind of loss function is tolerant under symmetric label noise"; this question was to be revealed as nontrivial, since Long Philip and Rocco [296] proved that the potential function of boosting algorithms can lead to a solution that performs no better than random guess under arbitrary noise level, which highlights the importance of loss selection when learning with label noise.

Subsequently, Aritra et al. [7] proved that, in binary classification, if the chosen loss function is symmetric and bounded, then minimizing this loss can yield a classifier whose expected error equals that of the classifier trained on clean data. Formally, the loss function needs to satisfy

$$\ell(f(\mathbf{x}), +1) + \ell(f(\mathbf{x}), -1) = K, \quad \forall \mathbf{x}, \forall f \in \mathcal{H}, \tag{2.2}$$

where K is some positive constant.

Overall, learning under symmetric label noise has been shown to be relatively well understood, both theoretically and empirically. Under mild assumptions and without requiring knowledge of the exact noise rate, strong theoretical guarantees such as asymptotic consistency can still be achieved. However, the assumption of symmetric noise is highly restrictive, as it imposes a uniform corruption pattern across classes. In the following sections, we shift our focus to more realistic scenarios that impose weaker and more flexible assumptions on the label noise generation process.

2.1.2 *Learning with Class-Conditional (Asymmetric) Label Noise*

Unlike symmetric label noise, where the off-diagonal entries of the transition matrix are uniformly distributed, class-conditional label noise assumes that the transition process is class dependent. This means the off-diagonal entries of the transition matrix can vary across different classes and are not necessarily equal to one another. Such a setting introduces additional complexity, both from a theoretical standpoint and in practical implementation. Under asymmetric label noise, the transition matrix can be denoted as

$$T = \begin{bmatrix} 1-e_1 & e_{1,2} & \cdots & e_{1,C} \\ e_{2,1} & 1-e_2 & \cdots & e_{2,C} \\ \vdots & \vdots & \ddots & \vdots \\ e_{C,1} & e_{C,2} & \cdots & 1-e_C \end{bmatrix} \in \mathbb{R}^{C\times C},$$

where $e_i = \sum_{j\neq i}^{C} e_{i,j}$.

Typically, when handling class-dependent label noise, one common assumption needed to obtain theoretical guarantee is knowing the noise rate. Notably, in binary classification, under the assumptions that the noise rates are known, Natarajan et al [274] proved that using re-weighted margin loss can lead to risk-consistent minimizer of the clean loss function, that is, if we use re-weighted margin loss as the proxy loss to be optimized, given sufficient training samples, the minimizer of the proxy loss is asymptotically equivalent to the loss function defined over clean data.

However, knowing the noise rate could be unrealistic in real-world practices, to address that problem, Liu et al [239] proposed the **anchor point assumption** for this problem. More specifically, assume that we are able to locate at least one sample per class, that we know for sure it belongs to certain clean class ($P(Y|X = \mathbf{x}_{\text{anchor}}) \approx 1$) (anchor point), then, if the classifier we have is perfectly calibrated, we are able to obtain the noisy class posterior for a certain class directly. And because we are operating under the class-conditional label noise assumption, the noisy class posterior for $\mathbf{x}_{\text{anchor}}$ of ith-class is effectively the ith-row of the transition matrix. This enables us to obtain a theoretically consistent estimation of the transition probability in closed form, providing asymptotic consistency guarantee as well as PAC-style generalization guarantee.

2.1.3 *Learning with Instance-Dependent Label Noise*

Learning with instance-dependent label noise, also known as learning with data(feature)-dependent label noise, is generally perceived as imposes minimal assumption on the generation process of the label noise, since it has the highest degree of freedom among the noise generation assumptions. Concretely, random label noise can be seen as a special case of class-dependent label noise, and class-dependent label noise can be seen as a special case of instance-dependent label noise. In data-dependent label noise, the transition matrix can be denoted as a function in the form of

$$T(\mathbf{x}_k) = \begin{bmatrix} 1-e_1 & e_{1,2} & \cdots & e_{1,C} \\ e_{2,1} & 1-e_2 & \cdots & e_{2,C} \\ \vdots & \vdots & \ddots & \vdots \\ e_{C,1} & e_{C,2} & \cdots & 1-e_C \end{bmatrix} \in \mathbb{R}^{C \times C},$$

where $e_i = \sum_{j \neq i}^{C} e_{i,j}$.

Under class-dependent label noise assumption, we have concluded that if the noise rate is known, one can provably solve this problem, and if the noise rate is unknown, but the anchor point assumption is satisfied, we can consistently estimate the noise rate by fitting a model to noise first. However, under instance-dependent label noise assumption, anchor-point assumption alone can no longer solve the problem. To that extent, Xia et al. [413] take an alternative approach by proposing **part-dependent assumptions**, which assumes an instance can be decomposed into many parts, and different instances are essentially different weighted combination of the parts; therefore, by enforcing anchor-point assumption on parts, one can effectively approximate instance-dependent transition matrix by the weighted combination of the part-dependent transition matrices.

Another alternative approach is by Cheng et al. [68], which operates under **bounded error rate assumption**, specifically, under binary classification task, if the noise rates are bounded, and the upper bound of the noise rate is known; Cheng et al proposed to first select a group of **distilled examples**, whose labels are guaranteed to be the same to their label assigned by the Bayes optimal classifier. Therefore, it is straightforward to perceive, given sufficiently many distilled examples unbiasedly sampled from the underlying distribution, the learned classifier will converge to the Bayes optimal classifier asymptotically.

2.2 Deep Learning for Noisy Labels

In this section, we give an overview for the techniques to combat label noise, specifically in the deep learning context, and some unique phoneme emerging from deep learning models when trained with label noise.

What is unique about deep learning, when it comes to label noise? What are its fundamental differences when comparing with conventional statistical learning method? One notable difference observed by the seminal paper from Zhang et al [468] is the fact that deep neural network (DNN) approximation power is so strong, even when the label is assigned completely at random, DNNs can still "memorize" the mapping from features to random labels, raising questions to the conventional view on the generalization. To that end, more fascinating phenom of label noise were later revealed, such as the unique learning dynamics of DNNs known as the "memorization effect" [10], revealing there exists an implicit "preference" for data to be fitted by DNN.

2.2.1 Regularization Strategies

The most commonly used approach for combating overfitting under different learning scenario is arguably regularization; learning with label noise is no exception. The effectiveness of regularization has been especially pronounced for deep learning models; given the high flexibility and the approximation power to perfectly overfit even completely random labels, applying appropriate regularization technique to alleviate overfitting has been especially significant. The general formulation of regularization can be expressed as

$$\mathcal{L}_D(\theta, \lambda) = \frac{1}{n} \sum_{i=1}^{n} \ell(f_\theta(\mathbf{x}_i), y_i) + \lambda \Omega(\theta), \tag{2.3}$$

where $\mathcal{L}_D$ is the objective function defined over dataset D, ℓ is the loss function, and Ω refers to an arbitrary regularization term. In most cases, Ω is dependent on the function being learned.

Apart from using basic strategies such as ℓ-1 or ℓ-2 regularization, many research have contributed to advanced regularization techniques that are specifically tailored for learning with label noise. In this part, we will give a brief overview on *explicit regularization* and *implicit regularization* under the context of label noise.

2.2.1.1 Mixup

Mixup is a strong regularization technique introduced by Zhang et al in their seminal paper [460]. Motivated by vicinal risk minimization [57], Mixup aims to augment the support of the training distribution by reformulating the training objective as

$$\ell_{\text{mixup}} = \ell(f(\lambda \cdot \mathbf{x}_i + (1 - \lambda) \cdot \mathbf{x}_j), \lambda \cdot y_i + (1 - \lambda) \cdot y_j), \tag{2.4}$$

where $(\mathbf{x}_i, y_i)$ and $(\mathbf{x}_j, y_j)$ are two randomly sampled data points, λ is the mixing ratio in the range [0, 1] which is sampled at random, when $\lambda = 0$, and ℓ_{mixup} is equivalent to generic empirical risk minimization. Based on Eq. (2.4), it is clear that Mixup is essentially performing linear interpolation between randomly sampled data points; one straightforward advantage of this is that this implicitly creates more "virtual" sample for the training of DNNs, like other data augmentation techniques; another advantage has been attributed as it regularizes the transition of decision boundary across classes to be linear, avoiding overcomplicated extrapolation for unseen data. In the original paper of Mixup [460], the authors have empirically showcased the performance gain when facing label noise by applying Mixup.

In addition, DivideMix [221] is a notable attempt that employs Mixup as part of its integrated strategies to battle label noise. Specifically, DivideMix maintains two separate network to avoid the confirmation bias caused by a single network and uses Gaussian mixture model (GMM) to cluster the prediction loss into confident and unconfident part, and then, a refined Mixup is applied to the confident; this refined Mixup is mixing the observed label $\tilde{y}_i$ and the predicted label $\hat{y}_i$, so that when the observed label is incorrect, this could be remedied by the predictions. Lastly, for both confident and unconfident data, they will go through a MixMatch [32] process, which utilizes generic Mixup as data augmentation and consistency regularization. This approach has showcased notable empirical performances.

2.2.1.2 Early Stopping as Implicit Regularization

Early stopping is a crucial technique for preventing overfitting in machine learning models. Its core principle is to halt the training process when the model's generalization performance begins to deteriorate. Under the *empirical risk minimization* (ERM) paradigm, early stopping typically relies on an independent validation set to monitor changes in the model's generalization performance during training, which serves as the basis for determining the optimal time to stop training [449]. Specifically, traditional early stopping follows these steps:

1. The complete dataset is divided into three parts: training set, validation set, and test set. The training set is used for model training, the validation set for performance evaluation and hyperparameter tuning, and the test set for final model assessment.
2. At the end of each training epoch, the current model makes predictions on the validation set, and its loss function value on the validation set is calculated as an estimate of the model's generalization performance.
3. When the validation loss curve shows an upward trend, indicating that the model's performance on the validation set begins to decline, the training process is stopped, and the current model is selected as the final model. This moment is considered the optimal early stopping point.

Recent theoretical and empirical research findings, such as *deep double descent* [276] and *benign overfitting* [26], have challenged traditional machine learning paradigms and early stopping techniques. These phenomena suggest that under certain conditions, moderate overfitting of training data can actually improve model generalization. Nevertheless, early stopping remains an indispensable regularization method for preventing overfitting and improving generalization performance. Particularly in challenging scenarios with insufficient data quality or severe noise, application of early stopping strategies is crucial for ensuring model robustness and reliability.

In standard deep learning practice, early stopping method typically assumes that the validation set accurately represents the true data distribution and that its labels are entirely correct. However, when label noise is present—especially at high noise levels—creating a truthworthy enough validation set becomes a significant challenge.

In many paradigms of noisy learning, training data consists of a small clean subset and a large noisy subset. Allocating valuable clean samples to validation is unreasonable as it would further reduce clean samples available for training—which are crucial for learning genuine patterns. However, creating a validation set from noisy samples means noisy labels could mislead performance assessment and impact early stopping reliability. Although validation set noise may not completely obscure generalization trends [63], it reduces the reliability of validation feedback and affects the model's performance on clean test sets.

Method To address challenges, we propose leveraging the dynamic changes in model predictions on the training data, eliminating the need for a separate validation set [445]. The core idea is to monitor the fluctuations in model predictions across consecutive training epochs. By analyzing these fluctuations, we can accurately identify the critical turning point when the model transitions from learning generalizable patterns to overfitting noisy labels. We term this method *label wave*.

Specifically, we observe that during the initial stages of training, the model primarily learns patterns from correctly labeled samples, which is reflected in relatively stable predictions across epochs. As training progresses, the model attempts to fit samples with noisy labels, causing significant instability in its predictions. By tracking the rate of prediction changes in real time, we can detect this transition and determine the optimal stopping point for training.

Our method calculates a key metric called *prediction changes (PC)*, which quantifies the number of label predictions that differ between two consecutive epochs for the training samples. Formally, for each epoch t, we calculate the following statistic:

$$\mathrm{PC}_t = \sum_{i=1}^{n} \mathbb{1}_{\hat{y}_i^t \neq \hat{y}_i^{t-1}}, \tag{2.5}$$

where n denotes the total number of training samples, $\hat{y}_i^t$ is the predicted label for sample i in epoch t, and $\mathbb{1}$ is an indicator function that returns 1 when the condition is met and 0 otherwise.

During training, we monitor how PC_t changes across epochs t. We define the optimal stopping point as the first local minimum of PC_t followed by an upward trend, which typically indicates that the model has begun overfitting samples with noisy labels. The method is simple to implement and can be seamlessly integrated into standard neural network training workflows. The specific steps are as follows:

1. Initialize the model parameters and set up the training loop as usual.
2. For each training epoch, after updating the model parameters using the training data, perform the following:
   - Calculate the current predicted labels $\hat{y}_i$ for all training samples.
   - Compute the PC_t metric by comparing predictions for each sample between the current epoch t and the previous epoch $t-1$.
3. Identify the first local minimum in PC_t as the optimal early stopping point.
4. Terminate training at the identified stopping point to prevent the model from overfitting noisy labels.

To further enhance robustness, smoothing techniques like moving averages can be applied to the PC_t calculations to filter out minor fluctuations. Additionally, introducing a patience parameter allows the algorithm to continue training for a few more epochs after detecting the minimum, ensuring the stopping point is not overly sensitive to transient fluctuations.

The foundation of the *label wave* method lies in understanding of the learning dynamics in deep neural networks. Numerous empirical studies suggest that neural networks initially tend to learn simpler, more generalizable patterns in the data [10] before moving on to memorizing specific noisy labels [468]. These phenomena are often referred to in the literature as *learning simple patterns* and the *memorization effect*, respectively.

As shown in Fig. 2.1, in the early stages of training, the model tends to prioritize fitting the majority of the training data, which typically corresponds to correctly labeled samples. As a result, the model's predictions on these samples quickly stabilize, leading to a sustained decrease in the (PC_t) metric. However, as training progresses, the model adjusts its parameters to fit samples with noisy labels,

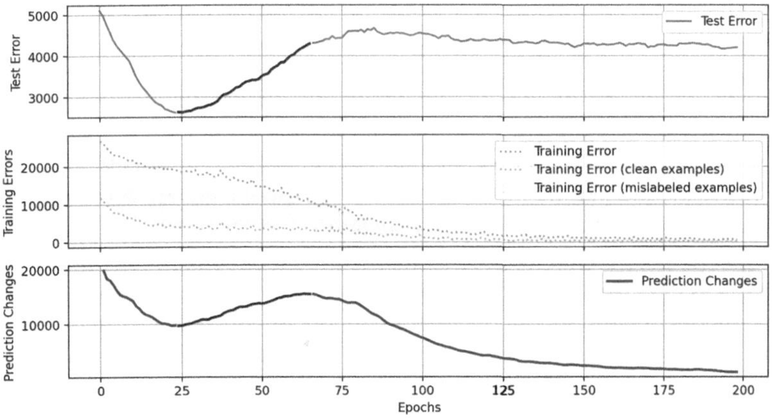

Fig. 2.1 Visualization of model training dynamics showing test error (top), training errors (middle), and prediction changes (PC_t) (bottom). The purple segment indicates the transition phase where the model learning confusion patterns

Table 2.1 Differences among the selection methods on CIFAR-10 (with 40% Sym. label noise)

Noise (Sym.)	*20%*	*30%*	*40%*	*50%*	*60%*	*70%*	*80%*
Global Max (%)	84.1±0.1	81.7±0.4	80.5±0.7	75.5±0.5	68.6±0.7	57.6±1.0	37.5±0.9
Label wave (%)	83.4±0.3	81.5±0.4	80.2±0.7	75.2±0.7	68.2±1.0	56.6±1.6	37.1±1.0
Difference	**0.7**	**0.2**	**0.3**	**0.3**	**0.4**	**0.9**	**0.4**
Noise (Ins.)	*20%*	*30%*	*40%*	*50%*	*60%*	*70%*	*80%*
Global Max (%)	84.3±0.5	82.8±0.7	77.7±0.7	64.7±0.8	46.8±1.5	29.5±1.0	20.6±0.7
Label wave (%)	84.0±0.7	82.7±0.5	76.8±0.8	63.9±0.7	45.9±2.0	28.9±0.7	20.1±1.1
Difference	**0.3**	**0.2**	**0.9**	**0.8**	**0.9**	**0.6**	**0.5**

Table 2.2 Test accuracy (mean±std) of each method on CIFAR-10 (with 40% Sym. label noise)

Method	*Val.* 5%	*Val.* 10%	*Val.* 20%	*Val.* 30%	*Label wave*
CE	80.3±0.7	79.9±0.5	78.9±0.5	77.4±0.6	**81.6±0.4**
Taylor-CE	83.9±0.6	83.3±0.3	82.3±0.4	81.7±0.1	**85.1±0.3**

causing increased prediction fluctuations and a subsequent rise in the (PC_t) metric. Therefore, by identifying the point at which the (PC_t) metric begins to rise, we can accurately determine when the model starts to significant learn from mislabeled samples. Stopping training at this point maximizes the model's generalization ability while minimizing the negative impact of label noise.

We term the stage where the (PC_t) metric rises as *learning confusion patterns*. During the *learning confusion patterns* stage, the model begins attempting to fit noisy labels, causing fluctuations or even a decline in its fitting ability on the training data. The emergence of this stage indicates that the model has started learning incorrect and overly complex patterns from the data, which ultimately impairs its generalization performance.

Experiments To evaluate the practical effectiveness of our *label wave* method, we conducted a series of experiments on CIFAR datasets with different noise settings.

Table 2.1 presents the differences between the test accuracy of models selected by the *label wave* method and the global maximum test accuracy during the training process under both symmetric and instance-dependent label noise, with noise ratios ranging from 20% to 80%. The results show that the *label wave* method consistently selects models that are close to the global maximum in test accuracy, with the differences ranging from 0.2% to 0.9%, regardless of the noise type or level.

We further compared the method with traditional validation-based early stopping methods on CIFAR-10 and CIFAR-100 with 40% symmetric label noise. We varied the size of the validation set from 5% to 30% of the training data. Tables 2.2 and 2.3 compare the test accuracy of models selected by the *label wave* method and traditional validation-based methods with different validation set sizes. On both CIFAR datasets with 40% symmetric label noise, the *label wave* method consistently outperforms all validation-based methods, achieving the highest test accuracy with both the cross-entropy (CE) [317] and Taylor cross-entropy (Taylor-CE) [110] loss functions.

Table 2.3 Test accuracy (mean±std) of each method on CIFAR-100 (with 40% Sym. label noise)

Method	*Val.* 5%	*Val.* 10%	*Val.* 20%	*Val.* 30%	*Label wave*
CE	49.7±0.9	48.4±0.8	46.0±1.2	43.7±0.5	**51.0±0.3**
Taylor-CE	56.7±0.4	56.5±0.2	55.2±0.5	54.4±0.3	**57.6±0.3**

These experimental results demonstrate that the *label wave* method can effectively select models with good generalization performance, outperforming traditional validation-based early stopping methods. By leveraging the prediction dynamics on the training data, our method eliminates the need for a separate validation set, allowing all available labeled data to be used for model training. Key advantages include:

- Elimination of validation set dependence: The method allows all available labeled data to be used for training without setting aside a validation set, particularly useful when training data is scarce.
- Adaptivity to noise levels: The method can adapt to different levels of label noise without requiring specific tuning of algorithm parameters.
- Ease of integration: The method can be easily integrated into existing model training workflows, with minimal additional computational steps per epoch.

However, the method also has certain limitations:

- Computational overhead: Calculating the PC_t metric requires additional storage and comparison of predictions for each training sample after each epoch, which may impose some memory overhead. However, in practice, only storing predictions for the previous two epochs limits additional memory requirements.
- Dependence on assumptions: The method is based on the assumption that models typically follow a pattern where they first learn simple and generalizable patterns. This is followed by a phase where they begin to learn confusion and incorrect patterns. In the final stage, models tend to memorize noisy labels. While this learning progression holds true in most scenarios in learning with noisy labels, we acknowledge that there may be exceptions in certain specific cases.

2.2.2 Memorization Effect of Deep Learning

Deep neural networks (DNNs) can learn almost any function, no matter how complicated, as long as they are large enough [79, 125]. This means they can even memorize all data, including examples with wrong labels. What's interesting is that, despite this ability to memorize, DNNs often perform very well on real-world data, which is a bit of a mystery. This has led to a lot of research into how memorization and generalization work together in deep learning.

Interestingly, DNNs do not learn correctly labeled and mislabeled data simultaneously. Instead, they tend to prioritize simpler patterns early in training before progressing to more complex ones [10, 461]. In noisy datasets, the majority of data is typically correctly labeled and exhibits consistent patterns, while mislabeled data is less common and more irregular. As a result, DNNs initially focus on learning the patterns of the correctly labeled data. As they encounter and fit increasingly complex patterns, they do begin to absorb the mislabeled examples.

Arpit et al. [10] show that deep neural networks (DNNs) can memorize random noise data, demonstrating their ability to fit even arbitrary patterns. To better understand this behavior, a series of experiments analyzing “easy” and “hard” examples are conducted. They define that easy examples are those that the model correctly classifies after just a few training epochs, while hard examples take much longer to learn. They found that clean datasets contain both easy and hard examples, whereas mislabeled data lack this property. This reinforces the idea that DNNs learn meaningful patterns from clean data rather than memorizing it. Additionally, they discovered that increasing the size of the dataset slows down the memorization process, especially for noisy data. Larger datasets provide more opportunities for models to learn generalizable patterns, reducing its tendency to rely on memorization.

Zhang et al.’s research [461] upends traditional views of generalization in deep learning, showing that DNNs can memorize random data while still excelling on real-world tasks. Generalization depends not just on model capacity or explicit regularization but also on implicit regularization from optimization algorithms and data properties. They swapped true labels in CIFAR10 and ImageNet with random ones and found DNNs achieved zero training error despite no meaningful input-output link. With noisy data, DNNs are still able to fit the data perfectly, learning patterns from clean examples while memorizing mislabeled ones. Yet, as noise ratio increased, generalization weakened, nearing random guessing. This dual capacity reveals an intricate balance between memorization and generalization in DNNs.

2.2.2.1 The Use of Memorization Effect

After the memorization effect found, many researchers take advantage it and develop a new kind of methods. One common strategy is to stop training early, before the network starts overfitting to mislabeled examples. After stopping, the method identifies examples that are likely to be correctly labeled from the noisy dataset and uses them as reliable data for further training.

A popular criterion for selecting these reliable examples is the small-loss trick. Han et al. [136] found that correctly labeled examples usually have smaller losses compared to mislabeled ones. This approach works even better when multiple networks are used together [192, 221]. Another effective method is to use the network’s own predictions [182, 355]. Here, predictions are compared with the original noisy labels, and examples where they match are considered confident. Although these methods lack theoretical guarantees, they have shown practical success in real-world applications, largely because they don't require assumptions about the noise distribution [15, 20, 245].

2.2.2.2 Data Selection

Deep learning models training with label noise commonly employ sample selection as a primary solution [448], where high-confidence clean samples are selected from noisy datasets for training [136, 182]. This approach typically leverages the memorization effect of deep neural networks [10], where networks tend to learn dominant patterns first before gradually adapting to and memorizing rare samples with noisy labels. To utilize this effect, a typical strategy begins training with a small but relatively clean subset of data and then gradually expands the training set to prevent overfitting to noisy labels. Existing sample selection methods primarily follow two approaches. The first directly filters high-confidence clean samples for classifier training, commonly known as the "small-loss trick" [136, 182, 414]. The second controls the training process to ensure the classifier primarily learns from high-confidence clean samples in the noisy dataset, typically referred to as the "early stopping" approach [18, 313]. Specifically, sample selection for learning with noisy labels typically follows these steps:

1. Initialize the model with a small but relatively clean subset of data from the noisy dataset. This can be achieved by selecting samples with the lowest loss values or by using a predefined clean subset.
2. Train the model on the current subset for a certain number of epochs or until a specified criterion is met, such as a threshold for the validation loss.
3. Use the trained model to evaluate the confidence or reliability of the remaining samples in the noisy dataset. Common evaluation metrics include the loss value, prediction consistency, or uncertainty estimation.
4. Expand the training subset by selecting additional high-confidence clean samples based on the evaluation metric. The selection can be based on a fixed threshold, a percentage of the total samples, or a dynamic criterion that considers the model's performance.
5. Repeat steps 2–4 until a stopping criterion is met, such as a maximum number of iterations, or a desired dataset size.

While typical strategy in sample selection methods has demonstrated impressive performance across various tasks, their focus on preventing overfitting to noise simultaneously limits the model's ability to learn from *clean hard examples* (CHEs) within noisy datasets.

CHEs are clean samples that lie close to decision boundaries and exhibit high loss values, with a significant portion representing samples from nondominant subclasses [106]. Distinguishing CHEs from truly mislabeled samples presents a significant challenge [15], as their loss distributions often substantially overlap [414]. To maintain the purity of selected high-confidence samples, existing methods often must exclude numerous potential CHEs [15]. However, these CHEs are crucial for achieving near-optimal generalization performance, particularly in learning underrepresented subgroups in natural datasets through the memorization effect [468].

Method To address the difficulty of effectively selecting CHEs in existing methods, we propose a new method called *late stopping* [446]. Different from traditional early stopping strategies, we utilize a longer training process and design a novel sample selection criterion, namely, *First k-epoch Learning* (FkL), to distinguish between mislabeled samples and clean samples.

Specifically, we consider a model f_i trained on a noisy training set D_i, which contains n samples $(\mathbf{x}_j, y_j)_{j=1}^n$, where y_j is the given label for sample $\mathbf{x}_j$ (which may not be consistent with the true label). After the t-th training epoch, the model's predicted label for sample $\mathbf{x}_j$ is $\hat{y}_j^t$. We define a binary indicator variable $\mathrm{acc}_j^t = \mathbf{1}(\hat{y}_j^t = y_j)$, indicating whether the model predicts sample $\mathbf{x}_j$ as the given label y_j in the t-th epoch. Based on this, the FkL value for sample $\mathbf{x}_j$ is defined as

$$\mathrm{FkL}_j = \underset{t \in \mathcal{T}}{\mathrm{argmin}} \left(\mathrm{acc}_j^t \wedge \cdots \wedge \mathrm{acc}_j^{(t-k+1)} = 1 \right).$$

Intuitively, if $\mathrm{FkL}_j = t$, it indicates that classifier f_i "learned" sample $\mathbf{x}_j$ only after the t-th epoch. If a sample has a small FkL value, it means it was learned early in training; conversely, it indicates it was learned late in training. Through analyzing the distribution of FkL values of training samples, we found that the vast majority of samples with larger FkL values (i.e., learned late in training) are mislabeled samples. Therefore, we can use the FkL criterion to screen out high-confidence mislabeled samples.

Based on the FkL criterion, we propose the *late stopping* algorithm framework. The entire algorithm employs a hard-to-easy positive feedback loop, improving classifier performance by continuously refining the training set. Unlike most sample selection methods, our method starts with a larger noisy training set and maximally retains CHEs in the training set while gradually removing high-confidence mislabeled samples through extended training and reduced selection criteria.

The *late stopping* algorithm takes as input a noisy training dataset D_1, along with several control parameters: an iteration ratio $m\%$ that controls the sample selection process, a noise ratio $n\%$ representing the estimated proportion of noisy labels, and two iteration limits—$T_{\max}$ for the maximum number of epochs in each training cycle and $I_{\max}$ for the maximum number of iterations. Through iterative training and sample selection, the algorithm outputs a refined training set and its corresponding trained classifier.

1. For each iteration I from 1 to $I_{\max}$, create a new training set D_i and initialize the sample counter $s_{F_i} = 0$ and empty sample collection $D_{F_i} = \emptyset$.
2. Train classifier f_i on the current training set D_i for up to $T_{\max}$ epochs, continuously updating the sample collection D_{F_i} and counter s_{F_i}.
3. During training, if the sample counter s_{F_i} exceeds $S_{F_{i-1}} \times (1 - m\%)$, stop the current iteration and use D_{F_i} as the training set D_{i+1} for the next iteration.

4. After each iteration, check if $m \times i > n$. If true, terminate the algorithm and return the current classifier f_i along with the sample collection D_{F_i}.

Let us further delve into the "from hard to easy" positive feedback loop employed in *late stopping* and how it overcomes the limitations of existing methods to maximize the learning of CHEs. According to statistical learning theory [272], higher-quality training data yields classifiers with better performance. Therefore, we can improve the classifier's generalization performance in learning with noisy labels by constructing a positive feedback loop. Most methods relying on confident example selection adopt such a loop, aiming to ensure that each new classifier obtained after each training step exhibits improved generalization performance compared to the previous step, thus more robustly guiding the subsequent sample selection to acquire higher-quality training data.

To obtain a classifier with strong generalization performance in the initial stage, these methods typically impose strict constraints on the memorization effect to exclude all potential mislabeled examples, even if this entails removing a considerable number of CHEs. By suppressing the memorization of mislabeled examples, they can achieve better generalization performance than directly training on the original noisy training set. Subsequently, based on this stronger classifier, they iteratively expand the training set size or enhance the classifier's robustness. This strategy is effective when the intra-class sample homogeneity is high, as the classifier can generalize well from simple examples to hard ones in such cases.

However, the class distributions in natural datasets are often imbalanced. In this scenario, the aforementioned methods based on typical positive feedback loops struggle to achieve near-optimal generalization performance. This is because CHEs, as a nondominant minority pattern, usually have large losses, and their effective learning primarily relies on the memorization effect, occurring in the late stage of training [15, 414]. Conventional positive feedback loops, in an effort to prevent overfitting to noisy labels, often strictly limit this memorization effect, causing the model to focus on learning typical classes and subclasses while continuously neglecting the learning of minority classes and subclasses (i.e., CHEs) [106].

To address this issue, we adopt a novel "from hard to easy" positive feedback loop in *late stopping*. On one hand, we prolong the training process in each iteration to fully learn CHEs by leveraging the memorization effect. On the other hand, we use each new classifier's robust learning ability in the current iteration (i.e., FkL) as the basis for sample selection, focusing only on removing high-confidence mislabeled examples without pursuing the classifier's generalization performance. As shown in Fig. 2.2, as the iterations proceed, the training set is continuously purified, the noise ratio gradually decreases, and the number of clean examples containing CHEs remains stable. Benefiting from this, each new classifier obtained from the training in each iteration can memorize more CHEs while more accurately distinguishing between clean examples and mislabeled examples, further enhancing the quality of the subsequent training set.

Discussion We assume that the intermediate stage of training may be the key to the noise robustness of deep learning. Although the generalization performance may be

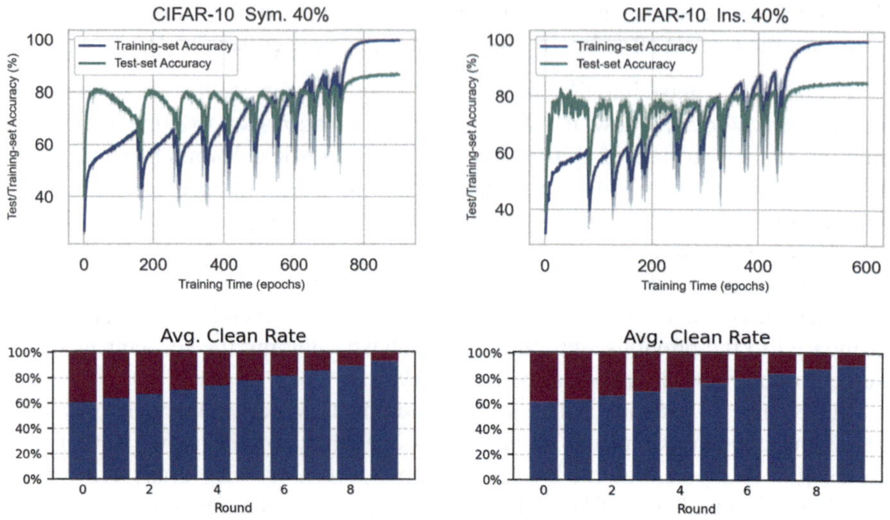

Fig. 2.2 Illustration of the model training dynamics in the late stopping framework, showing the test accuracy and training accuracy of the classifier, as well as the clean ratio of the training set in each iteration. The classifiers are re-initialized at the beginning of each new iteration, resulting in ten peaks in the accuracy curves, corresponding to the ten iterations set in this example

overfit to noisy labels, deep neural networks might exhibit a certain spontaneous noise-robust learning ability throughout the entire training process. Therefore, we attempt to separate this robust learning ability from the generalization performance by prolonging the training until the generalization performance significantly declines (i.e., *late stopping*) and extracting the results of robust learning from it (i.e., FkL).

Experiments We validated the effectiveness of *late stopping* on both CIFAR-10 and CIFAR-100 datasets under both symmetric and instance-dependent label noise. Table 2.4 presents a comparison of the sample selection performance between the FkL criterion and the loss criterion within the same training process. It can be observed that under the *late stopping* framework, FkL significantly outperforms the loss criterion in both tasks of selecting clean examples and mislabeled examples, especially in the range of $40k$–$45k$. In terms of classification performance, as shown in Tables 2.5 and 2.6, our method achieves superior results compared to the baseline (early stopping) methods under various noise types and levels. This superiority is particularly evident on CIFAR-100, which has a larger number of classes.

Key advantages include:

- Maximizing the retention of clean examples: By prolonging the training process and using a looser sample selection criterion, the method ensures that the majority of clean examples, especially the hard ones, are retained in the training set throughout the learning process.

Table 2.4 Label precision of the selected examples (45 k in total, with 27 k clean examples)

		Selecting clean examples			Selecting mislabeled examples		
Dataset	Criterion	0–10 k	0–15 k	0–25 k	30–45 k	35–45 k	40–45 k
CIFAR-10	*loss*	92.7	89.6	82.9	75.7	83.5	91.5
(Sym. 40%)	FkL(ours)	**95.5**	**95.6**	**93.6**	**96.4**	**99.8**	**99.9**
CIFAR-10	*loss*	75.6	72.3	71.4	60.3	68.7	77.3
(Ins. 40%)	FkL(ours)	**81.9**	**81.7**	**81.5**	**78.5**	**88.9**	**99.3**
CIFAR-100	*loss*	88.4	84.7	78.5	68.0	75.7	84.7
(Sym. 40%)	FkL(ours)	**97.5**	**96.8**	**95.1**	**94.6**	**97.9**	**98.9**
CIFAR-100	*loss*	71.3	68.2	65.3	45.9	44.2	42.6
(Ins. 40%)	FkL(Ours)	**89.3**	**88.6**	**84.4**	**78.2**	**86.3**	**91.8**

Table 2.5 Test performance (mean±std) of each approach using ResNet-18 on CIFAR-10

Method	Sym. 20%	Sym. 40%	Ins. 20%	Ins. 40%
Early stopping	85.0±0.4	79.6±1.3	85.9±1.1	79.9±1.4
Late stopping (ours)	**91.1±0.2**	**88.9±0.4**	**91.1±0.2**	**87.4±0.4**

Table 2.6 Test performance (mean±std) of each approach using ResNet-34 on CIFAR-100

Method	Sym. 20%	Sym. 40%	Ins. 20%	Ins. 40%
Early Stopping	57.6±2.6	45.7±2.6	59.9±1.6	43.7±1.5
Late Stopping (Ours)	**68.7±0.7**	**64.1±0.4**	**68.6±0.7**	**59.3±0.5**

- Adaptivity to different noise conditions: The method can adapt to various noise types and levels without requiring specific tuning of algorithm parameters.

However, *late stopping* also has certain limitations:

- Potential misselection: A small number of clean examples might be mistakenly removed, while some mislabeled examples that are close to the dominant classes may be incorrectly retained.
- Imperfect removing: To limit the computational resources consumed, the method employs a relatively large removal rate in each iteration, which may result in an excessive removal of CHEs from certain rare subclasses.
- Longer training overhead: The method involves training the classifier on a larger dataset for an extended period for multiple iterations, which may not be advantageous in terms of efficiency.

2.3 Overparameterized Models for Noisy Labels

Modern deep learning models are overparameterized, meaning they have far more parameters than training examples. This high capacity allows them to fit complex patterns but also to memorize noisy labels. In fact, if trained naively on corrupted data, an overparameterized network can eventually fit the noise perfectly, leading to poor generalization [152]. For example, extending training on a dataset with random label flips causes the model to overfit the incorrect labels and degrade in performance [152]. To address this, researchers have developed strategies that exploit the power of overparameterized models while mitigating their tendency to learn noise. Two prominent families of methods are (1) the ***pretraining paradigm***, which leverages prior training on additional data or tasks to immunize the model against noise, and (2) the ***deep k-NN method***, which uses deep networks in conjunction with k-nearest neighbors to identify and handle mislabeled samples. We discuss each in turn below.

2.3.1 Pretraining Paradigm

With the emergence of the pretraining paradigm, increasing research has started to utilize the pretrained model to mitigate the presence of potentially incorrect supervision in various downstream tasks. Conceptually, this can be viewed as utilizing the marginal distribution $P(\mathbf{X})$ to remedy the misspecified conditional distribution $P(\tilde{Y}|\mathbf{X})$.

2.3.1.1 Overview and Mechanism

The pretraining paradigm involves first training a model on a large auxiliary dataset or task (often with clean labels or in a self-supervised manner) and then fine-tuning it on the target dataset with noisy labels. In practice, this often means starting with a network pretrained on a high-quality dataset (e.g., ImageNet for images or unsupervised language modeling for text) and then adapting it to the noisy data. The intuition is that pretrained models have learned generalizable features that make them less reliant on the possibly corrupted labels in the target data. When fine-tuned, such a model can latch onto real patterns quickly and is less prone to fitting noise compared to training from scratch. Empirically, this approach has been shown to markedly improve robustness: Hendrycks et al. [152, 153] demonstrated that pretraining improves a model's resilience to label noise on large-scale datasets [152]. In their study, even though pretraining did not always boost standard accuracy, it significantly reduced overfitting to wrong labels and improved metrics related to robustness and uncertainty [152]. In essence, the pretrained model begins in a favorable parameter regime that acts as a strong prior, so it requires fewer drastic weight updates based on noisy data, thereby limiting the impact of mislabeled examples.

2.3.1.2 Theoretical Justification

From a theoretical perspective, one can view pretraining as a form of regularization or informed initialization [432]. Overparameterized neural networks can represent many functions, including ones that fit random noise [152]. By constraining the model's initial weights to encode useful representations (learned during pretraining), we effectively restrict the hypothesis space to functions that already perform well on a related task. This restriction makes it harder for the fine-tuning process to diverge toward pathological solutions that fit noise. While a full theoretical account of why pretraining helps with noisy labels is complex (due to the nonlinear and high-dimensional nature of deep networks), empirical evidence strongly supports its benefit.

2.3.1.3 Effectiveness in Different Domains

The pretraining paradigm has proven effective across computer vision (CV), natural language processing (NLP), and other domains. In CV, it is now common to fine-tune ImageNet-pretrained models on noisy-label datasets. For instance, the Clothing1M dataset (with $\sim$1 million clothing images labeled through online shopping tags, which include many errors) is a standard benchmark for noisy labels. Almost all high-performing methods on Clothing1M initialize the model with weights pretrained on ImageNet [482]. The rich diversity and correctness of ImageNet labels provide a strong starting point, so the network only needs to adjust slightly to the clothing data, largely resisting the urge to overfit the noisy labels [482]. As a concrete comparison, a ResNet trained from scratch on Clothing1M performs markedly worse than the same ResNet fine-tuned from an ImageNet initialization [482]. This trend extends to other real-world noisy data: for example, the WebVision dataset (a noisy web-crawled version of ImageNet) and noisy versions of CIFAR have all been shown to benefit from models pretrained on cleaner data [482]. When an external clean dataset is not available, researchers have turned to self-supervised pretraining on the noisy data itself. By learning representations through tasks like contrastive learning (which do not require labels), the model can still obtain a good feature extractor prior to looking at the noisy labels [482]. Li et al. (2021) follow this approach and report that self-supervised pretraining on the raw images, followed by a robust fine-tuning procedure, can match the performance of using a supervised ImageNet pretrained model on Clothing1M [482].

NLP sees a parallel story. Large language models (e.g., BERT, GPT) are pretrained on massive text corpora, learning rich linguistic features. Fine-tuning such models for classification tasks with noisy labels (e.g., classifying sentiment from tweets, where user-provided hashtags or emojis serve as imperfect labels) tends to be far more robust than training a smaller network from scratch on the same data. In fact, even BERT can eventually overfit noise if overtrained, but its pretrained knowledge provides a head start in learning the signal before the noise [301]. Recent research has proposed robust fine-tuning frameworks that take a pretrained BERT and adapt it to noisy labels with additional regularization, achieving superior results to naive fine-tuning [301]. This indicates that the benefits of pretraining for noisy labels are not limited to vision; they are ubiquitous across domains.

2.3.2 Deep k-NN Method

The deep k-nearest neighbors (kNN) method is a technique that combines deep neural networks with a classic instance-based learning approach to tackle noisy labels. Rather than relying solely on the end-to-end learning of the network, this method introduces an explicit data cleaning step using kNN in the model's learned feature space [14]. The approach can be summarized as follows [14]:

1. ***Preliminary Training:*** Train an initial deep model M on the noisy dataset. This model is overparameterized and will learn an intermediate feature representation of the data (e.g., the activations from the penultimate layer or the logit layer).
2. ***Neighbor Analysis:*** For each training sample, compute its feature representation using M. Then, find its k nearest neighbors in the feature space (usually measured by Euclidean distance or cosine similarity). Examine the labels of these neighbors.
3. ***Noisy Label Detection:*** Identify samples that are label inconsistent with their neighborhood. In practice, if a sample's given label disagrees with the majority of its k nearest neighbors' labels, it is flagged as likely mislabeled. The assumption is that correctly labeled examples tend to cluster together in representation space, whereas incorrectly labeled ones will be surrounded by neighbors from a different class.
4. ***Filtering/Correction:*** Remove the flagged noisy examples from the training set (or optionally correct their labels if there is reason to believe the neighbors' majority label is the true label). This yields a filtered dataset.
5. ***Retraining:*** Train a new deep model (or fine-tune M) on the filtered dataset, which should now have significantly reduced label noise. The final model can be the same architecture A as M and will presumably generalize better now that the training set is cleaner [14].

By incorporating this data-driven filtering step, the deep kNN method explicitly cleans the training data before the final learning phase. Importantly, even though the initial model M was trained on noisy data, its learned features are often good enough to distinguish most clean vs. corrupted labels. Bahri et al. (2020) introduced this method and found that the preliminary network's representation is surprisingly reliable: despite being trained with some wrong labels, it clusters data by their true underlying classes to a large extent, enabling kNN to detect out-of-place examples [14]. This aligns with the memorization dynamics of deep networks; they tend to fit easy patterns first (grouping similar inputs) and only later memorize noise. The kNN filtering intercepts the process by using the network's intermediate features after an initial training pass, catching many of the mislabeled samples that haven't been fully assimilated into the clusters of correct samples [14].

Empirically, deep kNN approach has been shown to be highly effective [14]. Moreover, a notable practical benefit is that this method does not require a hold-out set of clean data; it can work in a fully self-contained manner using only the noisy dataset [14]. This is in contrast to some other approaches that rely on a small trusted subset of data to guide the training or validation (e.g., using clean data to estimate noise parameters or early stopping), which may not always be available. Deep kNN is also relatively insensitive to the exact choice of k (the number of neighbors): it was found to be robust across a range of k values [14], which is desirable because tuning hyperparameters in noisy-label settings can be tricky without clean validation data.

2.4 Learning with Noisy Labels to Weakly Supervised Learning

Learning with noisy label is closely related to several distinct topics in weakly-supervised learning, for instance, one paramount challenge in semi-supervised learning arises from noisy pseudo-labels generated during training. Similarly, in contrastive self-supervised learning, overly aggressive data augmentation can result in incorrectly specified positive samples, hence creating noisy supervision as well. In this section, we explore three representative cases to illustrate the interplay between noisy labels and other domains: (1) semi-supervised learning [227], (2) contrastive self-supervised learning [20], and (3) out-of-distribution generalization [165].

2.4.1 Semi-Supervised Learning with Instance-Dependent Thresholds

Semi-supervised learning (SSL), closely related to learning with label noise, addresses the problem of training models using a combination of labeled and unlabeled data points. Denoting the size of labeled data as n and unlabeled data as m, typically we have $m \gg n$. Consequently, to develop an effective model from this type of data mixture, SSL leverages the unlabeled data, predominantly employing a technique called *self-training*. Self-training initially trains a weak model using only labeled data, subsequently assigning pseudo-labels to the unlabeled data based on the predictions of this model. If the model's prediction confidence for an unlabeled instance exceeds a predefined threshold, that instance is included in the training set. By iteratively repeating this process, the training dataset gradually expands, progressively enhancing the capabilities of the SSL model.

However, a fundamental issue arising during self-training is that pseudo-labels may not always be accurate, and incorrect pseudo-labels can lead the model to overfit these erroneous signals. Thus, devising effective pseudo-label thresholds becomes critical for successful SSL.

Nevertheless, many existing methods overlook this challenge, typically treating the threshold as a fixed hyperparameter that remains uniform for all training data points and constant throughout the training process [32, 339].

Subsequent research has attempted to address this limitation by introducing dynamic thresholds [426], which adjust the pseudo-label threshold during training based on predefined rules, typically following a monotonic pattern. Building on dynamic threshold methods, AdaMatch proposed (self-)adaptive thresholding [33], which further incorporates the training dynamics by adapting the threshold based on the current average prediction confidence of the model. However, these approaches, while considering training dynamics, still apply a uniform threshold across all unlabeled data points, thus neglecting more granular information. Addressing this limitation, methods such as FlexMatch [467] propose class-dependent pseudo-labeling strategies, improving upon dynamic thresholds by setting thresholds individually for each class based on class-specific curriculum criteria.

In the context of semi-supervised learning (SSL), we examine the scenario where there is a limited set of labeled examples $\mathbf{X}_l := \{\mathbf{x}_1, ..., \mathbf{x}_n\}$ and their associated labels $\mathbf{y}_l := \{y_1, ..., y_n\}$. Additionally, we have a larger group of unlabeled examples $\mathbf{X}_u := \{\mathbf{x}_{n+1}, ..., \mathbf{x}_{n+m}\}$, whose labels $\mathbf{y}_u := \{y_{n+1}, ..., y_{n+m}\}$ remain hidden, typically with m significantly greater than n. We also assume that both labeled and unlabeled examples are drawn from the same distribution, i.e., $\mathbf{X} := \{\mathbf{X}_l, \mathbf{X}_u\} \in \mathcal{X}$, $\mathbf{y} := \{\mathbf{y}_l, \mathbf{y}_u\} \in \mathcal{Y}$, following the independent and identically distributed (i.i.d) principle.

The model, denoted as f and parameterized by θ, utilizes both clean and pseudo-labels for training. Due to the inherent inaccuracies in f_θ, the pseudo-labels $\hat{Y}$, it is likely to include errors. Consequently, the softmax outputs from f_θ essentially estimate the class posterior of the noisy pseudo-labels. We define the true noisy class posterior distribution at the t-th iteration as $P(\hat{Y}|X)$, while our estimated noisy class posterior from f_θ is denoted as $P_\theta(\hat{Y}|X)$. Since the actual clean labels for the unlabeled set remain unknown, our goal is to determine the *Bayes optimal label* for these instances, which is the label that assigned by the Bayes optimal classifier.

As previously discussed, label noise presents a significant hurdle in semi-supervised learning (SSL). Additionally, the noise class posterior $P(\hat{Y}|X)$ varies with $\mathbf{x}$, indicating that label noise is dependent on individual instances in SSL. Specifically, the probability $P(\hat{Y}|X = \mathbf{x})$ is described by the equation $P(\hat{Y}|X) = T(\mathbf{x})^\top P(Y|X = \mathbf{x})$, where $T(\mathbf{x})$ denotes the **instance-dependent transition matrix**. This matrix accounts for the generation process of label noise. Here, $P(Y|X = \mathbf{x})$ signifies the underlying clean class posterior for unlabeled instances. The i, j-th element of $T(\mathbf{x})$ is defined as

$$T_{i,j}(\mathbf{x}) = P(\hat{Y} = j|Y = i, X = \mathbf{x}). \tag{2.6}$$

This definition implies that the ij-th entry of $T(\mathbf{x})$ estimates the likelihood that an instance $\mathbf{x}$, originally labeled with the clean label $y = i$, is misclassified as $\hat{y} = j$.

Method Estimating instance-dependent transition matrix $T(\mathbf{x})$ has been known as a challenging problem in learning with label noise. Fortunately, in SSL, additional information are available to facilitate the estimation process of $T(\mathbf{x})$. Specifically, as we know, the quantity of the unlabeled data is usually dominant over the amount of the labeled data; this means, as the training progresses, the SSL classifier usually tends to overfit to the noisy pseudo-labels generated during the SSL training. Here, we are implicitly operating under two assumptions:

Assumption 1 (Stabilization of Pseudo-Label Distribution) *In semi-supervised learning, as training iteration $t \to \infty$, the posterior distribution of pseudo-labels for unlabeled data converges:*

$$\lim_{t\to\infty} P_{\theta_t}(\hat{Y}|X_u) \approx P(\hat{Y}|X),$$

where $P_{\theta_t}(\hat{Y}|X_u)$ denotes the model's predicted pseudo-label distribution at iteration t and $P(\hat{Y}|X)$ is the underlying true noisy class posterior.

This assumption is mild, since this has already been empirically validated by many existing works, also, one can prove this to be true by using the fact that the probability distribution $P_{\theta_t}(\hat{Y}|X_u)$ is parameterized by θ, and as θ converges, $P_{\theta_t}(\hat{Y}|X_u)$ will also converges correspondingly.

Assumption 2 (Pseudo-Label Dominance in SSL) *In semi-supervised learning, as $m \to \infty$, the posterior probability distribution produced by the classifier converges to the pseudo-label posterior distribution:*

$$P_\theta(Y|X) \longrightarrow P(\hat{Y}|X) \quad as \quad m \to \infty.$$

This assumption states as we have asymptotically more unlabeled datapoints, the influence of the predicted pseudo-label posterior will eventually override the clean class posterior. Under these two assumptions, we can now summarize the noise class posterior in SSL is essentially tractable and will be reflected on the labeled data. This means, we can collect n observations of $P(\hat{Y}|X = \mathbf{x}_l, Y = y_l)$, which is equivalent to the label noise transition probability in the context of SSL. This allows us to be able to directly model the $T(\mathbf{x})$ in SSL, and, if the $(\mathbf{X}_l, Y_l)$ is representative enough to $P(X, Y)$, the estimation should be of a small error, but in reality, such estimation could require the supplement of confident sample in order to be more reliable.

Two primary benefits for knowing the transition matrix are evidently clear under SSL, the first of which is approximating the clean class posterior by inverting the noise generation process described in Eq. (2.6), that is, if the transition matrix is invertible, then the clean class posterior can be directly approximated by $T(\mathbf{x})^{-1}P(\hat{Y}|X = \mathbf{x})$. If $T(\mathbf{x})$ is not invertible, one can apply forward correction [289] to obtain an equivalent minimizer of the noisy loss as the one defined over the clean loss; specifically, the forward correction can be defined as

$$\ell_{\psi}^{\rightarrow}(f_{\theta}(\mathbf{x}_u), \hat{y}_u) = \ell(\hat{T}(\mathbf{x}_u)^{\top}\psi^{-1}f_{\theta}(\mathbf{x}_u), \hat{y}_u), \tag{2.7}$$

where ψ is an invertible link function such as sigmoid and $\hat{T}$ is our estimated transition matrix. The intuition behind forward correction has been discussed in the former sections, so we will not elaborate here.

Yet, another purpose of $T(\mathbf{x})$ is estimating the instance-dependent thresholds for unlabeled datapoints. Knowing that, if the $T(\mathbf{x})$ is known, the thresholds can be stated as

$$\tau_t(x_u) = \hat{T}_{k,k}(\mathbf{x}_u)P_{\theta_t}(Y = s|X = \mathbf{x}_u) + \sum_{i \neq k}^{|Y|} \hat{T}_{i,k}(\mathbf{x})P_{\theta_t}(Y = i|X = \mathbf{x}_u). \tag{2.8}$$

Intuitively, Eq. (2.8) measures the the probability that the predicted pseudo-label came from a class other than itself through the noise transition process, which can be understood as the probability of unreliability. If the predicted probability outweighs the probability of unreliability, this suggests that we have more reason to accept this pseudo-label prediction as correct; otherwise, we should reject this pseudo-label at this iteration and not add that to the training data, hence fulfilling the purpose of instance-dependent thresholds. This threshold function is based on the derivation of Zheng et al. [484], where they proved that under the context of learning with label noise, classifiers trained with such noise-filtering threshold function satisfy bounded error rates, thus theoretically motivating the instance-dependent threshold function we adapt here.

Experiment InstanT is evaluated on three widely used image classification datasets, namely, CIFAR-10, CIFAR-100, and STL-10—each featuring different numbers of labeled and unlabeled examples. To study varying degrees of data scarcity, we follow recently introduced more stringent settings [394] by assigning 1, 4, 25 labeled samples per class on CIFAR-10, 2, 4, 25 on CIFAR-100 and 1, 4, 10 on STL-10. All experiments, including comparisons with other methods, are conducted within the USB (Unified SSL Benchmark) framework to promote fair evaluation and reproducibility. We employ the pretrained ViT as backbone model for evaluation, adopt AdamW with a learning rate of 5×10^{-4} for CIFAR-10/100 and 1×10^{-4} for STL-10, and train for 204,800 iterations with a batch size of 8 across all datasets.

Results are summarized into Table 2.7, showing that InstanT generally secures leading performance across most configurations. Notably, in setups with extremely few labeled samples, InstanT offers the most pronounced gains, including an average accuracy improvement of 2.17% on CIFAR-10 (10) relative to the previous best. Although accurately estimating the transition matrix may be challenging under severe label scarcity, one practical workaround is to integrate highly confident predictions and their pseudo-labels into the labeled pool to refine its estimation.

Table 2.7 Top-1 accuracy with pretrained ViT. The best performance is bold and the second best performance is underlined. All results are averaged with three random seeds {0,1,2} and reported with a 95% confidence interval

Dataset	CIFAR-10			CIFAR-100			STL-10		
# Label	10	40	250	200	400	2500	10	40	100
PL	37.65±3.1	88.21±5.3	95.42±0.4	63.34±2.0	73.13±0.9	84.28±0.1	30.74±6.7	57.16 ±4.2	73.44±1.5
MT	64.57±4.9	87.15±2.5	95.25±0.5	59.50±0.8	69.42±0.9	82.91±0.4	42.72±7.8	66.80±3.4	77.71±1.8
MixMatch	65.04±2.6	97.16±0.9	97.95±0.1	60.36±10.3	72.26±0.1	83.84±0.2	10.68±1.1	27.58 ±16.2	61.85±11.3
VAT	60.07±6.3	93.33±6.6	97.67±0.2	65.89±1.8	75.33±0.4	83.42±0.4	20.57±4.4	65.18±7.0	80.94±1.0
UDA	78.76±3.6	97.92±0.2	97.96±0.1	65.49±1.6	75.85±0.6	83.81±0.2	48.37±4.3	79.67 ±4.9	89.46±1.0
FixMatch	66.50±15.1	97.44±0.9	97.95±0.1	65.29±1.4	75.52±0.1	83.98±0.1	40.13±3.4	77.72 ±4.4	88.41±1.6
FlexMatch	70.54±9.6	97.78±0.3	97.88±0.2	63.76±0.9	74.01±0.5	83.72±0.2	60.63±12.9	78.17 ±3.7	89.54±1.3
Dash	74.35±4.5	96.63±2.0	97.90±0.3	63.33±0.4	74.54±0.2	84.01±0.2	41.06±4.4	78.03±3.9	**89.56±2.0**
AdaMatch	85.15±20.4	**97.94±0.1**	97.92±0.1	73.61±0.1	78.59±0.4	**84.49±0.1**	68.17±7.7	83.50±4.2	89.25±1.5
InstanT	**87.32±10.2**	97.93±0.1	**98.08±0.1**	**74.17±0.3**	**78.80±0.4**	84.28±0.5	**69.39±7.4**	**85.09 ±2.8**	89.35 ±1.9

However, under the scenarios with more labeled data, InstanT's dominance becomes less apparent. This could attribute to the increase of the classifiers' performance benefited by more labeled data, hence less noisy labels in the predicted pseudo-labels, and reduces the need for noise-resistant thresholds. Nevertheless, when compared to its closest peer AdaMatch, InstanT still achieves higher accuracy in nearly every case, underscoring the meaningful performance gain InstanT offers.

2.4.2 *Robustness for Strong Data Augmentation*

Recently, self-supervised learning methods have been a significant milestone. Representative contrastive approaches like SimCLR and MoCo have demonstrated strong results by using the power of data augmentation to enhance model performance [59, 146, 403]. However, these methods also introduce a critical challenge: the use of aggressive augmentations, which can alter the semantics of images in unpredictable ways. Aggressive augmentations refer to operations that have the potential to change the semantic meaning of an image, such as applying grayscale, color jittering, or Gaussian blurring. Other used augmentations such as random cropping and flipping are referred as weak augmentations.

Aggressive augmentations can lead to improved model performance in certain cases, but they also introduce a significant semantic shift problem for training images. This is evident when comparing original and augmented views side by side. As demonstrated in Fig. 2.3, the first row shows pristine images from ImageNet and CIFAR-100 datasets. Then, the second row shows corresponding augmented views, showcasing the widely adopted composition of augmentations. A closer examination reveals that the augmented views struggle to maintain semantic consistency with their original counterparts. This raises concerns about pushing these images toward similar representations during training, as it can lead to decreased model performance. Therefore, these semantically inconsistent pairs can be regarded as noisy pairs in the training set.

To mitigate the semantic shift problem caused by aggressive augmentations, RSA leverages the memorization effect of deep neural networks [16, 18]. By creating pairs of aggressively augmented images and relatively clean images, it's possible to exploit the fact that weak augmentations induce fewer semantic shifts. At the

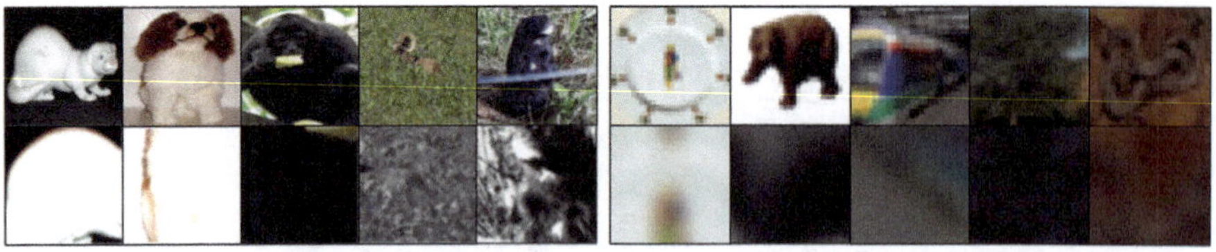

Fig. 2.3 In the first row, the original images are shown side by side with noisy samples resulting from aggressive data augmentation in the second row

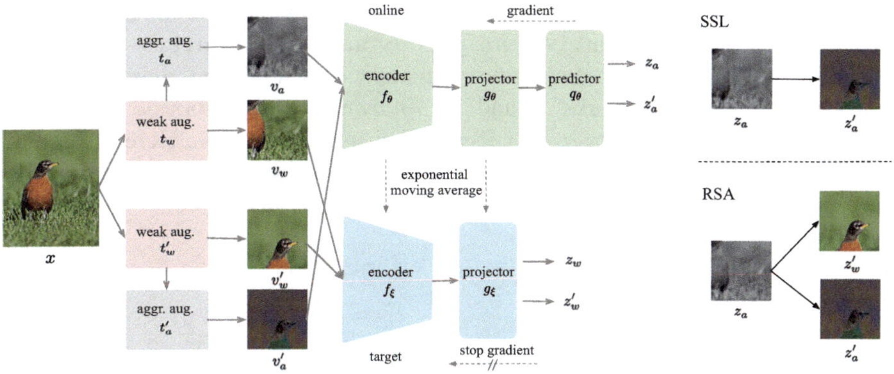

Fig. 2.4 RSA utilizes an asymmetric-style framework combining online and target networks. Weak augmentations are first applied to generate v_w and v'_w, followed by aggressive augmentations to produce v_a and v'_a. On the right, RSA is compared with classical SSL methods. Unlike traditional approaches, RSA enforces a balance between weak- and aggressive-augmented views

beginning, we set larger weights to aggressively augmented data pairs. Over time, the weights of pairs of strong and weak augmentations increase, and the weights of aggressively augmented data pairs decrease. This adaptive approach, as shown in Fig. 2.4, helps reduce the harm caused by semantic shifts and improves model performance.

Method To address noisy pairs in self-supervised learning, a novel method named **RSA** is proposed to mitigate semantic shift caused by aggressive augmentations. The approach leverages the *memorization effect*, where DNNs first fit clean pairs of training data during the early learning phase and later overfit to noisy pairs. Since noisy pairs constitute a minority of the dataset, their initial impact is minimal but grows as training progresses. To address this, RSA assigns different weights to clean and noisy pairs, gradually decreasing the weights of noisy pairs as their impact rises.

To reduce the increased computational costs, RSA addresses two key issues:

1. **Augmentation overhead:** Generating augmented images demands high CPU, disk, and storage resources.
2. **Increased processing demand:** Adding more instances leads to extra backpropagation steps, raising GPU processing time.

RSA introduces a **multistage augmentation strategy** for the first challenge. This approach generates double instances with minimal extra computational cost by mixing various transformation techniques and arranging them in a queue. For the second challenge, RSA employs an innovative design where aggressive and weakly augmented views are processed by separate networks, ensuring each instance undergoes a single forward pass. Despite this optimization, RSA achieves backpropagation twice, maintaining parity with many state-of-the-art SSL methods that employ a symmetrized loss function [53, 59, 130, 292].

Multistage augmentation strategy processes images using both aggressive t_a and weak t_w augmentations. An image x is transformed through two distinct augmentation instances, producing four augmented views. Then, these images are processed by neural networks f_θ and f_ξ, followed by projectors g_θ and g_ξ, and finally through the predictor q_θ to generate their representations. Here, f_θ denotes the parameters of the online network, while f_ξ corresponds to the parameters of the target network. This process is systematically described as follows::

$$\begin{aligned} v_w &= t_w(\mathbf{x}) & v'_w &= t'_w(\mathbf{x}), \\ v_a &= t_a(v_w) & v'_a &= t'_a(v'_w) \\ z_a &= q_\theta(g_\theta(f_\theta(v_a))) & z_w &= g_\xi(f_\xi(v_w)), \\ z'_a &= q_\theta(g_\theta(f_\theta(v'_a))) & z'_w &= g_\xi(f_\xi(v'_w)). \end{aligned} \tag{2.9}$$

A key step in this method is to form four pairs of representations. To maintain consistency between corresponding views in the embedding space, aggressive-augmented views are preserved alongside their weak and aggressive counterparts. For aggressive-augmented views, instead of encoding them on the target network, online network's representations with a stop-gradient operation are used. This approach enables balancing the loss terms for weak- and aggressive-augmented views. The total loss function can be written as

$$\begin{aligned} \mathcal{L} = &(1-\beta)\ell_{\text{MSE}}(z_a, z'_w) + \beta\ell_{\text{MSE}}(z_a, z'_a) \\ &+(1-\beta)\ell_{\text{MSE}}(z'_a, z_w) + \beta\ell_{\text{MSE}}(z'_a, z_a), \end{aligned} \tag{2.10}$$

where β is a learnable parameter that controls the trade-off between weak- and aggressive-augmented views. By using this loss function, the method encourages networks to learn representations that achieve a balance between weak- and aggressive-augmented views, reducing overfitting to noisy pairs. As training progresses, noisy impacts increase, requiring an adjustment of β. To address this, the memorization effect is used, which decreases β with a cosine decay equation:

$$\beta = \beta_{\text{base}} \times \frac{1}{2}(\cos(\pi\frac{k}{K}) + 1), \tag{2.11}$$

where β_{base} is fixed at the start of training and k and K are the current and total training steps, respectively.

Experiments Table 2.8 shows a comparison of RSA with baselines on ImageNet-1K. We can find that RSA demonstrates superior performance over other methods after 100 epochs, surpassing BYOL's results even at a longer training time of 200 epochs. RSA with 200 epochs performs well, showing a 2.5% improvement over BYOL. These linear evaluation results support the idea that RSA consistently improves generalization across different scenarios.

Table 2.8 Linear classification on ImageNet-1K: All methods use a standard ResNet-50 backbone without multi-crop strategy

Method	Supervised	SimCLR	MoCo v2	ReSSL	W-MSE 4	SimSiam	SwAV	BYOL	RSA	RSA
Batch size	256	4096	256	256	4096	256	4096	2048	2048	2048
Epochs	100	1000	200	200	400	200	200	300	100	200
Accuracy	76.2	69.3	67.5	69.6	72.6	70.0	69.1	72.4	71.4	**73.1**

Table 2.9 **Transfer learning:** Object detection and instance segmentation. All models were pretrained on ImageNet-1K for 200 epochs and then fine-tuned on the MS COCO dataset [231] using a 1x schedule

	Object detection			Instance segmentation		
Method	AP^{bb}	AP^{bb}_{50}	AP^{bb}_{75}	AP^{mk}	AP^{mk}_{50}	AP^{mk}_{75}
random init.	32.8	50.9	35.3	29.9	47.9	32.0
supervised	39.7	59.5	43.3	35.9	56.6	38.6
InstDis [403]	38.8	58.4	42.5	35.2	55.8	37.8
PIRL [270]	38.6	58.2	42.1	35.1	55.5	37.7
MoCo [146]	39.4	59.1	42.9	35.6	56.2	38.0
MoCo V2 [60]	40.1	59.8	44.1	36.3	56.9	39.1
InfoMin Aug. [366]	40.6	60.6	44.6	36.7	57.7	39.4
RSA (Ours)	**41.1**	**61.4**	**45.1**	**37.3**	**58.6**	**40.1**

Table 2.10 **Transfer learning:** Object detection and instance segmentation. All models were pretrained on ImageNet-1K for 200 epochs and then fine-tuned on the MS COCO dataset [231] using a 2x schedule

	Object detection			Instance segmentation		
Method	AP^{bb}	AP^{bb}_{50}	AP^{bb}_{75}	AP^{mk}	AP^{mk}_{50}	AP^{mk}_{75}
Random	38.4	57.5	42.0	34.7	54.8	37.2
Supervised	41.6	61.7	45.3	37.6	58.7	40.4
InstDis [403]	41.3	61.0	45.3	37.3	58.3	39.9
PIRL [270]	41.2	61.2	45.2	37.4	58.5	40.3
MoCo [146]	41.7	61.4	45.7	37.5	58.6	40.5
MoCo v2 [60]	41.7	61.6	45.6	37.6	58.7	40.5
InfoMin Aug. [366]	42.5	62.7	46.8	38.4	59.7	**41.4**
RSA (Ours)	**42.7**	**62.9**	**47.1**	**38.5**	**60.0**	**41.4**

To further test the quality of RSA learned representations, we evaluate more downstream tasks. Object detection tasks use Mask R-CNN [145]; instance segmentation tasks employ FPN [232]. The results on object detection and instance segmentation in Tables 2.9 and 2.10 demonstrate RSA's superiority over all baselines. RSA outperforms the strong baseline InfoMin Aug [366], which utilizes strong augmentation techniques like RandAugment [78]. These findings confirm that RSA not only achieves state-of-the-art performance on downstream tasks but also improves the overall quality of the learned representations.

Above discussed learning under instance-dependent label noise under trustworthy scenarios, below a realistic noisy distribution, is considered. To solve this problem, a robust training framework is proposed to enhance the robustness of machine learning models.

2.4.3 Robust Generalization

Learning under noise and corruption has been a vital challenge in the practical applications of learning models, as they are much more fragile to subtle noises than human perception systems [124, 150]. During training, the encountered noise corruptions are essentially perceived as distribution shifts, significantly hindering the prediction results [229, 249, 374, 395]. Therefore, to mitigate the performance degradation, enhancing robust generalization to noisy data distributions has drawn lots of attention [8, 319].

In real-world scenarios, noise corruption is often associated with photon-limited imaging problems [253, 368], which occur due to the arrival of numerous photons at an image sensor. The varying number of noise photons results in different levels of corruption severity, leading to multiple data distributions and challenging the performance of learning models [149]. Particularly, photon-limited corruption $\mathcal{E}$ is formed by cumulative noise photons u, which are caused by discrete elements with specific probabilities during a period. As a result, the corruption severity s follows a Poisson distribution, *i.e.*, $s \sim P(s;\lambda) = \frac{e^{-\lambda}\lambda^s}{s!}$, as illustrated in Fig. 2.5. Therefore, realistic data is not completely clean; instead, they contain part of corrupted samples, and the proportion goes less as severity increases.

To solve the challenge, SharpDRO is proposed to learn a flat loss landscape for dealing with worst-case data, hence, mitigating the generalization drawback of distributionally robust optimization (DRO). Specifically, sharpness [111, 404] is leveraged to evaluate loss values. A sharp loss landscape is sensitive to noise and generalizes poorly, whereas a flat loss landscape is robust to input changes and generalizes well (Fig. 2.6b). However, applying sharpness minimization directly on multiple distributions is suboptimal [54, 166], as the major data distribution could skew the learning result, leading to failures on weakly corrupted data. As a solution, SharpDRO aims to penalize worst-case sharpness. As illustrated in

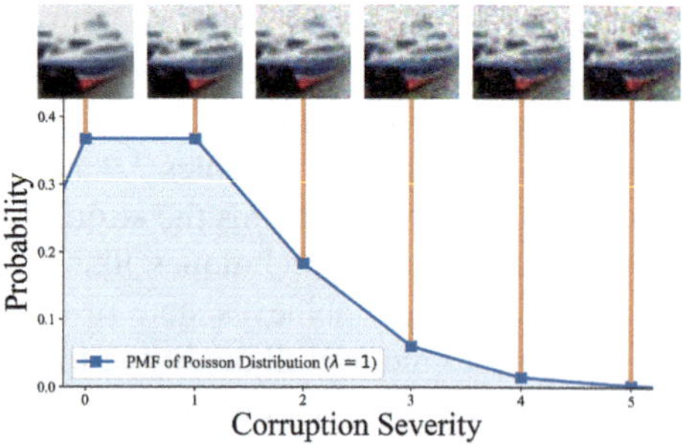

Fig. 2.5 Illustration of photon-limited corruptions.

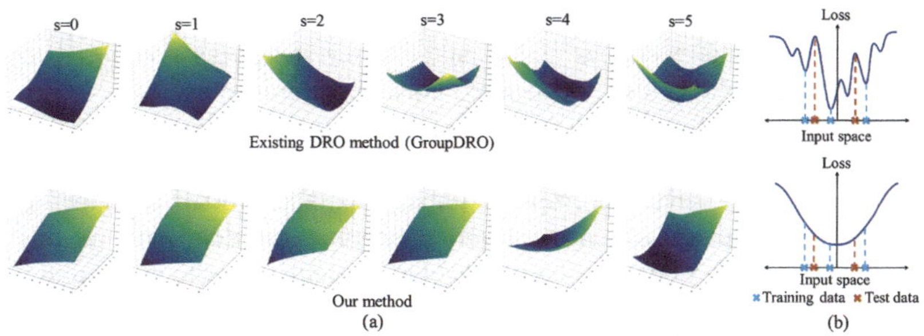

Fig. 2.6 Motivation of SharpDRO. (**a**) Loss surface visualization of various methods. The columns show noisy distributions varied from $s = 0$ to 5. (**b**) Relationship between sharpness and generalization

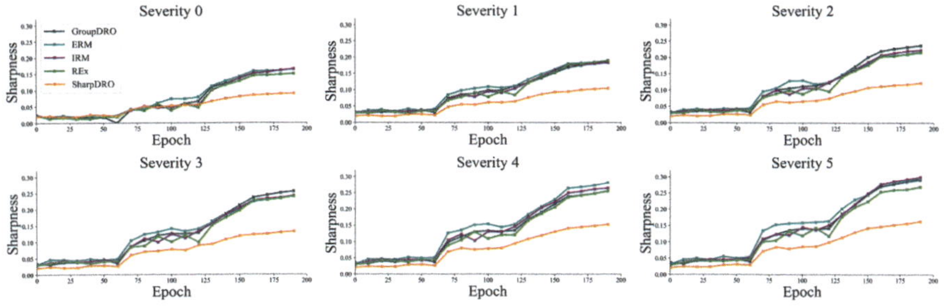

Fig. 2.7 Sharpness during training under clean ($s = 0$) and noisy distributions ($s = 1$ to 5)

the lower part of Fig. 2.6a, SharpDRO successfully achieves smooth loss curves, ensuring generalization with robustness across multiple noise severities.

Method In robust generalization, we are given a training set D^{train} with n image instances. Each $\mathbf{x} \in \mathcal{X}$ is assigned a label $y \in \mathcal{Y} = \{1, 2, \ldots, C\}$. Moreover, the training dataset is influenced by noise types, where the severity s follows a Poisson distribution $P(s; \lambda)$, where $\lambda = 1$ by assumption. Hence, 1 noise photon u is expected to occur within a time interval. As a result, the general training distribution p is composed of S sub-distributions p_s related to severity level $s \in \{1, 2, \ldots, S\}$. Our target is to develop a robust model parameterized by $\theta \in \Theta$ that generalizes well under challenging scenarios where s is large.

The goal of SharpDRO is designed as

$$\min_{\theta} \left\{ \mathcal{L}_{\text{SharpDRO}} := \mathbb{E}_{(\mathbf{x},y)\sim Q} \left[\ell(\theta; (\mathbf{x}, y)) \right] + \mathbb{E}_{(\mathbf{x},y)\sim Q} \left[\mathcal{R}(\theta; (\mathbf{x}, y)) \right] \right\}, \tag{2.12}$$

the first term is the risk minimization using the loss function ℓ, and the worst-case distribution Q is decided according to predictions. The second term $\mathcal{R}$ denotes sharpness minimization, which focuses on enhancing generalization on Q. As shown in Fig. 2.7, sharpness increases when severity goes stronger. Therefore, to enhance robust generalization, we only stress the worst-case distribution. Consequently, our approach produces significantly lower sharpness compared to other baselines, particularly on highly corrupted noise distributions.

A vital challenge here is the imbalance in the training data (Fig. 2.5). While performance on clean data is acceptable, robustness against noisy distributions is poor because of the severe corruption impact and the limited amount of noisy data. To enhance generalization, we leverage sharpness to deal with various noise severities properly. Particularly, sharpness [111] is defined as the maximum change in loss when parameter θ is deviated by perturbation ϵ:

$$\mathcal{R} := \max_{\|\epsilon\|_2 \leq \rho} \{\ell(\theta + \epsilon; (\mathbf{x}, y)) - \ell(\theta; (\mathbf{x}, y))\}. \tag{2.13}$$

$$\mathbb{E}_{(\mathbf{x},y)\sim Q}\mathcal{R} := \ell(\theta + \rho\, \mathrm{sign}(\nabla \ell(\theta; (\mathbf{x}, y)))) - \ell(\theta; (\mathbf{x}, y)). \tag{2.14}$$

The key intuition is that perturbations in the direction of the gradient norm significantly increase the loss value. When training on corrupted distributions, noisy data points are sparsely scattered in the high-dimensional space. As a result, the neighborhood of each data point is not sufficiently explored, leading to a sharp loss landscape. During testing, unseen noisy data are likely to fall in unexplored regions with high loss values, causing inaccurate model predictions.

To address this, instead of applying sharpness minimization across the entire dataset—which can degrade generalization performance [54]—we focus on minimizing sharpness over the worst-case distribution Q. By performing worst-case sharpness minimization, we improve the classifier's loss landscape smoothness. Consequently, when predicting unseen data, a flatter loss surface is more likely to yield a lower loss than a sharp one, enabling SharpDRO to generalize better than other DRO methods. However, the model's robustness heavily depends on the selection of the worst-case distribution Q, which we discuss next.

In general, worst-case data selection involves identifying the most uncertain data distribution Q from an uncertainty set $\mathcal{Q}$, defined as an f-divergence ball around the training distribution P [29, 99, 161]. When annotation labels are available to indicate different levels of corruption, each image $\mathbf{x}$ is associated with a class label y and a distribution index s. The worst-case distribution Q is then determined by selecting the sub-distribution $p_s \subset p$ that produces the highest loss. We optimize accordingly:

$$\min_{\theta}\Bigg\{ \max_{\substack{\omega_S;\\ Q=\{\omega_S p_S\}_{s=1}^{S}}} \Big\{ \sum_{(\mathbf{x}_i, y_i)\in Q} [\ell(\theta, \omega_s; (\mathbf{x}_i, y_i))] \Big\} + \sum_{(\mathbf{x}_i, y_i)\in Q} [\mathcal{R}(\theta, \omega_s; (\mathbf{x}_i, y_i))] \Bigg\}. \tag{2.15}$$

The variable ω_s lies within an $(S-1)$-dimensional probability simplex. The first term aligns with the learning objective of GroupDRO [161, 319], aiding in identifying the worst-case distribution Q. The second term, by emphasizing the selected p_s, serves as a sharpness regularizer, promoting a flatter loss landscape

Table 2.11 Comparisons with various baseline methods

Data	Type	Method	Corruption severity					
			0	1	2	3	4	5
CIFAR-10	Gaussian	ERM	90.9	89.2	86.4	85.9	83.5	78.8
		IRM	91.8	90.3	89.5	86.7	81.8	80.0
		REx	91.3	89.5	88.1	86.7	83.3	80.5
		GroupDRO	90.2	89.1	88.4	84.3	83.0	78.2
		SharpDRO	**93.1**	**92.2**	**91.4**	**89.2**	**87.1**	**84.5**
	Shot	ERM	**92.5**	91.1	89.9	85.6	85.7	78.8
		IRM	90.4	90.3	89.4	86.3	84.3	79.1
		REx	91.1	90.6	90.2	86.8	84.7	80.5
		GroupDRO	92.2	91.4	89.4	84.0	84.7	78.3
		SharpDRO	91.7	**93.3**	**92.5**	**91.0**	**88.2**	**83.9**
CIFAR-100	Gaussian	ERM	68.2	64.8	60.6	56.9	53.9	50.2
		IRM	64.7	64.7	62.2	54.5	53.4	50.4
		REx	68.0	65.1	61.8	56.8	53.2	51.5
		GroupDRO	66.1	61.7	59.3	53.6	54.0	50.6
		SharpDRO	**72.3**	**72.2**	**71.5**	**62.4**	**59.7**	**55.7**
	Shot	ERM	67.6	65.1	62.9	56.0	55.1	47.3
		IRM	67.5	65.7	62.7	59.5	55.8	48.3
		REx	65.7	63.8	61.9	59.3	53.8	48.1
		GroupDRO	67.0	65.8	63.1	58.9	57.5	49.3
		SharpDRO	**71.3**	**70.5**	**68.3**	**63.6**	**60.5**	**53.4**
ImageNet30	Gaussian	ERM	87.5	84.6	81.9	76.5	71.2	65.3
		IRM	86.6	84.4	80.6	75.2	70.7	64.8
		REx	86.3	83.8	81.1	75.6	71.5	66.1
		GroupDRO	85.1	84.2	81.2	76.3	72.0	66.3
		SharpDRO	**89.6**	**88.3**	**85.5**	**82.8**	**77.5**	**70.2**
	Shot	ERM	86.9	84.8	83.6	79.7	75.4	64.6
		IRM	86.8	85.1	81.5	73.5	68.5	62.5
		REx	83.8	86.3	82.5	73.9	70.6	64.0
		GroupDRO	86.7	85.6	84.5	80.7	76.2	65.4
		SharpDRO	**88.3**	**88.1**	**86.4**	**85.3**	**78.2**	**68.2**

on the worst-case data. Consequently, SharpDRO enhances generalization beyond GroupDRO.

Experiment Table 2.11 demonstrates that SharpDRO consistently outperforms other methods, with larger performance gains as corruption severity increases. Notably, on the CIFAR-10 dataset with "Gaussian noise" corruption, SharpDRO surpasses the second-best method by 2.2% at severity level 0, and this margin grows to 5.7% at severity level 5. This highlights SharpDRO's superior generalization under severe corruption. Additionally, SharpDRO consistently outperforms other approaches across various corruption types, underscoring its robustness.

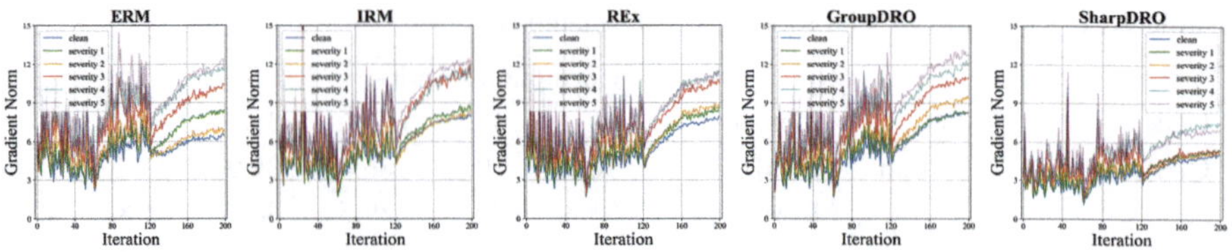

Fig. 2.8 Gradient norm comparisons across various methods over different corrupted distributions

To demonstrate the stability of our method across challenging distributions, we evaluate the gradient norm on a validation set with corruption severities ranging from 0 to 5. As shown in Fig. 2.8, SharpDRO achieves the lowest gradient norm while maintaining nearly uniform values across all corrupted distributions. This indicates that SharpDRO is the most distributionally stable method among those compared.

Chapter 3
Trustworthy Machine Learning with Adversarial Examples

It is well established that machine learning models are vulnerable to adversarial examples (AEs), which introduce imperceptible perturbations into the original image yet can significantly degrade the performance of well-trained classifiers, often resulting in nearly 100% misclassification rates. As machine learning models become more essential to society, AEs present significant safety risks to their dependability. To prevent various failure modes in real-world deployments, it is essential to develop robust defense strategies against adversarial attacks, thereby ensuring the reliability of machine learning systems.

Currently, various techniques have been proposed to enhance the robustness of machine learning models throughout their life cycle, from preprocessing to development and deployment. In this chapter, we focus on three primary adversarial defense frameworks: statistical learning, deep learning, and diffusion-based models, which align closely with different stages of the machine learning pipeline. Statistical learning-based methods focus on modeling adversarial noise by learning the transition relationship between adversarial and natural labels. In the deployment stage, erroneous predictions caused by adversarial perturbations can be corrected in post-processing using the learned transition matrix, thereby inferring the natural label. In the development phase, deep learning-based defenses generate adversarial examples (AEs) on the fly to expose vulnerabilities in the training model. These AEs are incorporated into the training process, encouraging the model to better capture the underlying data distribution and thereby enhancing its robustness. In the preprocessing stage, diffusion-based approaches aim to either purify or detect adversarial examples before they are passed to the decision model. Adversarial purification techniques effectively reverse the distribution shift caused by adversarial perturbations, mapping adversarial examples back to the natural data manifold. On the other hand, adversarial detection methods identify anomalies by quantifying deviations from the natural data distribution. Through an in-depth discussion of how each framework contributes to enhancing model robustness and security under

B. Han, T. Liu, *Trustworthy Machine Learning under Imperfect Data*,
https://doi.org/10.1007/978-981-96-9396-2_3

adversarial settings, we provide insights into strengthening the reliability and resilience of machine learning models against worst-case adversarial scenarios.

3.1 Statistical Learning for Adversarial Examples

Adversarial examples rely on the injected adversarial noise to manipulate the predictions of the target model. Protecting training data and identifying the label flips caused by the adversarial noise is a key foundation for correcting erroneous predictions. As a tool for modeling and understanding complex data, statistical learning can naturally assist in purifying data and obtaining the probability distribution of label flips from the input adversarial examples. Thus, in this section, we introduce adversarial poisoning and kernel learning for adversarial robustness. Besides, we model the adversarial noise based on statistical learning to construct the transition relationship between the adversarial and natural labels, which is concretized as a probability matrix. Further, a deep neural network is introduced to learn the respective transition matrices for different input adversarial examples. In this way, we are able to infer the natural label by multiplying the predicted adversarial label with the transition matrix, thus mitigating the interference of adversarial noise on the target model. Meanwhile, the transition matrix can assist in detecting the adversarial noise in the input sample, thus providing additional safeguards for trustworthy machine learning at the data level.

3.1.1 Adversarial Poisoning Against SVM

With the development of machine learning and deep learning technologies, the requirements of these technologies for data and computational resources are becoming more and more enormous, and simply introducing private resources may not meet the need of these complex systems for applications [206, 393]. To address these problems, researchers are seeking for improving the artificial intelligence (AI) systems using various resources, containing outsourced data and third-party platforms where they can be conveniently used for reducing their costs [456]. However, such a method which explicitly utilizes unprotected resources makes the system easy to be leaked to attackers, posing significant threats to them [287]. Among these threats, poisoning attacks have received a lot of attention due to its destructiveness on these resources. Poisoning attacks seek to disrupt the model by contaminating the training data or model parameters and are performed usually on the training phase [303]. For instance, these attacks maliciously manipulate open-source data, thereby poisoning any models trained on them, where the training examples or the corresponding labels are all possible to be the goal of attacks [21]. Besides, some strategies upload the poisoned weights on a server for disrupting distributed system like federated learning models [391].

Among various machine learning models, the support vector machine (SVM) is crucial for various applications. It is a classical data-driven machine learning model, which learns specific support vectors to hold a decision hyperplane for classifications. SVM has been proven to be vulnerable to poisoning attacks, and even a few poisoned examples can reduce the prediction accuracy of SVM [299, 435]. The training procedure, loss function or the model weights of SVM, and so on are all possible to be the goal of poisoning attack. Therefore, studying the attacks and defenses of SVM in the field of poisoning attacks is considered to be meaningful for trustworthy machine learning systems.

Poisoning Attacks Poisoning attacks against SVM mainly focus on disrupting the decision boundary dominated by support vectors, so that the classification accuracy is going to be decreased a lot. They can be categorized into two strategies, namely, data poisoning attacks and model poisoning attacks. Among the data poisoning attacks, a gradient-based approach [36] aims at finding malicious gradients using the properties of optimal solution, which successfully finds a local maximum value in the error space to craft poisoned examples. They focus on a hinge loss optimization strategy to craft poisoned examples by an indiscriminate attack against the integrity of the model. Optimization-based attacks [262] generate poisoned examples through representing optimal attacks on the training sets as a bi-level optimization problem. Koh et al. [196] perform poisoning attacks also by the bi-level optimization, focusing on the highest test loss for attacking. Some strategies [131] introduced some poisoned examples with specific poisoning patterns, which cause wrong predictions of SVM when facing with these patterns.

In addition, label flipping attacks [23, 461] seek to change the labels of selected training examples for misleading SVM systems, building a mismatch between data and labels to inject wrong knowledge. Xiao et al. [415] select some mislabeled data using an optimization framework to maximize the classification error of SVM systems. Zhang et al. [458] explored the label flipping attacks on distributed SVM systems, showing the effectiveness of label poisoning attacks on distributed systems. Among the model poisoning attacks, the malicious model weights can be uploaded on the distributed systems through optimizing the stealthiness and attack performances to mislead the global model, which also verifies the vulnerability of federated learning systems [35]. These attacks show that the vulnerability of SVM mainly comes from its dependency on the support vectors and training data, promoting the defense mechanisms against poisoning attacks.

Defenses Against Poisoning Attacks Various poisoning attacks drive the development of defenses against them. Since poisoning samples can be considered as outliers, this leads to two natural defense approaches: (1) data purification defense, i.e., which firstly removes outliers and then uses the cleaned data for machine learning [75], or (2) optimization defense, which directly optimizes the algorithm to enhance robustness to make it resilient to outliers [72, 178].

Steinhardt et al. [343] study two basic methods of data purification defense, which eliminate the points outside a specified sphere or slab. They show that high dimensionality gives attacker more rooms for constructing attacks to evade outlier removal. Besides, Laishram and Phoha [209] apply the seminal DBSCAN (density-based spatial clustering of applications with noise) method to eliminate outliers for SVM and show that it can successfully identify most of the poisoning data. However, DBSCAN and similar clustering-based defenses lack theoretical robustness guarantees. Several other outlier removal methods for fighting poisoning attacks have also been studied recently [290].

In addition, in this field, designing robust optimization algorithms has also received sustained attention. A substantial part of robust optimization algorithms relies on the idea of regularization. For example, the relation between robustness and regularization of SVM is studied by Xu et al. [423], and the 1-norm SVM is proposed by Zhu et al. [489] to enhance the robustness against noise. Additionally, some other robust SVM algorithms [326] have been proposed for reliable performance. However, as discussed in [178, 262], due to the fact that the attacker can arbitrarily locate outliers in the feature space, the effectiveness of these methods for defending against poisoning attacks is not very satisfactory. Another strategy to achieve the robustness is to supplement strong assumptions about the data distribution or the attacker's behavior [108, 397], but these assumptions are usually not well satisfied in practice.

3.1.2 Adversarial Awareness via Deep Kernel

Kernel-based methods are widely used in machine learning and pattern recognition. This kind of method aims to map data into a high-dimensional feature space so that linear methods can be applied in such space to solve nonlinear problems. Kernel functions are the key components of kernel-based methods, which define the measure of similarity between input data. By leveraging appropriate kernel functions, we can directly compute the inner product of the input data at a high-dimensional feature space without explicitly performing feature mapping, thereby effectively handling high-dimensional data while avoiding the massive computational complexity of high-dimensional feature mappings. In recent years, kernel-based methods are also introduced in modern deep learning methods. By incorporating kernel functions into nonlinear neural networks, we can effectively capture intricate nonlinear data relationships under complex data distributions and model architectures. Recent researches introduce deep kernel-based methods to explore the vulnerability of deep neural networks against adversarial attacks. Carlini et. al [50] systematically evaluate the effectiveness of ten adversarial example detection methods, which includes some early-stage kernel-based methods, such as local singular value decomposition (SVD) and principal component analysis (PCA). The authors demonstrate that relying solely on kernel-based detection may not offer sufficient defense against carefully crafted adversarial examples.

Kernel-Based Adversarial Detection For mitigating the interference caused by adversarial noise, Feinman et. al [105] propose a novel method to detect adversarial samples by exploiting the artifacts that are produced in feature space. Specifically, the authors leverage the Bayesian uncertainty estimation and the dropout mechanism in neural networks to calculate the model's prediction uncertainty for the input sample. They additionally perform kernel density estimation in the deep feature space of the model and found that there exists a difference in the distribution of normal samples and adversarial samples in this space, which can be used to detect adversarial samples. The proposed method does not rely on any specific attack algorithm, thereby ensuring its general applicability. Besides, Das et. al [85] also utilize the kernel density estimation to detect adversarial samples in medical images. According to the characteristics of medical images, the authors adjust the kernel density feature extractor, which further improves the accuracy and reliability of detection.

Kernel-Based Adversarial Robustness In addition to detecting adversarial samples, kernel learning is also applied to enhance the model's robustness. Gaussian process (GP) is a statistical method that is applied to describe random processes, which leverages kernel functions to measure the similarity between input data points. Bradshaw et. al [43] combine GP with deep neural networks (DNN) and propose the hybrid GPDNN model, which enhance the model's robustness against adversarial examples and the uncertainty estimation capability.

Besides, Gao et. al [397] leverage the neural tangent kernel (NTK) theory to analyze the adversarial training process in neural networks. NTK connects the training dynamics of neural networks with kernel methods, allowing the training process of neural networks can be approximated as the training of a linear model under specific conditions. The authors further theoretically improve that adversarial training in neural networks can converge to a nearly optimal robust model, which provides a new perspective for analyzing and improving adversarial robustness. Loo et. al [250] compare the evolution of the empirical NTK under standard training and adversarial training and reveal how adversarial training affects the feature learning and robustness of neural networks through empirical studies. Moreover, Tsilivis et. al [372] exploit NTK to generate adversarial examples in a training-free fashion and reexamine the roles of robust and non-robust features in standard classification tasks via kernel spectral decomposition.

Additionally, Han et. al [138] propose a nonparametric kernel density estimator to replace the self-attention mechanism in conventional transformer-based models and introduce robust kernel density estimation that can improve the adversarial robustness of transformer models. Based on the proposed methods, the authors design multiple transformer architectures that are capable to maintain model performance under adversarial attacks and data contamination.

3.1.3 Model Adversarial Noise for Adversarial Defense

Deep neural networks have been found to be susceptible to carefully designed small perturbations, known as adversarial noise, which can mislead the model into making evidently incorrect predictions. The vulnerability of deep neural networks poses a significant threat to their applications across various fields, such as facial recognition, autonomous driving, and medical image analysis.

In recent years, research on defending against adversarial noise has gained increasing attention in academia, with adversarial training being a core method that improves the robustness of the target model by incorporating adversarial samples into the training process. However, the relationship between adversarial noise and natural data remains underexplored in current research.

As shown in Fig. 3.1, the research focus has shifted to the potential relationship between adversarial samples and natural samples, which can be formalized as a modeling representation of adversarial noise, providing new solutions to mitigate the negative impact of adversarial samples. Specifically, this modeling approach uncovers the inherent connections between the two types of data, paving an effective path for constructing more robust machine learning models. If this latent relationship can be effectively learned, it may be possible to infer natural data (*e.g.*, clean images or labels) from adversarial data. While denoising-based strategy can incorporate this concept by transforming adversarial examples into natural samples, it may encounter issues at the pixel dimension, such as observable image quality

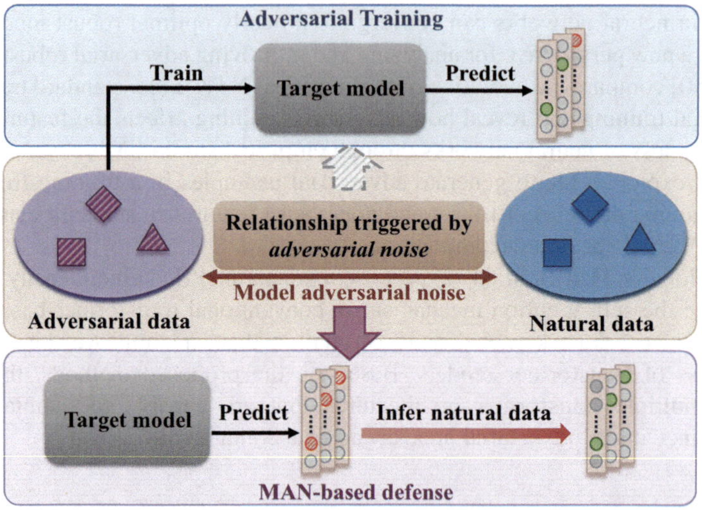

Fig. 3.1 The characteristics of adversarial noise modeling. Adversarial noise triggers an inversion relationship from natural data to adversarial data (e.g., from true natural labels to incorrect predicted labels). Modeling adversarial noise by learning this relationship can help infer natural data (such as natural labels) from adversarial data, thereby correcting erroneous predictions

degradation [424] and error amplification effects due to residual adversarial noise [230].

To address the issues mentioned above, considering that the number of pixels in generated images is typically much larger than the number of labels, this section focuses on modeling adversarial noise within the low-dimensional label space. That is, it characterizes adversarial noise by modeling the transition relationship between adversarial labels and natural labels. It is noteworthy that adversarial labels are not merely passive noise outcomes; in certain specific scenarios (such as in targeted attacks), they can also actively guide the generation process of adversarial noise. This bidirectional relationship inherently imbues adversarial labels with rich feature information, which is crucial for accurately capturing the essential characteristics of adversarial noise.

From the perspective of adversarial attack mechanisms, the generation of adversarial noise is influenced by multiple factors, including the feature distribution characteristics of natural samples, the specific implementation of the attack algorithm, and the parameter configuration of the target model. By systematically modeling adversarial noise in the label space, a comprehensive analytical framework can be provided for the development of defense strategies.

Based on this theoretical analysis, this study proposes a novel adversarial defense framework—Multidimensional Adversarial Noise (MAN) Modeling. Specifically, a *label transition network* is introduced to learn the *label transition matrix*, modeling the relationship between adversarial labels and natural labels. Additionally, a joint optimization strategy for the model and the network is adopted, resulting in a significant improvement in the overall system's robustness. This dual optimization mechanism not only enhances the model's ability to recognize adversarial samples but also improves its stability in complex attack environments.

In summary, the main contributions of this section are as follows:

1. This section introduces a prediction bias correction strategy that enhances the model's robustness by modeling adversarial noise to learn the relationship between adversarial data and natural data.
2. This section explicitly presents the transition relationship in the form of a label transition matrix, which is learned from the input sample via a label transition network.
3. The proposed strategy can detect adversarial examples to identify potentially malicious interference before inputting data into the target model, thereby enhancing the reliability of model predictions.

Method: Adversarial Noise Modeling Studying the characteristics of adversarial noise is crucial for effectively mitigating its destructive impact, modeling the relationship between adversarial samples and natural samples. This allows adversarial data to be input into this relationship to infer natural data, correct prediction biases, enhance the robustness of the target model, and promote trustworthy visual analysis. Denoising strategies that learn the pixel space mapping from adversarial samples to natural samples involve estimating this relationship; however, image denoising methods face limitations such as image quality loss and the error amplification effect of residual adversarial noise. To address this limitation, we model adversarial noise in the low-dimensional label space.

Adversarial noise has subtle and highly universal characteristics, making it effectively modeled. Specifically, the training of classification models typically focuses on improving the accuracy of natural data while also tending to utilize all available signals during the training process, including those that are somewhat robust but susceptible to attacks. Adversarial noise can be generated by interfering with these vulnerable features, leading to the transformation between adversarial labels and natural labels. Therefore, adversarial noise contains specific destructive characteristics, and by constructing a transfer relationship from adversarial labels to natural labels, it can be effectively modeled.

It is important to highlight that the formation mechanism of adversarial noise is multidimensional and complex, as it is influenced not only by the data distribution characteristics but also closely related to the structure and parameters of the target model. Therefore, when constructing adversarial noise models, a systematic modeling approach must be adopted, fully considering the interactions between these key factors. Specifically, since the label transfer process is directly driven by adversarial samples, both the sample feature distribution and the dual influence of adversarial noise on label flipping probabilities must be considered in the modeling process. This comprehensive consideration helps accurately identify and correct the prediction bias caused by adversarial noise.

It is worth noting that the learned label transfer relationship can be organically integrated with the target model. This integration method shows significant advantages in the face of strong adaptive attacks: by jointly training the label transfer probability matrix with the target model, deep learning models can effectively resist high-intensity interference, significantly enhancing the overall robustness of the system. For more complex label transfer scenarios, a deep neural network architecture can be further introduced. This architecture autonomously learns from instance-related adversarial samples, precisely extracting the corresponding label transfer patterns by deeply analyzing misleading features and remaining target features in the samples. This multilayered, multi-perspective modeling approach provides a reliable theoretical foundation and technical support for building robust adversarial defense systems.

Label Transition Matrix To effectively implement the above method, the key lies in establishing an explicit mapping relationship between adversarial labels and natural labels. This study innovatively proposes a probabilistic transformation framework, which quantifies the likelihood of adversarial label conversion to natural labels by constructing a specialized label transformation matrix. Unlike the transition matrices commonly used in traditional label noise learning, this matrix has unique properties and application scenarios.

Additionally, due to the sample-specific nature of adversarial noise, it is necessary to design a specific label transition matrix for each input sample. Besides, to ensure the natural accuracy of the model, we use a mixture of both as the input sample (referred to as mixed sample $\mathbf{x}'$) and use a transition matrix $T \in [0, 1]^{C \times C}$

to model the relationship between the mixed label y' and natural label y, which is defined as

$$T_{i,j}(\mathbf{x}') = P\left(y = j \mid y' = i, \mathbf{x}'\right), \tag{3.1}$$

where $T_{i,j}$ denotes the element located at (i, j) in the matrix $T(\mathbf{x}')$.

The basic idea of this method is that given the mixed class posterior probability (the class posterior probability of the mixed label y') $P(y' \mid \mathbf{x}') = [P(y' = 1 \mid \mathbf{x}'), \ldots, P(y' = C \mid \mathbf{x}')]^{\top}$, we can use $P(y' \mid \mathbf{x}')$ and $T(\mathbf{x}')$ to infer the natural class posterior probability (the class posterior probability of the natural label y) $P(y \mid \mathbf{x}')$:

$$P(y \mid \mathbf{x}') = T(\mathbf{x}')^{\top} P(y' \mid \mathbf{x}'). \tag{3.2}$$

Subsequently, the category corresponding to the maximum value in the inferred class posterior probabilities is chosen as the final prediction result.

Label Transition Network: This section utilizes a label transition network to estimate the label transition matrix from the mixed data $\{(\mathbf{x}'_i, y'_i, y_i)\}_{i=1}^{2n}$. The transition network g_ω, parameterized by ω, takes the mixed sample x' as input and outputs the label transition matrix $T(\mathbf{x}') = g_\omega(\mathbf{x}')$. Subsequently, the natural class posterior probability $P(y \mid \mathbf{x}')$ can be inferred based on Eq. (3.2), enabling the prediction of natural labels. This network is learned to minimize the discrepancy between the predicted and actual natural labels, with the loss function defined as

$$\mathcal{L}_T = -\frac{1}{2n}\sum_{i=1}^{2n} \ell_{\text{CE}}(y'_i \cdot T(\mathbf{x}'_i), y_i), \tag{3.3}$$

where ℓ_{CE} denotes a cross-entropy loss. This method conducts joint adversarial training on the classification model and the label transition network to enhance the overall performance. The framework is illustrated in Fig. 3.2.

Experiments: Performance Against General Attacks The experiments were conducted using PGD, AA, FWA, L_2 norm C&W, and DDN attacks for testing. Specifically, the number of iterations for PGD, FWA, and DDN attacks was 40 and was 200 for C&W, respectively. The perturbation budgets for L_2 norm attacks and L_∞ norm attacks were set to 0.5 and 8/255, respectively. The evaluation results of the proposed method, along with the separate testing results of the target model, are shown in Table 3.1. The results indicate that the method can effectively correct prediction bias when faced with general adversarial attacks.

Performance Against Adaptive Attacks Adaptive attacks are a powerful adversarial attack strategy, where the attacker has access to the target model's network architecture and model parameters, allowing them to design targeted attack algo-

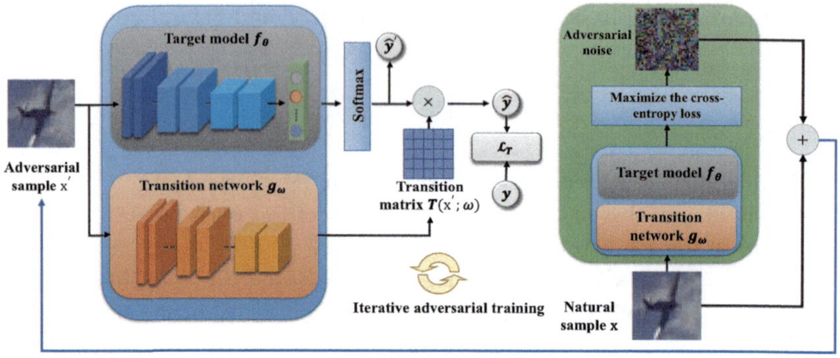

Fig. 3.2 Framework of the prediction bias correction method based on adversarial noise modeling

Table 3.1 The classification accuracy (percentage) against general attacks on CIFAR-10. The target model is ResNet-18

Defense	None	PGD	AA	FWA	C&W	DDN
MAN	89.01	81.07	79.90	80.02	77.89	77.82
Target model	88.98	0.00	0.00	0.00	0.00	0.00

rithms. This section will explore three adaptive attack scenarios to assess the effectiveness of the proposed method in enhancing model robustness.

1. **Disrupting the final output**: In this section, an attack is implemented on the combination of the target model and the label transition matrix to generate adversarial samples. In this scenario, the goal of the untargeted attack is to disrupt the final prediction results. The optimization objective for adversarial samples can be expressed as

$$\tilde{x} = \arg\max_{\tilde{\mathbf{x}}} \ell_{\mathrm{CE}}(\tilde{y} \cdot T(\tilde{\mathbf{x}}), y), \quad \text{s.t. } \|\mathbf{x} - \tilde{\mathbf{x}}\| \leq \epsilon, \tag{3.4}$$

where $\tilde{y}$ denotes the adversarial label corresponding to the output $f_\theta(\tilde{\mathbf{x}})$. Similarly, adversarial samples can be generated through a targeted attack, which, in this setting, solves the following optimization problem:

$$\tilde{\mathbf{x}} = \arg\max_{\tilde{\mathbf{x}}} -\ell_{\mathrm{CE}}(\tilde{y} \cdot T(\tilde{\mathbf{x}}), y^*), \quad \text{s.t. } \|\mathbf{x} - \tilde{\mathbf{x}}\| \leq \epsilon, \tag{3.5}$$

where the y^* is the target label set by the attacker.

Table 3.2 The classification accuracy (percentage) against white-box adaptive attacks. The target model is ResNet-18. Values in **bold** indicate the optimal results

Defense	None	PGD	AA	FWA	C&W	DDN
CIFAR-10						
AT	**83.39**	42.38	39.01	15.44	0.00	0.09
+MAN	82.72	**44.83**	**39.43**	**29.53**	**43.17**	**10.63**
TRADES	**80.70**	46.29	42.71	20.54	0.00	0.06
+MAN	80.34	**48.65**	**44.40**	**29.13**	**1.46**	**0.31**
MART	**78.21**	50.23	43.96	25.56	0.02	0.07
+MAN	77.83	**50.95**	**44.42**	**31.23**	**1.53**	**0.47**
Tiny-ImageNet						
AT	**48.40**	17.35	11.27	10.29	0.00	0.29
+MAN	48.29	**18.15**	**12.45**	**13.17**	**16.27**	**4.01**
TRADES	**48.25**	19.17	12.63	10.67	0.00	0.05
+MAN	48.19	**20.12**	**12.86**	**14.91**	**0.67**	**1.10**
MART	**47.83**	20.90	15.57	12.95	0.00	0.06
+MAN	47.79	**21.22**	**15.84**	**15.10**	**0.89**	**1.23**

The results in Table 3.2 indicate the proposed method (MAN) has superior performance in robust accuracy compared to AT. Moreover, the proposed method can be applicable to various adversarial training methods, such as TRADES and MART. As shown in Table 3.2, the evaluation results indicate that the proposed method can be applied to different adversarial training strategies to correct the prediction bias of the target model and further improve robustness accuracy.

2. **Disrupting the label transition matrix**: In this section, we take the label transition matrix as target. If the attack succeeds, the correction of prediction bias will fail. A counter-diagonal identity matrix (as shown in Fig. 3.3) is selected as the attack target. The optimization objective is

$$\begin{aligned} \tilde{\mathbf{x}} &= \arg\max_{\tilde{\mathbf{x}}} \ -\ell_{\text{MSE}}(T(\tilde{\mathbf{x}}), T^*), \\ &\text{s.t.} \ \left\|\mathbf{x} - \tilde{\mathbf{x}}\right\| \leq \epsilon, \end{aligned} \tag{3.6}$$

where ℓ_{MSE} denotes the mean squared error loss. This section uses L_∞ PGD with this strategy to evaluate the predictive bias correction effect of the method proposed in this section. Based on the experimental data above, the proposed method in this section demonstrates high robust accuracy, effectively resisting adaptive attacks and showcasing its superiority.

3. **Dual adversarial attack**: In this section, we investigate another adaptive attack scenario: dual adaptive attack, which disrupts both the final prediction result and the initial prediction result. Its optimization objective is expressed as

$$\begin{aligned} \tilde{\mathbf{x}} &= \arg\max_{\tilde{\mathbf{x}}} \, [\ell_{\text{CE}}(\tilde{y} \cdot T(\tilde{\mathbf{x}}), y) + \ell_{\text{CE}}(\tilde{y}, y)]. \\ &\text{s.t.} \ \left\|\mathbf{x} - \tilde{\mathbf{x}}\right\| \leq \epsilon. \end{aligned} \tag{3.7}$$

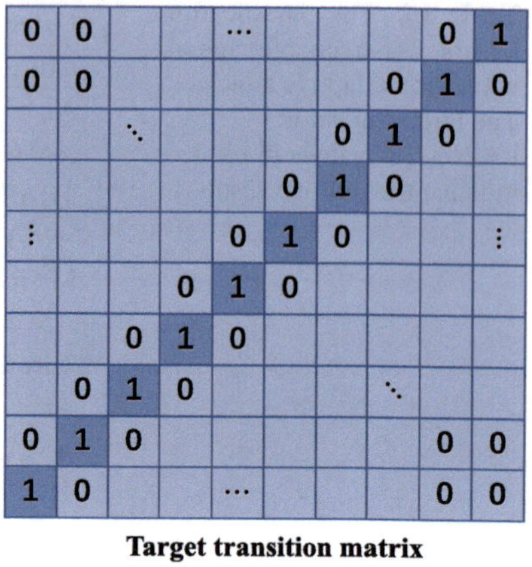

Fig. 3.3 Target transition matrix under target adversarial attacks T^*

Table 3.3 The classification accuracy (percentage) against dual white-box adaptive attacks on CIFAR-10. The target model is ResNet-18

Defense	None	PGD	AA	FWA	C&W	DDN
MAN	82.72	70.63	45.13	64.09	45.21	38.16
Target model	8.67	2.36	1.43	1.65	2.10	0.73

Similarly, adversarial samples can be generated through targeted attacks. In this scenario, the optimization objective for the targeted attack can be designed as

$$\tilde{\mathbf{x}} = \arg\max_{\tilde{\mathbf{x}}} [-\ell_{\mathrm{CE}}(\tilde{y} \cdot T(\tilde{\mathbf{x}}), y^*) + \ell_{\mathrm{CE}}(\tilde{y}, y)],$$
$$\text{s.t.} \ \left\|\mathbf{x} - \tilde{\mathbf{x}}\right\| \leq \epsilon, \tag{3.8}$$

where y^* is the target label in vector form set by the attacker. Five types of attacks are combined with the dual attack strategy to make evaluations. The results in Table 3.3 indicate that the method can effectively correct prediction bias, thereby improving the model's performance.

Detecting Adversarial Samples The proposed method can be applied to detect adversarial samples. The diagonal elements of the transition matrix can reflect whether the input is adversarial. If the element p corresponding to the predicted label is greater than the values of other elements on the diagonal, the input is natural; otherwise, it is adversarial. Thus, $1 - p$ is utilized as the prediction probability. The test samples from CIFAR-10 are employed to calculate the receiver operating characteristic (ROC) curve and the area under the curve (AUC) against attacks. The measured results are shown in Fig. 3.4, which indicates that the method achieves high AUC values for adversarial samples generated by both types of adversarial attacks. This further demonstrates the positive impact of the adversarial noise

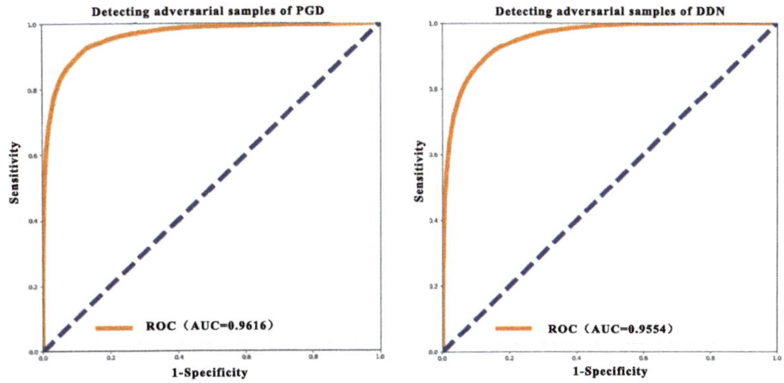

(a) (b)

Fig. 3.4 The performance of the proposed method in detecting adversarial samples. (**a**) PGD-40. (**b**) DDN-40

modeling strategy in addressing adversarial attacks and constructing reliable deep learning models.

3.2 Deep Learning for Adversarial Examples

Adversarial examples (AEs) exploit adversarial perturbations to manipulate the decisions of deep neural networks (DNNs). One of the foundational approaches to enhancing DNN robustness against such attacks is adversarial training (AT). The core idea of AT is to generate AEs during training and incorporate them into the learning process, compelling the model to capture the underlying distribution.

Mathematically, AT is formulated as a min-max optimization problem. The inner maximization seeks to generate AEs by maximizing the classification loss, while the outer minimization updates the model parameters to correctly classify the generated AEs. Based on this framework, research on AT primarily focuses on three key directions: (1) improving the inner maximization by refining AE generation, (2) enhancing the outer minimization by optimizing the model's learning process, and (3) mitigating overfitting arising from the imperfect min-max optimization.

3.2.1 Adversarial Training

3.2.1.1 Geometry-Aware Adversarial Training

The objective of AT [258] is to develop a deep model that is adversarially robust, and its predictions are locally consistent to a small neighbor of its inputs [124].

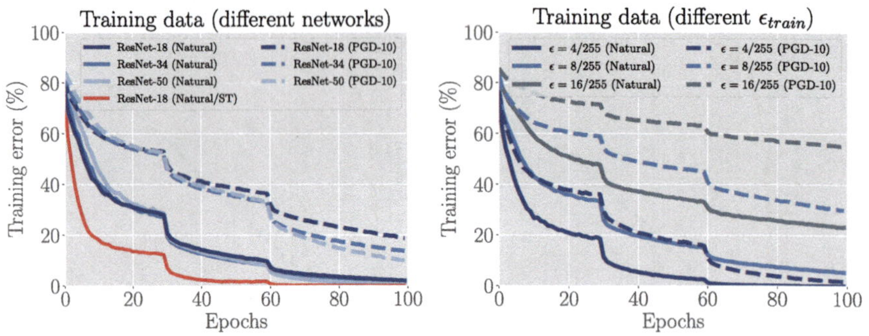

Fig. 3.5 Training error on PGD-10 and natural data over the training epochs of the standard AT on the CIFAR-10 dataset. AT with networks of varying capacity is presented in the left panel. The standard test accuracy by standard training (ST) is represented by the red line. AT on ResNet-18 under varying perturbation bounds ϵ_{train} is illustrated in the right panel

AT achieves *adversarial robustness* against adversarial data by smoothing the prediction within a small neighbor using adversarial data. However, this approach frequently results in the undesirable deterioration of *standard accuracy* on natural data [373, 462]. Nevertheless, there are two compelling concerns that exist with respect to the data in the general training scheme.

First, the over-parameterized deep networks [461] could still be afflicted by an insufficient *model capacity* as a result of the overwhelming normalizing effect of AT [258]. Fitting adversarial data is a challenging task for the model capacity: it requires a substantial number of trainable parameters or prolonged training epochs to achieve a near-zero error on adversarial training data (see Fig. 3.5). Compared to standard training, which involves fitting the natural data samples, adversarial training smooths the neighborhoods of natural data. Consequently, adversarial data occupy substantially more model capacity than natural data.

Secondly, the significance of the data and its adversarial variants is not consistent. Certain datasets are geometrically distant from the class boundary (refer to Fig. 3.6). These are relatively *guarded*. The adversarial variants are particularly difficult to misclassify. However, certain data are in close proximity to the class boundary. They are relatively vulnerable to attack. They are effortlessly misclassified in their adversarial variants. Consequently, AT should exercise caution in utilizing the restricted model capacity to match the localities of the critical data that facilitate the fine-tuning of the decision boundary.

As training progresses, an adversarially robust model produces an increasing amount of protected training data while reducing the proportion of vulnerable data. If all data is treated uniformly, the model may struggle with the sheer volume of adversarial variations of the protected data, potentially leading to undesirable robust overfitting [311] due to its limited capacity. Therefore, treating all data equally in adversarial training may be an overly cautious approach.

Method *Geometry-Aware Instance-Reweighted Adversarial Training* [469] is proposed to alleviate the previous issues. GAIRAT employs an alternative methodology

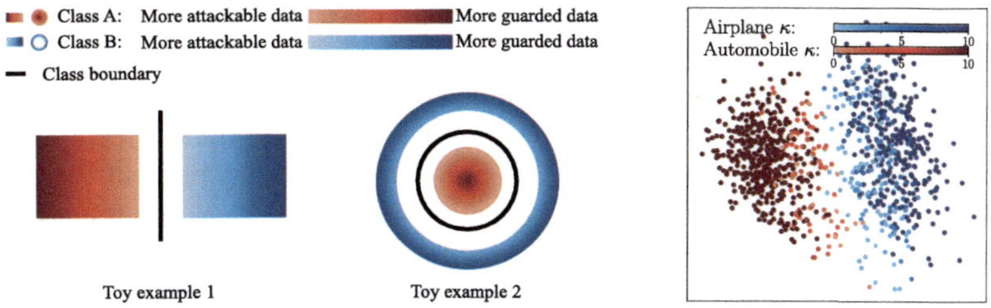

Fig. 3.6 The more attackable data (lighter red and blue) is located closer to the class boundary, while the more guarded data (darker red and blue) is located farther away from the class boundary. Two miniature examples are illustrated in the left panel. The output distribution of the model for two randomly selected classes from the CIFAR-10 dataset is depicted in the right panel. The degree of robustness of a data point (represented by the color gradient) is determined by the minimum number of iterations κ of PGD to identify its misclassified adversarial variant

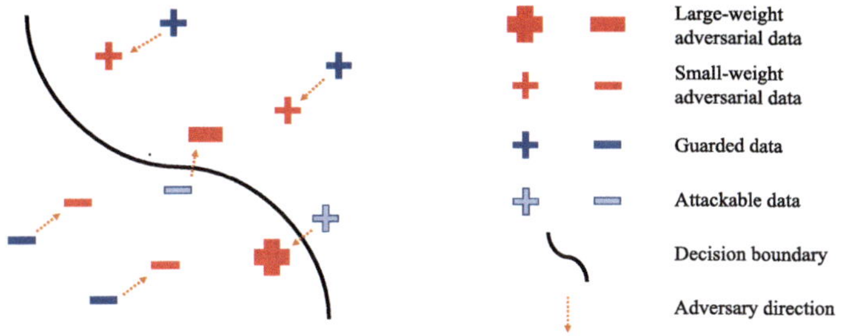

Fig. 3.7 An illustration of GAIRAT. The losses of adversarial data (larger red) are expressly assigned a higher weight in GAIRAT than their natural counterparts, which are closer to the decision boundary (lighter blue). GAIRAT explicitly assigns lower weights to the losses of adversarial examples (smaller red) whose natural counterparts are farther from the decision boundary (darker blue). See Fig. 3.6 for illustrations using the CIFAR-10 dataset and two synthetic datasets.

for adversarial training data, as illustrated in Fig. 3.7. GAIRAT dynamically adjusts the weight assigned to the loss of an adversarial variant of attackable or guarded data points during model updates, based on its importance in refining the decision boundary. A data point is considered attackable or guarded depending on its *geometric distance*, which measures its proximity to the decision boundary. This distance is estimated by the minimum number of iterations, κ, required for the projected gradient descent (PGD) method to generate a misclassified adversarial example. The function $\omega(\mathbf{x}, y)$, known as the *geometry-aware weight assignment function*, determines the weight applied to the loss of an adversarial variant $\tilde{\mathbf{x}}$. The process of generating $\tilde{\mathbf{x}}$ continues to follow the PGD method. The outer minimization is

$$\min_{\theta} \frac{1}{n} \sum_{i=1}^{n} \omega(\mathbf{x}_i, y_i) \ell(f_\theta(\tilde{\mathbf{x}}_i), y_i). \tag{3.9}$$

The constraint first guarantees that $y_i = \arg\max_i f_\theta(\mathbf{x}_i)$ and, secondly, that $\omega(\mathbf{x}_i, y_i)$ is a nonincreasing function with respect to the *geometric distance*. This distance is the distance from data $\mathbf{x}_i$ to the decision boundary, and it is defined as $\omega(\mathbf{x}_i, y_i) \geq 0$ and $\frac{1}{n}\sum_{i=1}^{n} \omega(\mathbf{x}_i, y_i) = 1$. When $y_i \neq \arg\max_i f_\theta(\mathbf{x}_i)$, there are no constraints. Outliers, which are x that are substantially away from the decision boundary, may be discarded, while those that are close to the decision boundary may be assigned high weights. In terms of weight calculation, GAIRAT heuristically develops various nonincreasing functions ω. For instance, one such function is as follows:

$$w(\mathbf{x}, y) = \frac{(1 + \tanh(\lambda + 5 \times (1 - 2 \times \kappa(\mathbf{x}, y)/K)))}{2}, \tag{3.10}$$

where $\kappa/K \in [0, 1]$, $K \in \mathbb{N}^+$, and $\lambda \in \mathbb{R}$. In addition, a burn-in period can be introduced to initialize AT process by setting $\omega(\mathbf{x}_i, y_i) = 1$ since the geometric distance in early training process may not be informative.

Experiments The primary experimental analyses are illustrated in Fig. 3.8. The training statistics are collected at each training epoch, which includes the geometry value $\kappa(\mathbf{x}, y)$ of each training data, standard/robust training and test error, and the flatness of loss with respect to adversarial test data. The geometry value κ of the training data of standard AT is displayed in the bottom-left panel. As the learning

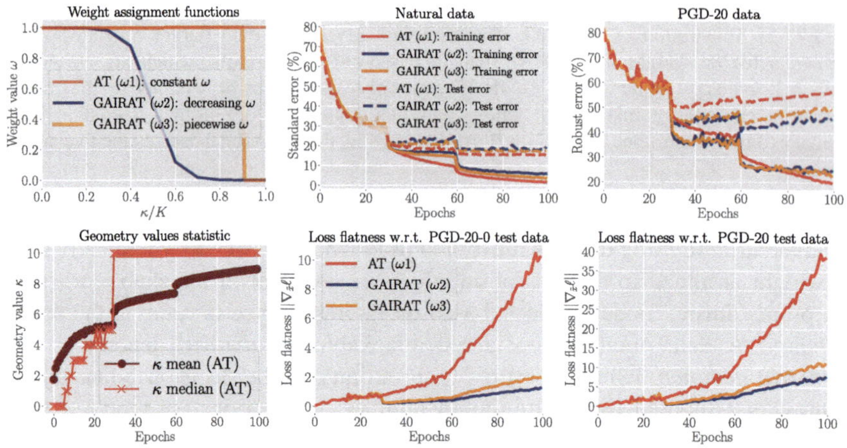

Fig. 3.8 The CIFAR-10 dataset was used to compare AT (ω_1, red lines) with GAIRAT (ω_2, blue lines and ω_3, yellow lines) using ResNet-18. With respect to the geometry value κ, *upper-left* displays several weight assignment functions ω. The standard AT's training statistic is reported by *bottom-left*, which also determines the mean (light red cross) and median (dark red circle) of all training data's geometry values for each epoch. Standard training/test errors are reported by *upper-middle*, whereas robust training/test errors are reported by *upper-right*. The loss flatness with regard to most adversarial test data and friendly adversarial test data is reported by *bottom-middle* and *bottom-right*, respectively

rate decreases to 0.01 at Epoch 30, there is a sudden surge in the amount of guarded training data. Following this, the model progressively generates more encrypted data whose adversarial variants are correctly classified. This reinforcement of existing knowledge—primarily from correctly classified adversarial examples (majority)—while paying minimal attention to misclassified ones (minority) leads to robust overfitting. The red (dashed and solid) lines in these figures illustrate this phenomenon.

To ensure that learning from the relatively rare attackable data is not overshadowed by the abundant guarded data, GAIRAT deliberately assigns lower weights to the losses of adversarial variants of the guarded data. Two varieties of weight assignment functions are represented by the blue (ω_2) and yellow (ω_3) lines in the upper-left. These functions assign instance-dependent weight to the loss based on the geometry value κ. In GAIRAT, the model is compelled to allocate sufficient attention to the infrequent attackable data. The first 30 epochs serve as a burn-in phase, after which the instance-dependent weight assignment ω is applied from Epoch 31 onward (represented by both blue and yellow lines). All other hyperparameters remain the same as in standard adversarial training (AT) (red lines). Compared to conventional AT (red lines), GAIRAT (both yellow and blue lines) achieves lower errors on adversarial test data while exhibiting higher errors on adversarial training data, as shown in the upper-right panel. In general, GAIRAT effectively mitigates robust overfitting and achieves superior performance in the context of adversarial robustness.

3.2.1.2 Adversarial Training with Complementary Labels

In the context of weakly supervised learning, complementary labels (CLs) [173, 174] offer imperfect supervision by assigning a label that is distinct from the ground truth. Recent research [71, 117, 173, 439] has demonstrated promising success in the training of a model to predict the true labels during inference using complementary loss $\bar{\ell}$. The general value in real-world applications is derived from cost-effectiveness and privacy preservation. A novel and challenging setting, *Adversarial training (AT) with CLs*, is currently being investigated in order to advance AT [258] to more practical scenarios.

Despite numerous endeavors, the direct integration of AT with existing CL methodologies [71, 173] consistently leads to failure. Conversely, the two-stage training baseline illustrated in Fig. 3.9 is effective. In order to align the learning process with the performance of ground-truth labels, complementary learning employs correction mechanisms to calculate complementary losses $\bar{\ell}$. The complementary losses $\bar{\ell}$ are employed in the majority of experiments that integrate CLs with AT to generate adversarial examples and perform outer minimization. Nevertheless, the direct interpretation typically leads to the failure training curves, with the exception of the two-stage baseline. This baseline consists of an initial complementary learning phase that predicts the pseudo-label, followed by an AT phase that employs the predicted labels generated in the first stage to produce

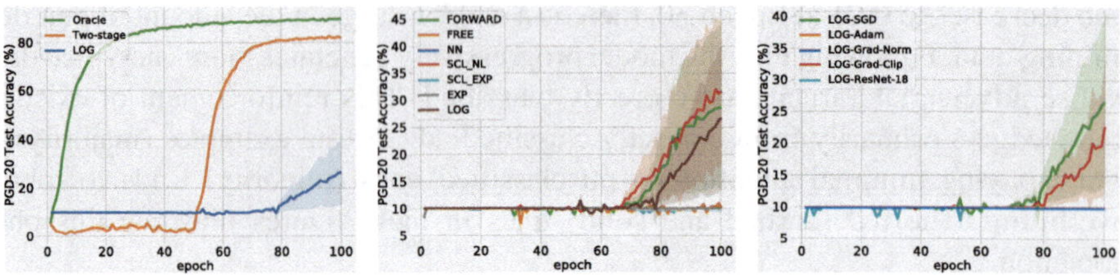

(a) Direct combination (b) Complementary Losses (c) Training Tricks

Fig. 3.9 (**a**) The direct combination of AT with CLs results in weak performance when compared to the two-stage baseline; (**b**) the use of a variety of supplementary loss functions results in *consistent experimental failure* in AT with CLs; (**c**) an array of training alternatives, including optimizers, was evaluated without success

robust models. The comparison emphasizes that AT with CLs encounters substantial obstacles, including *low-quality adversarial examples* and *intractable adversarial optimization.*

Method A new learning strategy [487] that employs incrementally informative attacks is proposed to resolve the problem of AT with CLs. Two critical components of this methodology are the *warm-up attacks (warm-up)*, which facilitate the transition from natural to adversarial training within a suitable budget, and the *pseudo-label attack (PLA)*, which improves adversarial generation by generating insightful predictions from the model's increasing discrimination. These components would be adjusted during training using a flexible learning scheduler.

Warm-Up Attack Adversarial optimization with CLs is more difficult than standard optimization, as evidenced by empirical studies in Fig. 3.10. Variability in variable gradients is the result of training with adversarial samples from multiple labels. Dynamic adversarial budgets can be implemented to gradually improve adversarial generation in order to mitigate variation. In order to achieve inner optimization, the warm-up attack progressively increases the epsilon ball from 0 to ϵ across e_s epochs while maintaining the constant step size and attack steps, taking into account the epsilon ball radius, attack steps, and step size α for implementing adversarial generation.

Pseudo-Label Attack At the initial training stage, the adversarial data within a small epsilon ball can be used to obtain a simple pattern. The original complementary label can be advantageous for the self-evolution of the model to become more discriminative. The model can attribute high confidence to the potential ordinary label and low confidence to others in addition to the training. This allows us to utilize the progressively informative predicted labels to generate high-quality adversarial data. Consequently, it is recommended that the convex combination outputs of partial labels and anticipated ordinary labels be enhanced. The assault is optimized by learning with CLs using corrective methods and subsequently using predicted labels as conventional labels. The specific loss function is defined as

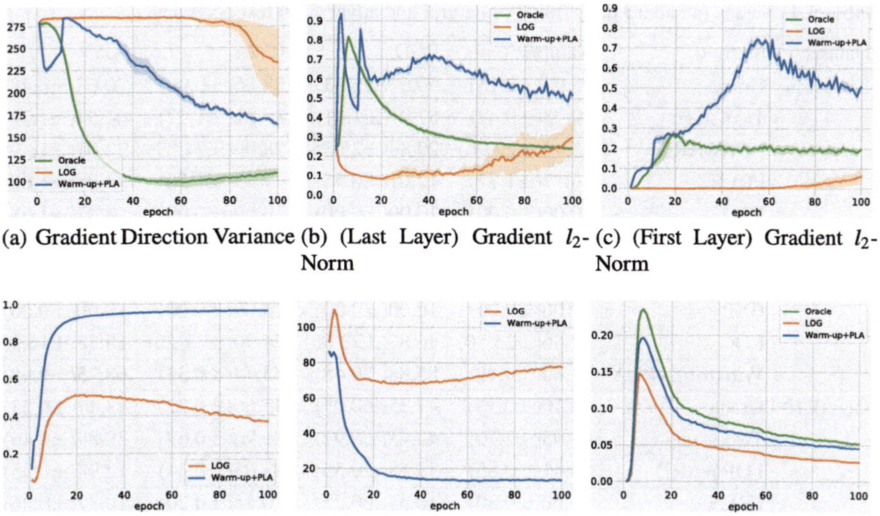

(a) Gradient Direction Variance (b) (Last Layer) Gradient l_2-Norm (c) (First Layer) Gradient l_2-Norm

(d) Cosine Similarity (e) l_1-Distance (f) $p_\theta(y|\mathbf{x}) - p_\theta(y|\tilde{\mathbf{x}})$

Fig. 3.10 [Intractable adversarial optimization] (**a**) The trace sum of the covariance matrix of gradient directions with respect to the first-layer parameters. (**b**) The l_2-norm of stochastic gradients in the last layer. (**c**) The l_2-norm of stochastic gradients in the first layer. [Low-quality adversarial generation] (**d**) the cosine similarity of gradient directions (i.e., $\text{sign}(\nabla_{\mathbf{x}}\bar{\ell}(\mathbf{x}, \bar{y}; \theta))$) with the oracle in the inner maximization; (**e**) the l_1-distance between the generated adversarial examples and the oracle. (**f**) The difference in model predictions between natural and adversarial data for the standard label

$$\bar{\ell}(\mathbf{x}, \bar{y}; \theta) = -(C-1)\log(\gamma \sum_{j \neq \bar{y}} p_\theta(j|\mathbf{x}) + (1-\gamma)p_\theta(\hat{y}|\mathbf{x})), \tag{3.11}$$

$$\hat{y} = \arg\max_{j \neq \bar{y}} p_c(j|\mathbf{x}),$$

where the learning transition is regulated by the hyperparameter $\gamma \in [0, 1]$. To prevent deterioration and stabilize optimization, a simple exponential moving average is used to gather model prediction dynamics. The label with the largest probability is then selected as the projected label. In particular, the model predictions $p_c(\mathbf{x})$ are updated as $p_c(\mathbf{x}) = \beta p_c(\mathbf{x}) + (1-\beta)p_\theta(\mathbf{x})$ given the model predictions $p_\theta(\mathbf{x})$ of the current epoch. $p_c(\mathbf{x})$ is initialized with $\frac{(\mathbf{1}-e_{\bar{y}})^T}{C-1}$, $p_c(\bar{y}|\mathbf{x}) = 0$, and $\beta = 0.9$.

Experiments The empirical findings are presented in Table 3.4, where *Oracle* represents the conventional AT [258] with ordinary labels. This comparison was made to assess the performance disparity between AT with CLs and ordinary labels. A number of baselines are employed to facilitate the comparison—1) *Two-stage method*: this method initially trains a model for 50 epochs using complementary learning methods (e.g., LOG [109]), which exhibits superior performance and stability in comparison to other methods. Then, the checkpoint with the greatest

Table 3.4 Means (standard deviations) of natural and adversarial test accuracy

Dataset	Method	Natural	PGD	CW	AA
Kuzushiji	Oracle	95.94(±0.15)	90.01(±0.43)	88.06(±0.96)	70.63(±0.48)
	Two-stage	89.75(±0.42)	82.91(±1.01)	80.21(±1.27)	64.57(±1.79)
	FORWARD	35.48(±27.96)	29.84(±25.55)	28.09(±24.37)	22.01(±18.98)
	FREE	16.17(±1.77)	12.01(±0.55)	9.33(±1.50)	4.08(±1.60)
	NN	10.00(±0.00)	10.00(±0.00)	10.00(±0.00)	8.87(±1.6)
	SCL_NL	40.83(±24.03)	32.82(±22.88)	29.93(±22.53)	20.86(±19.74)
	SCL_EXP	10.00(±0.00)	10.00(±0.00)	10.00(±0.00)	8.21(±2.54)
	EXP	10.00(±0.00)	10.00(±0.00)	10.00(±0.00)	10.00(±0.00)
	LOG	32.66(±25.50)	26.87(±22.66)	24.90(±21.20)	18.78(±16.95)
	Warm-up+PLA	**91.60(±0.49)**	**85.88(±0.48)**	**83.74(±0.35)**	**68.75(±0.68)**
CIFAR10	Oracle	78.10(±0.36)	47.35(±0.05)	45.66(±0.22)	43.47(±0.23)
	Two-stage	64.98(±2.70)	42.48(±0.92)	39.90(±0.67)	38.89(±0.46)
	FORWARD	15.41(±0.85)	13.58(±0.56)	13.03(±0.34)	12.94(±0.35)
	FREE	11.60(±0.30)	10.54(±0.23)	10.45(±0.20)	10.32(±0.26)
	NN	11.79(±0.61)	11.13(±0.56)	11.14(±0.56)	11.10(±0.53)
	SCL_NL	14.45(±1.07)	13.13(±1.03)	12.84(±1.17)	12.79(±1.18)
	SCL_EXP	13.49(±1.76)	12.55(±1.39)	12.45(±1.35)	12.38(±1.31)
	EXP	10.97(±1.01)	10.56(±0.60)	10.38(±0.36)	10.37(±0.34)
	LOG	11.70(±1.06)	11.04(±0.72)	10.49(±0.59)	10.44(±0.55)
	Warm-up+PLA	**65.88(±1.51)**	**43.29(±0.70)**	**41.30(±0.59)**	**40.28(±0.39)**

validation performance is used to relabel the complementary training set, and a new model is adversarially trained using the relabeled training set later. 2) *Direct adversarial training*: adversarial training is conducted directly with complementary objectives, including FORWARD [439], FREE [174], NN [174], SCL_NL [71], SCL_EXP [71], EXP [109], and LOG [109]. The results of the checkpoint with the highest PGD20 test accuracy are reported, and adversarial robustness is assessed using PGD20 [258], CW30 [51], and AutoAttack [76] (AA). The proposed warm-up assault and pseudo-label attack exhibit superior performance on both datasets, which serves as evidence of their general efficacy.

3.2.1.3 Reliable Adversarial Distillation with Unreliable Teachers

Most adversarial training methods [258, 387, 462] emphasize the acquisition of model robustness in adversarial data generation [124]. It is, however, difficult to optimize directly from training on large datasets due to the fact that robust accuracy is restricted by the use of hard labels to maximize loss [243]. Recent research indicates that robustness can be extracted from the adversarially pretrained model [64, 122]. To enhance the robustness of the model, an adversarial robust model is pretrained on the dataset, and, subsequently, *adversarial distillation* [122]

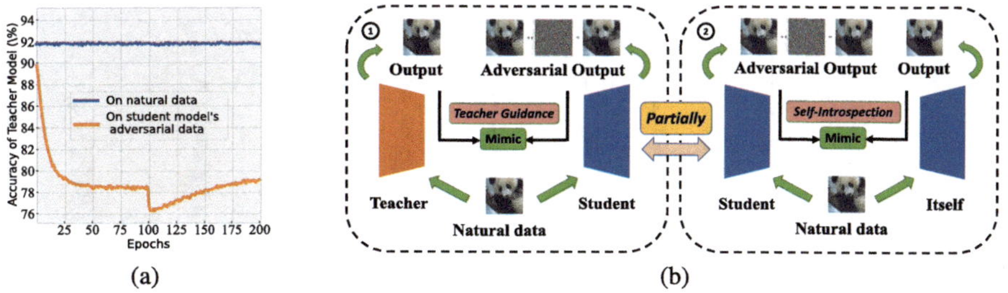

Fig. 3.11 Unreliability issue: A comparison of the accuracy of both teacher and student models on natural and adversarial data from the CIFAR-10. As opposed to preserving the natural data and the teacher model's typical performance, distillation reduces the teacher model's robust accuracy on the student model's adversarial data, rendering its counsel less reliable. The following is an overview of introspective adversarial distillation (IAD): the student partially trusts the teacher's guidance and self-introspection during hostile distillation. Within IAD, the student model generates adversarial data and partially replicates both the teacher model and itself. (**a**) Unreliability issue. (**b**) Overview of introspective adversarial distillation (IAD)

motivates a student model to replicate the outputs (soft labels) [156] of the adversarially pretrained instructors in order to extract robustness.

Nevertheless, there is a critical distinction in adversarial distillation: the instructor and pupil models can generate distinct adversarial training data based on their distinct model parameters, despite the fact that they share the same natural data during the distillation. The instructor model may progressively fail to provide the correct prediction for the adversarial data queried from the student model, as illustrated in the left panel of Fig. 3.11. The teacher model is unable to consistently develop the ability to provide completely reliable supervision as a result of the progressively increased complexity of the adversarial data generated by the robust pupil model.

Method Introspective adversarial distillation (IAD) [491] is proposed to resolve the issue of the unreliable teacher in adversarial distillation [122]. This method is illustrated on the right side of Fig. 3.11. The underlying intuition is to gradually incorporate self-prediction results as the supervision for adversarial training and partially trust the instructor model during the adversarial distillation. In particular, the rationality can be illustrated through the examination of the subsequent two viewpoints.

Fully Trust: Progressively Unreliable Guidance The teacher model is entirely trusted in the previous methods [64, 122] when attempting to extract robustness from adversarially pretrained models. The student model in adversarial robust distillation (ARD) generates adversarial data and optimizes prediction to align with the instructor model's output, as illustrated in the left portion of Fig. 3.11. Despite the fact that the instructor model is highly optimized for adversarial data queried independently, it may not perform as well on the more challenging adversarial data of the student model. The instructor model's accuracy on adversarial data decreases

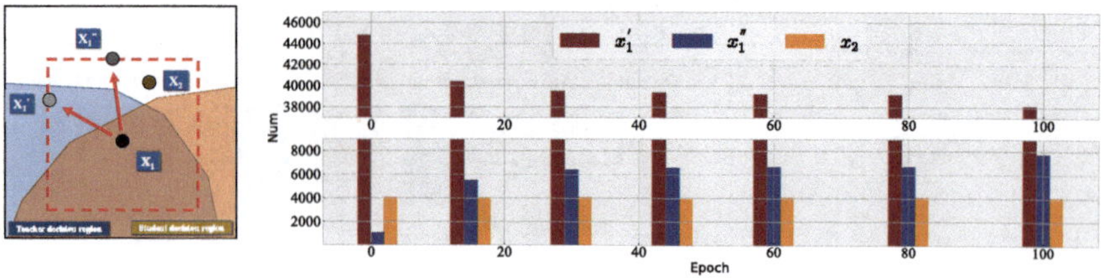

(a) Toy illustration (b) Changes on three data types during distillation

Fig. 3.12 (**a**) Toy illustration: Three educators simulate prediction scenarios. Red dashed sections denote AT's unit-norm ball, while blue/orange sections denote the teacher/student model's decision zone (where it can accurately predict). (1) x_1'. Teachers are capable of precisely predicting hostile situations based on blue natural examples x_1; (2) x_1''. The incorrect instructor model prediction is demonstrated by the adversarial example from x_1 outside the orange and blue sections; (3) x_2. Natural situations are beyond the capacity of the teacher model to predict. (**b**) Modifications to three data types during the distillation process: During adversarial distillation on the *CIFAR-10* dataset, three categories of data are monitored. In contrast to the constant number of examples like x_2, the number of instances like x_1' decreases while that of x_1'' increases during adversarial distillation

during distillation, despite its strong performance on natural data, as illustrated in Fig. 3.12. Ultimately, the guidance provided by the teacher model is unable to produce the appropriate results when the student model queries adversarial data.

Partially Trust: Construction of Reliable Guidance The guidance for adversarial distillation should be reevaluated due to the fact that unreliable instructors impede reliability. In natural instances, the predicted teacher model label is denoted by $\mathcal{T}(\mathbf{x})$ ($\mathcal{T}(\tilde{\mathbf{x}})$), and the targeted label is denoted by y. Adversarial data are categorized into distinct categories and analyzed individually using the toy representation (Fig. 3.12)(1) The teacher model performs well on both natural and adversarial data, as illustrated in Fig. 3.12, where $\mathcal{T}(\mathbf{x}) = y \cap \mathcal{T}(\tilde{\mathbf{x}}) = y$. Adversarial variants, such as x_1', are the most reliable among the three varieties. Nevertheless, the sample quantity in this type decreased during distillation, as shown in Fig. 3.12(2) $\mathcal{T}(\mathbf{x}) = y \cap \mathcal{T}(\tilde{\mathbf{x}}) \neq y$: the teacher model's outputs on the adversarial data in Fig. 3.12 are inconsistently accurate, which may result in the student model's guidance and expertise being potentially inaccurate. The student model may demonstrate reliability as a result of the adversarial robustness that was obtained during distillation, which may explain the decrease in the robust accuracy of the teacher model. (3) The teacher model is unreliable for data such as x_2 in Fig. 3.12 due to incorrect labels on genuine data. Specifically, $\mathcal{T}(\mathbf{x}) \neq y \cap \mathcal{T}(\tilde{\mathbf{x}}) \neq y$. In lieu of the instructor model's inaccurate prediction, student models may depend on their own outputs that are consistent with their own output. 4) $\mathcal{T}(\mathbf{x}) \neq y \wedge \mathcal{T}(\tilde{\mathbf{x}}) = y$: in the event that the initial prediction is incorrect, then the generation of $\tilde{\mathbf{x}}^*$ exacerbates the prediction error, as indicated by the adversarial data generation process, $\tilde{\mathbf{x}}^* = \arg\max_{\tilde{\mathbf{x}} \in \mathcal{B}_\epsilon(\mathbf{x})} \ell(f(\tilde{\mathbf{x}}), y)$. Consequently, this group is nonexistent.

Building on prior analyses of adversarial distillation, the objective of IAD can be introduced as follows:

$$\ell_{\text{IAD}} = \underbrace{O(AD_i;\alpha)}_{\text{Label \& Teacher Guidance}} + \gamma \underbrace{\ell_{\text{KL}}(S(\tilde{\mathbf{x}}|\tau)||S(\mathbf{x}|\tau))}_{\text{Student Introspection}}, \tag{3.12}$$

where $O(AD_i;\alpha)$ represents the prior adversarial distillation baseline, such as ARD or AKD^2, weighted by α and γ. $S(\cdot|\tau)$ is a Softmax operator with the temperature $\tau \in (0, +\infty)$, *e.g.*, $S(\mathbf{x}|\tau) = \frac{\exp(p_y(\mathbf{x})/\tau)}{\sum_{y' \neq y} \exp(p_{y'}(\mathbf{x})/\tau)}$, $S(\cdot)$ is the conventional Softmax with the temperature $\tau = 1$, $T(\cdot|\tau)$ is the tempered variant of the teacher output $T(\cdot)$, $\tilde{\mathbf{x}}$ is the adversarial data generated from the natural data $\mathbf{x}$, y is the hard label, and ℓ_{CE} is the cross-entropy loss and ℓ_{KL} is the KL-divergence loss. The annealing parameter $\alpha \in [0, 1]$, used to balance the teacher model in adversarial distillation, is defined as $\alpha = (P_T(\tilde{\mathbf{x}}|y))^\beta$, where $P_T(\cdot|y)$ represents the teacher model's prediction probability for the label y, whereas β sharpens the prediction. Specifically, IAD can be implemented as two types to involve previous advanced designs [64, 122] for adversarial distillation, which we present their loss functions as

$$\begin{aligned}\ell_{\text{IAD-I}} =& \lambda \ell_{\text{CE}}(S(\mathbf{x}), y) + (1-\lambda) \cdot \tau^2 \cdot \\ & (\ell_{\text{KL}}(S(\tilde{\mathbf{x}}|\tau)||T(\mathbf{x}|\tau)) + \gamma \ell_{\text{KL}}(S(\tilde{\mathbf{x}}|\tau)||S(\mathbf{x}|\tau))),\end{aligned} \tag{3.13}$$

$$\begin{aligned}\ell_{\text{IAD-II}} =& \lambda_1 \ell_{\text{CE}}(S(\tilde{\mathbf{x}}), y) + \lambda_2 \cdot \tau^2 \cdot (\alpha \cdot \ell_{\text{KL}}(S(\tilde{\mathbf{x}}|\tau)||T_{\text{AT}}(\tilde{\mathbf{x}}|\tau)) \\ & + \lambda_3 \tau^2 \ell_{\text{KL}}(S(\tilde{\mathbf{x}}|\tau)||T_{\text{ST}}(\tilde{\mathbf{x}}|\tau)) + \gamma \ell_{\text{KL}}(S(\tilde{\mathbf{x}}|\tau)||S(\mathbf{x}|\tau))).\end{aligned} \tag{3.14}$$

Experiments The results are presented in Table 3.5, where the first and fifth rows, respectively, demonstrate AT [258] and TRADES [462], while the other techniques employ these models as teacher models in the distillation process. In the context of *CIFAR-10* and *CIFAR-100*, IAD-I and IAD-II exhibit consistent enhancements in adversarial robustness, as measured by PGD-20 [258], CW∞ [51], and AA [76] accuracy, in comparison to student models distilled by ARD [122] or AKD^2 [64], as well as adversarially pretrained teacher models [258]. Additionally, when the teacher model is trained on AT [258], IAD-II frequently outperforms IAD-I, suggesting that AKD^2 may be a more favorable starting point than ARD in this context. Nevertheless, the robustness of IAD-I exceeds that of IAD-II when the teacher model is trained using TRAEDS. Based on their distillation philosophy, which is represented by $\ell_{\text{KL}}(S(\tilde{\mathbf{x}}|\tau)||T(\mathbf{x}|\tau))$ for IAD-I and $\ell_{\text{KL}}(S(\tilde{\mathbf{x}}|\tau)||T(\tilde{\mathbf{x}}|\tau))$ for IAD-I, IAD-II may be contingent upon the more informative of the two options, $T(\mathbf{x}|\tau)$ or $T(\tilde{\mathbf{x}}|\tau)$, in the context of adversarial distillation from a variety of teachers.

Table 3.5 Test accuracy (%) on *CIFAR-10/CIFAR-100* datasets using *ResNet-18*

Datasets		CIFAR-10					CIFAR-100				
Teacher	Method	Natural	FGSM	PGD-20	CW_∞	AA	Natural	FGSM	PGD-20	CW_∞	AA
AT	AT	83.06%	63.53%	50.21%	49.22%	46.70%	56.21%	33.94%	24.60%	23.35%	21.55%
	ARD	83.13%	**64.11%**	51.36%	50.37%	48.05%	56.65%	35.82%	27.04%	25.30%	23.34%
	AKD[2]	**83.52%**	63.91%	51.36%	50.36%	48.08%	56.46%	35.76%	27.18%	25.33%	23.47%
	IAD-I	82.09%	63.20%	**52.14%**	**50.74%**	**48.66%**	55.53%	35.69%	27.40%	**25.80%**	23.66%
	IAD-II	83.21%	63.54%	51.85%	50.67%	48.58%	**57.26%**	**35.94%**	**27.50%**	25.68%	**24.06%**
TRADES	TRADES	81.26%	63.12%	52.98%	50.29%	49.47%	54.01%	35.23%	28.00%	24.26%	23.46%
	ARD	81.50%	63.38%	53.38%	51.27%	49.92%	55.82%	**37.56%**	30.21%	26.76%	25.36%
	AKD[2]	83.49%	64.00%	51.89%	50.25%	48.52%	**57.46%**	37.32%	28.62%	25.76%	24.36%
	IAD-I	80.63%	63.14%	**53.84%**	**51.60%**	**50.17%**	55.19%	37.47%	**30.60%**	**27.24%**	**25.84%**
	IAD-II	**83.76%**	**64.17%**	52.16%	50.59%	48.91%	57.08%	36.94%	29.08%	25.83%	24.45%

3.2.1.4 Defense Transformations for Counterattacking Adversarial Examples

Existing defense techniques frequently focus on particular classifications of adversarial examples, so limiting their efficacy in practical situations. In practice, we face several forms of attacks, and the specific characteristics of adversarial cases may remain ambiguous. This study investigates a novel approach to fighting against adversarial examples, based on the finding that such examples typically exist at decision boundaries and are susceptible to specific changes, by remapping them to the original clean data distribution. We parameterize affine transformations and utilize deep neural network (DNN) boundary information to develop defense mechanisms against adversarial perturbations. These learned transformations effectively revert a substantial fraction of adversarial examples to the clean data distribution. As a result, our defense strategy exhibits strong efficacy against various adversarial examples, including non-noise-based assaults such as stAdv [416] and EOT [12].

Method Given a pretrained classification model h, our aim is to build a transformation function $\mathcal{T}$ that maps adversarial examples $\tilde{\mathbf{x}}$ back to the clean data distribution, therefore assuring that $\mathcal{T}(\tilde{\mathbf{x}})$ faithfully predicts the label y. To accomplish this, we employ a spatial transformer network (STN) [177] to parameterize affine transformations. We then combine it with a U-Net [314] for strong feature representation. The U-Net generates adversarial samples to support the STN in learning efficient affine transformations. Figure 3.13 shows the general framework. The defense transformation utilizes an affine matrix $\Theta \in \mathbb{R}^{2\times 3}$ with a localization network g, to convert coordinates (u, v) of $\tilde{\mathbf{x}}$ into (u', v'):

$$\begin{bmatrix} u' \\ v' \end{bmatrix} = \begin{bmatrix} \theta_{11} & \theta_{12} & \theta_{13} \\ \theta_{21} & \theta_{22} & \theta_{23} \end{bmatrix} \begin{bmatrix} u \\ v \\ 1 \end{bmatrix}, \tag{3.15}$$

where θ_{ij} is the (i, j)-th element of the affine matrix Θ and $\Theta = g(\tilde{\mathbf{x}})$. We use bilinear interpolation to obtain the transformed sample $\mathcal{T}(\tilde{\mathbf{x}})$.

To train $\mathcal{T}$, we minimize the following objective function:

$$\min_{w} \frac{1}{mn} \sum_{i=1}^{n} \sum_{j=1}^{m} \ell\left(h\left(\mathcal{T}_w\left(\tilde{\mathbf{x}}_i^{(j)}\right)\right), y_i\right) + \frac{1}{2}\lambda \|w\|^2, \tag{3.16}$$

where ℓ denotes the cross-entropy loss, w signifies the parameters of $\mathcal{T}$, λ indicates a regularization hyperparameter, and m varieties of adversarial instances are produced for each input. This optimization utilizes the classifier's boundary knowledge to direct the acquisition of effective defensive transformations.

Experiments From Table 3.6, the proposed defense strategy surpasses existing approaches in handling both linear and nonlinear adversarial instances (AE), showcasing its capacity to enhance robustness in scenarios where random transformations

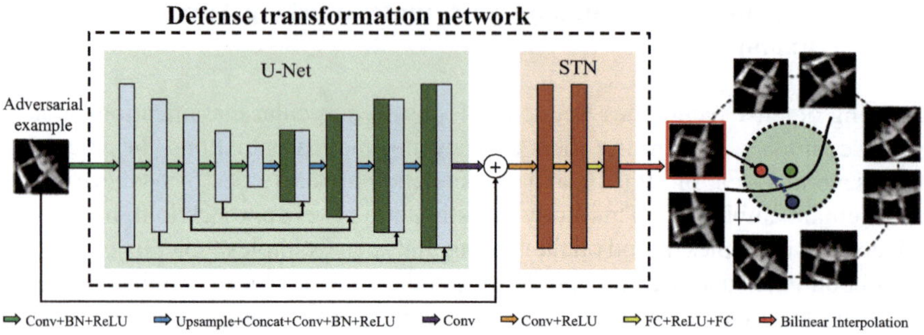

Fig. 3.13 Overview of the proposed defense method comprises a U-Net and a spatial transformer network (STN). During training, the defense transformation network learns an affine transformation to map adversarial examples back to the clean data distribution

and denoising techniques such as HGD, feature denoising, and CAFD are inadequate. Our solution significantly alleviates the deterioration induced by random affine transformations on clean data, diminishing the accuracy decline (e.g., from 93.51% to 36.98%) while attaining slightly inferior performance compared to HGD. Our technique demonstrates superior performance in acquiring effective affine transformations for various AEs on CIFAR-100. Notably, for FGSM, it enhances adversarial accuracy beyond clean accuracy (75.54% vs. 71.08%), possibly by converting certain misclassified clean samples into the right distribution.

We also examine the transferability of our defense transformation network by utilizing a pretrained network from model A to model B (referred to as "A $\rightarrow$ B"). Table 3.7 illustrates that our technique greatly surpasses HGD in generalization, particularly with nonlinear AE, showing its enhanced transferability between ResNet-56 and VGG-16-BN. Moreover, employing U-Net as a feature representation module mitigates the shortcomings of shallow architectures such as STN in acquiring effective defensive affine transformations. Table 3.8 indicates that the integration of U-Net with STN yields superior performance, exceeding STN alone by an average of 23.88%, underscoring the advantages of employing U-Net for resilient feature representation.

3.2.2 *Adversarial Overfitting*

3.2.2.1 Understanding Robust Overfitting of Adversarial Training

Adversarial noise, often invisible to humans, can manipulate deep neural networks, posing security risks in real-world applications. Protection mechanisms have been developed to address this vulnerability. Among these defenses, adversarial training (AT) [124, 258] stands out as one of the most robust empirical approaches [11].

Table 3.6 Comparisons (top-1 accuracy (%)) of defending against various adversarial examples over ResNet-56 on CIFAR-10 and CIFAR-100. "AE" denotes the adversarial examples

			Linear AE						Non-linear AE	
Datasets	Methods	Clean	FGSM	I-FGSM	PGD-10	PGD-100	DeepFool	CW-L2	stAdv	ADef
CIFAR-10	Natural training	**93.51**	21.03	0.00	0.00	0.00	5.08	3.66	0.39	4.25
	Random transformation	36.98	25.21	18.08	23.39	22.11	37.10	31.06	32.88	33.73
	Cropping-ensemble	86.31	32.56	0.46	6.47	5.12	72.89	50.08	42.08	44.04
	Mixup inference	81.47	42.38	16.58	28.40	25.89	79.51	66.31	64.94	74.23
	Feature denoising	82.32	78.68	79.13	80.48	80.77	82.00	80.56	80.06	79.90
	BaRT	81.48	79.53	82.15	83.35	82.91	82.27	81.01	81.34	81.21
	HGD	93.23	90.24	87.79	89.27	90.51	90.21	90.21	44.71	82.57
	CAFD	90.10	61.85	86.28	88.35	89.34	90.26	**90.70**	15.94	65.87
	Ours	90.41	**92.34**	**89.97**	**91.06**	**90.59**	**90.54**	90.44	**85.59**	**86.94**
CIFAR-100	Natural training	**71.08**	9.42	0.07	0.09	0.03	13.95	3.13	0.68	9.88
	Random transformation	22.42	10.28	6.58	8.76	8.78	20.55	16.16	17.44	17.86
	Cropping-ensemble	57.47	13.52	0.21	1.64	0.99	48.36	33.45	27.94	29.12
	Mixup inference	58.42	22.50	14.06	22.48	19.65	58.11	50.59	47.08	53.95
	Feature denoising	51.51	48.19	47.92	49.45	49.59	51.44	49.48	50.19	50.23
	BaRT	50.75	47.08	47.80	49.97	49.93	51.18	49.50	50.86	50.31
	HGD	70.48	59.66	46.61	57.00	58.60	61.22	65.03	25.94	55.73
	CAFD	65.46	29.94	53.00	59.27	60.67	65.82	**66.73**	6.74	49.01
	Ours	67.22	**75.54**	**63.49**	**67.33**	**66.55**	**66.80**	65.34	**59.01**	**61.76**

Table 3.7 Performance comparisons (top-1 accuracy (%)) of defending against the adversarial examples generated by different models on CIFAR-10. "AE" denotes the adversarial examples

Process	Methods	Linear AE		Nonlinear AE	
		FGSM	PGD-10	stAdv	ADef
ResNet-56 → ResNet-56	HGD	90.24	89.27	44.71	82.57
	Ours	**92.34**	**91.06**	**85.59**	**86.94**
ResNet-56 → VGG-16-BN	HGD	65.06	68.96	31.62	38.56
	Ours	**76.74**	**81.01**	**80.64**	**75.56**
VGG-16-BN → VGG-16-BN	HGD	90.21	90.38	73.70	68.89
	Ours	**92.74**	**91.49**	**81.63**	**76.94**
VGG-16-BN → ResNet-56	HGD	88.82	88.63	75.64	84.11
	Ours	**90.69**	**88.90**	**82.20**	**84.70**

AT introduces adversarial examples during training, framing it as a min-max optimization problem [258].

Robust overfitting (RO) is a common issue in AT. After early learning rate decay, accuracy on test data declines as training progresses [311]. We designed data ablation experiments and discovered that RO results from small-loss data. We introduce minimum loss constrained adversarial training (MLCAT) as a solution to this problem. MLCAT is a method used in AT to distinguish between small- and large-loss data, aiming to maximize classification loss under worst-case scenarios. By training with a minimum loss restriction, it reduces RO and offers an alternative viewpoint on AT.

The achievement of the double descent phenomenon in AT generally requires a significantly larger dataset [52, 453], which can be costly in terms of data collection and training time. Unlike these methods, we aim to eliminate RO without adding data. Removing small-loss data that contribute to RO can lower sample size, which is crucial for generalization [325]. Thus, we implement additional techniques to adjust the loss of such data, maximizing the value of each adversarial sample.

Method: Non-overfit AT vs. Overfitted AT the Data Distribution Perspective By adjusting the adversary strength through changes in the perturbation budget ϵ, we analyze the data distribution between overfit and non-overfit AT. In a L_∞ threat model, we train models with different ϵ values and use the appropriate ϵ in each scenario to assess the accuracy of the trained model. The test robustness for AT across a range of adversary strengths is displayed in Fig. 3.14a. Interestingly, RO is evident in strong adversaries but insignificant in weak adversaries. The distribution of training data for each setup over various loss ranges is displayed in Fig. 3.14b, c.

The data distributions under overfitted and non-overfitted AT differ dramatically, as shown in Fig. 3.14b, c: Unlike non-overfit AT, overfitted AT often incorporates both small- and large-loss data. This observation gives rise to (Q1): is it possible to eliminate RO in overfitted AT by suppressing large-loss data?

RO happens when the adversary gets stronger. For training with a powerful adversary, large-loss data are expected and beneficial, whereas small-loss data

Table 3.8 Performance comparisons (top-1 accuracy (%)) of defending against various adversarial examples over ResNet-56 with different network architectures on CIFAR-10. "AE" denotes adversarial examples

Methods	Architectures	Clean	Linear AE						Nonlinear AE	
			FGSM	I-FGSM	PGD-10	PGD-100	DeepFool	CW-L2	stAdv	ADef
Ours	STN	78.81	57.18	54.35	64.18	64.49	76.79	70.53	73.76	52.89
	Auto-Encoder + STN	70.06	63.82	62.25	65.47	65.09	69.43	66.44	67.10	67.82
	U-Net + STN	**90.41**	**92.34**	**89.97**	**91.06**	**90.59**	**90.54**	**90.44**	**85.59**	**86.94**

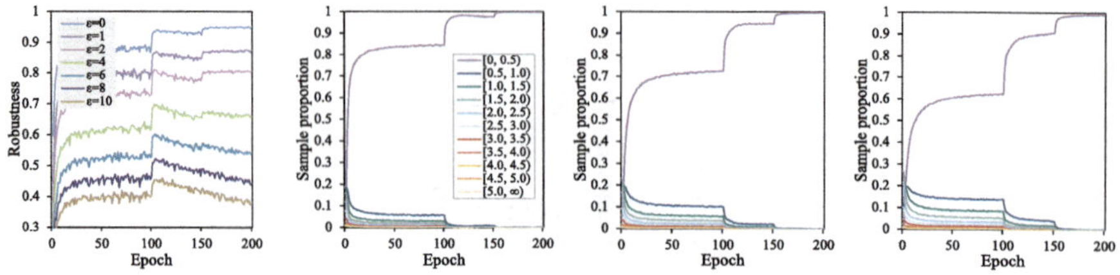

(a) (b) Data distribution across perturbation sizes of 0, 1, and 2 (shown sequentially from left to right)

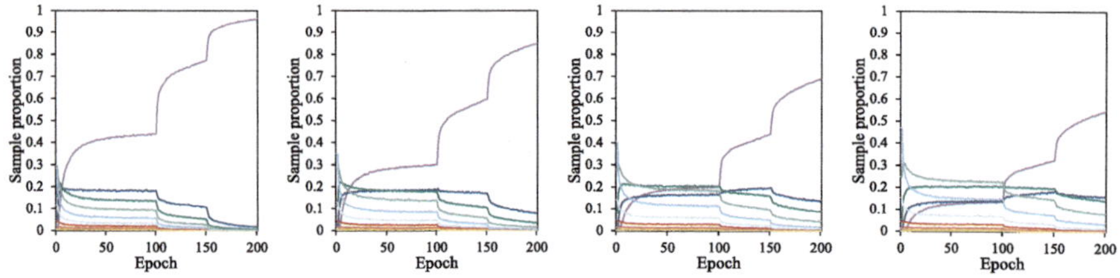

(c) Data distribution across perturbation sizes of 4, 6, 8, and 10 (shown sequentially from left to right)

Fig. 3.14 (**a**) Test robustness of adversarial training across different perturbation sizes ϵ; (**b**) and (**c**) Training data distribution across various loss ranges under different perturbation sizes ϵ

might not be as desirable. This viewpoint raises a second issue (Q2): could RO be eliminated if we suppress the small-loss data that do not match the adversary's strength?

Causes of Robust Overfitting: To address the two problems, we perform AT with $\epsilon = 8$ using a data ablation mode. We remove data from different loss ranges. We remove large-loss data prior to RO sets for Q1. For Q2, we remove small-loss data during AT. As demonstrated in Fig. 3.15a, b, AT without large-loss data still exhibits significant RO. Nevertheless, RO is eliminated when training without small-loss data. These results provide empirical evidence that RO is mostly caused by small-loss data.

Figure 3.14b, c indicate that there are two sources of small-loss data. One source is transformed from different loss ranges after the 100th epoch, whereas the other source is the original small-loss data. Two small-loss data subsets contribute uniquely to RO, according to additional research (see Fig. 3.15c). The results show that the transformed small-loss data is the main source of RO. In the following section, we suggest a novel AT technique to tackle this problem.

A Prototype of MLCAT: Small-loss data cause RO in AT. To overcome this, we propose MLCAT, which involves training the model on adversarial data with a minimum loss constraint. MLCAT seeks to prevent RO while keeping the training sample size.

Algorithm 1 outlines the MLCAT prototype, designed as a flexible framework where both the loss adjustment strategy S and the minimum loss threshold ℓ_{min}

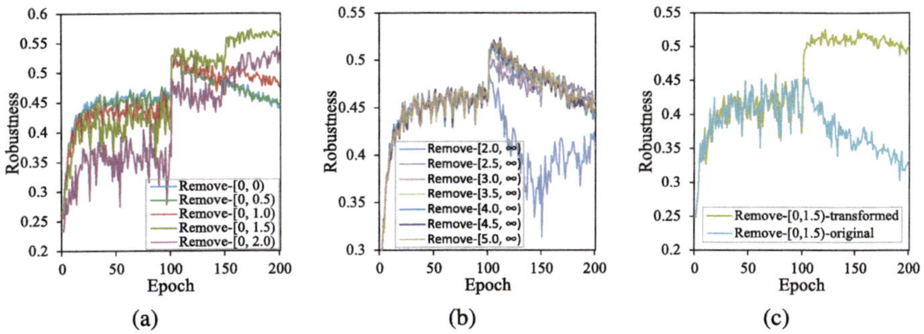

Fig. 3.15 Test robustness of adversarial training: (**a**) excluding small-loss data across various loss ranges; (**b**) excluding large-loss data across various loss ranges; and (**c**) excluding both original small-loss data and transformed small-loss data

---

Algorithm 1 MLCAT-prototype (in a mini-batch)

Require: base adversarial training algorithm A, optimizer O, network f_θ, training data $D = \{(\mathbf{x}_i, y_i)\}_{i=1}^n$, mini-batch B, batch size m, minimum loss conditions ℓ_{min} for A, loss adjustment strategy S

1: Sample a mini-batch $B = \{(\mathbf{x}_i, y_i)\}_{i=1}^m$ from D
2: $\tilde{B} = A$.inner_maximization(f_θ, B)
3: $\{\ell_i\}_{i=1}^m \leftarrow \ell(f_\theta, \tilde{B})$
4: $\ell_{\tilde{B}} \leftarrow 0$ # initialize loss accumulator
5: **for** $i = 1, ..., m$ **do**
6:     **if** $\ell_i \geq \ell_{min}$ **then**
7:         $\ell_{\tilde{B}} = \ell_{\tilde{B}} + \ell_i$
8:     **else**
9:         $\ell_i^S \leftarrow S(f_\theta, \tilde{\mathbf{x}}, \ell_{min})$ # adjust loss
10:         $\ell_{\tilde{B}} = \ell_{\tilde{B}} + \ell_i^S$ # accumulate adjusted loss
11:     **end if**
12: **end for**
13: $\ell_{\tilde{B}} \leftarrow \ell_{\tilde{B}}/m$ # average accumulated loss
14: $\nabla_\theta \leftarrow A$.outer_minimization($f_\theta$, $\ell_{\tilde{B}}$)
15: O.step(∇_w)

---

can be adapted based on the base AT algorithm A. The procedure is as follows: given a mini-batch B, the inner maximization pass of A is executed in Line 2. Lines 3–13 adjust the loss values, reducing them to a scalar for backpropagation. The loss accumulator $\ell_{\tilde{B}}$ is initialized in Line 4. In the loop, if adversarial example $\tilde{\mathbf{x}}_i$ is classified as large-loss data, its loss ℓ_i is added to $\ell_{\tilde{B}}$ in Line 7, enabling standard gradient descent via O in Line 15. For small-loss data, an adjusted loss ℓ_i^S is computed using S in Lines 9–10 and added to $\ell_{\tilde{B}}$, leading to an adjusted gradient descent in Line 15. Total accumulated loss $\ell_{\tilde{B}}$ is averaged by dividing by batch size m in Line 13 after the loop. After calling the outer minimization pass of A in Line 14, the optimizer O updates the model f_θ in Line 15.

Finally, it is important to highlight the versatility of MLCAT. If the strategy S is an identity mapping, MLCAT reduces to conventional AT. If it continuously returns 0, it becomes data ablation AT. Since sample size significantly affects RO in AT [325], we leverage S to transform small-loss data into large-loss data. This translation lets the model learn from all training data without RO, a key MLCAT notion of "turning waste into treasure." If ℓ_{min} is less than 0, MLCAT becomes standard AT. If it is greater than ∞, all data undergoes adjustment through S. In MLCAT, ℓ_{min} is directly related to perturbation size ϵ and task details. Therefore, balancing RO prevention and robustness improvement requires careful calibration of ℓ_{min}.

MLCAT through loss scaling: We use simple scaling to increase the loss of small-loss data. To prevent RO, we use the minimum loss threshold ℓ_{min} as a constraint. The loss ℓ_i is scaled according to the following rule:

$$\ell_i^S = \frac{\ell_{min}}{\ell_i} \cdot \ell_i = \ell_{min}. \tag{3.17}$$

Here, $\frac{\ell_{min}}{\ell_i}$ acts as the scaling factor, which is always greater than 1. The smaller the original loss ℓ_i, the larger the scaling factor and vice versa. Our main goal is to investigate RO, and MLCAT's loss-scaling approach is tailored for this. We denote this loss-scaling-based method as $\mathrm{MLCAT_{LS}}$.

MLCAT through weight perturbation. To increase the loss of small-loss data, we apply a weight perturbation technique to generate targeted perturbation noise for these data points. In line with our goal of mitigating RO, we use the minimum loss threshold ℓ_{min} as a constraint to guide the generation of adversarial weight perturbation v, calculated as follows:

$$v = \nabla_\theta \sum_i \mathbb{1}(\ell_i \le \ell_{min})\, \ell_i. \tag{3.18}$$

Here, $\mathbb{1}(\ell_i \le \ell_{min})$ is an indicator function that outputs 1 if $\ell_i \le \ell_{min}$ and 0 if $\ell_i > \ell_{min}$. The adjusted loss for small-loss data can then be represented as

$$\ell_i^S = \ell(f_{\theta+v}(\tilde{\mathbf{x}}), y_i). \tag{3.19}$$

The weight perturbation method prevents RO by directly modifying the model parameters, distinguishing it from loss-scaling method and making it orthogonal in implementation. We denote this weight perturbation-based approach as $\mathrm{MLCAT_{WP}}$.

Experiments: Setup We perform extensive experiments on three benchmark datasets (CIFAR10 [203], SVHN [278], and CIFAR100 [203]). Following Rice et al. [311], we use PreAct ResNet-18 [144] and Wide ResNet-34-10 [451] as network architectures. Model robustness is assessed by testing accuracy on adversarially perturbed data, using both 20-step PGD (PGD-20) [258] and AutoAttack (AA) [76]. The extent of RO is determined by tracking the robust accuracy gap over training.

Table 3.9 Test robustness (%) of MLCAT. Standard deviations across 5 runs are omitted due to their minimal values (<0.6%)

Network	Dataset	Method	PGD-20			AA		
			Best	Last	Diff	Best	Last	Diff
PreAct ResNet-18	CIFAR10	AT	52.29	44.43	−7.86	47.99	42.08	−5.91
		$MLCAT_{LS}$	56.90	56.87	**−0.03**	28.12	26.93	−1.19
		$MLCAT_{WP}$	**58.48**	**57.65**	−0.83	**50.70**	**50.32**	**−0.38**
	SVHN	AT	52.88	45.29	−7.59	45.09	40.36	−4.73
		$MLCAT_{LS}$	**64.28**	**62.30**	**−1.98**	34.48	32.33	−2.15
		$MLCAT_{WP}$	60.34	57.79	−2.55	**51.90**	**49.76**	**−2.14**
	CIFAR100	AT	28.01	20.39	−7.62	23.61	18.41	−5.20
		$MLCAT_{LS}$	20.09	18.14	−1.95	13.41	11.35	−2.06
		$MLCAT_{WP}$	**31.27**	**30.57**	**−0.70**	**25.66**	**25.28**	**−0.38**
Wide ResNet-34-10	CIFAR10	AT	55.57	47.37	−8.20	52.13	46.09	−6.04
		$MLCAT_{LS}$	**64.73**	**63.94**	−0.79	35.00	34.51	−0.49
		$MLCAT_{WP}$	62.50	61.91	**−0.59**	**54.65**	**54.56**	**−0.09**
	SVHN	AT	55.72	50.44	−5.28	48.00	42.41	−5.59
		$MLCAT_{LS}$	**78.96**	**77.38**	−1.58	34.66	34.21	−0.45
		$MLCAT_{WP}$	63.18	61.71	**−1.47**	**54.29**	**53.93**	**−0.36**
	CIFAR100	AT	30.74	24.89	−5.85	26.98	23.07	−3.91
		$MLCAT_{LS}$	22.86	22.18	−0.68	14.61	14.05	−0.56
		$MLCAT_{WP}$	**34.97**	**34.64**	**−0.33**	**29.49**	**29.25**	**−0.24**

Performance Evaluation: The evaluation results of MLCAT are presented in Table 3.9, where "Best" indicates the highest robustness achieved during training, "Last" represents the test robustness at the final checkpoint, and "Diff" shows the gap in robust accuracy between "Best" and "Last." First, we observe that both $MLCAT_{LS}$ and $MLCAT_{WP}$ demonstrate enhanced robustness under the PGD-20 attack. For the AA attack, $MLCAT_{WP}$ also boosts adversarial robustness, while $MLCAT_{LS}$ shows lower robustness compared to AT, due to the sensitivity introduced by the loss-scaling technique to logit scaling attacks [157]. Most notably, both $MLCAT_{LS}$ and $MLCAT_{WP}$ substantially reduce the robustness gap under both PGD-20 and AA attacks, indicating their effectiveness in mitigating RO across various network architectures and datasets in AT.

3.2.2.2 Abnormal Adversarial Examples in Catastrophic Overfitting

Multistep *adversarial training* (AT) is the *de facto* standard for enhancing the adversarial robustness of *deep neural networks* (DNNs) [11, 77]. However, this training paradigm notably increases the computational overhead because it requires multistep backward propagation to generate adversarial examples [258, 404], thereby limiting its real-world scalability and practicality. Naturally, single-step

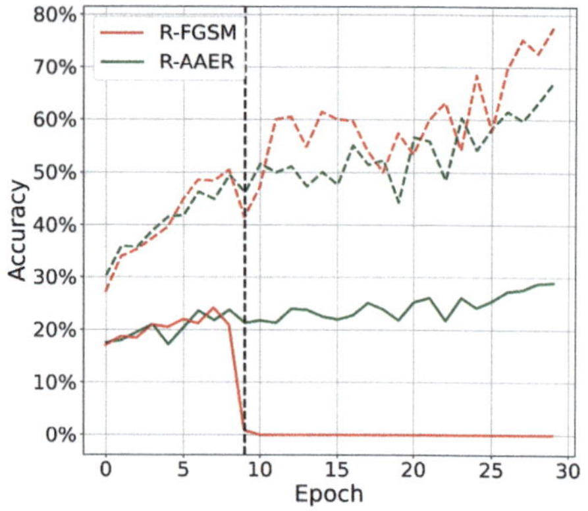

Fig. 3.16 The test accuracy of R-FGSM (red line) and R-AAER (green line) with 16/255 noise magnitude. The dashed and solid lines denote natural and robust (PGD-7-1) accuracy, respectively. The dashed black line corresponds to the 9th epoch, which is the point that R-FGSM occurs CO

AT becomes a promising training paradigm [96, 236, 401], as it can enhance the robustness of DNNs with minimal computational overhead.

Regrettably, as illustrated in Fig. 3.16, single-step AT suffers from the *catastrophic overfitting* (CO), which causes the model to abruptly lose its ability to defend against multistep adversarial attacks. Recent studies have indicated that classifiers trained with single-step AT often exhibit highly distorted decision boundaries [6, 191]. However, it is still unknown what mechanisms cause the classifier to display distortion so quickly.

To delve deeper into the CO process, this study identifies *abnormal adversarial examples* (AAEs) that exist in CO-affected DNNs displaying anomalous behavior [237]. It is worth noting that, although these AAEs are generated through the direction of gradient ascent, the loss associated with them anomalously decreases. As a result, we introduce *Abnormal Adversarial Examples Regularization* (AAER) method to mitigate CO by obstructing the generation of AAEs. In detail, AAER simultaneously penalizes the number and the output variation of AAEs, thereby effectively avoiding the classifier distortion.

Method: Definition and Counting of AAE AT utilizes the worst-case scenario, *w.r.t.* training on adversarial examples, to enhance the robustness of DNNs to the input. In contrast, this study reveals that CO-affected DNNs show an unexpected reduction in loss linked to specific training samples after adding adversarial perturbation. This behavior is illustrated in Fig. 3.17, where normal adversarial examples (NAEs, blue) successfully increase the classification loss. Conversely, the AAEs (red) are anomalously positioned further from the decision boundary after adding the adversarial perturbation. We introduce the following formula to define AAEs:

$$
\begin{aligned}
\delta &= \operatorname{sign}\left(\nabla_{\mathbf{x}+\eta}\ell(\mathbf{x}+\eta, y; \theta)\right), \\
x^{\mathrm{AAE}} &\stackrel{\text{def}}{=} \ell\left(\mathbf{x}+\eta, y; \theta\right) > \ell\left(\mathbf{x}+\eta+\delta, y; \theta\right),
\end{aligned}
\tag{3.20}
$$

where δ is adversarial perturbation, η is the random initialization, and $\ell(\mathbf{x}, y; \theta)$ is the loss function parameterized by θ.

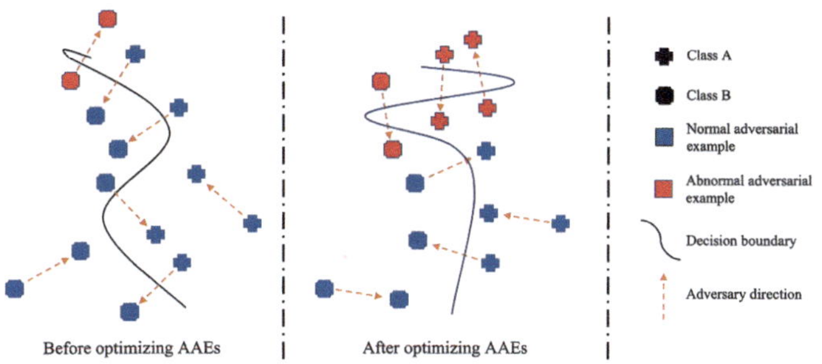

Fig. 3.17 This conceptual diagram illustrates the decision boundary of a DNN and its training samples. The training samples categorized as NAEs (blue) are effectively in misleading, whereas AAEs (red) are not. The left panel displays the decision boundary prior to optimizing AAEs, showing a slight distortion. The middle panel depicts the decision boundary after optimizing AAEs, highlighting a vicious cycle between increased distortion and optimizing AAEs

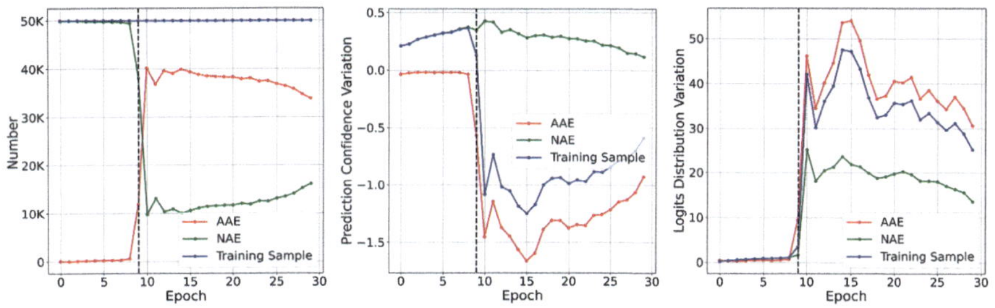

Fig. 3.18 From left to right, the figure illustrates the sample count, prediction confidence variance, and logit distribution for NAEs, AAEs, and training examples under R-FGSM with a noise level of 16/255. The dashed black line represents the 9th epoch, marking the onset of CO

Using AAEs as indicators, we track the underlying CO process, with associated changes in the number of AAEs displayed in Fig. 3.18 (left). It shows that before CO, a small amount of AAEs are observable, indicating minor initial distortions in DNNs. Examining the loss surfaces of both AAEs and NAEs, it is confirmed that before CO, DNNs had a sharper loss surface around AAEs compared to NAEs, as depicted in Fig. 3.19 (left and middle). These results indicate that the formation of AAEs is directly affected by decision boundary distortion.

During the occurrence of CO, the number of AAEs (red line) exhibits a significant increase, as illustrated in Fig. 3.18 (left). From Fig. 3.19 (right), we observe that the number of AAEs dramatically rises, while the accuracy defends against multistep attacks quickly drops to almost 0% in a few iterations, which forms a negative cycle between AAE optimization and CO deterioration. The number of AAEs still maintains a high level throughout the whole training process.

Outputs Variation of NAE and AAE The previous section highlights a significant correlation between the occurrence of CO and the number of AAEs. This section

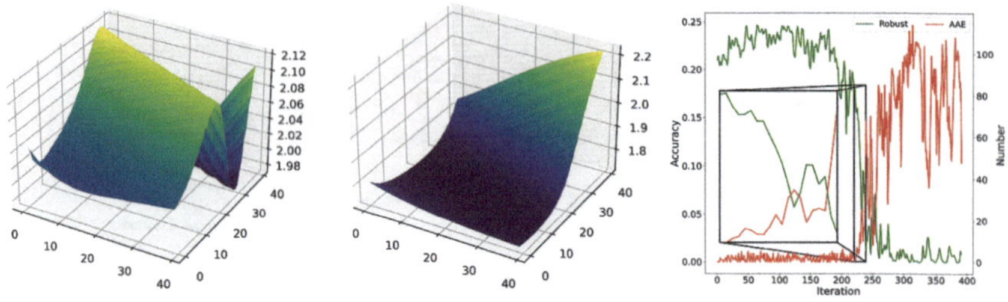

Fig. 3.19 Left/middle panel: The loss surfaces for AAEs/NAEs before CO (8th epoch). Right panel: Tracks the number of AAEs and test robustness through each iteration at the time of CO (9th epoch). The green line illustrates robust accuracy, while the red line indicates the number of AAEs

examines the prediction confidence and the distribution of logit of AAEs. We utilize cross-entropy (CE) to evaluate the variation in prediction confidence by assessing the loss change during the inner maximization process, which can be defined as follows:

$$\ell\left(\mathbf{x}+\eta+\delta, y ; \theta\right)-\ell\left(\mathbf{x}+\eta, y ; \theta\right). \tag{3.21}$$

We examine the changes in prediction confidence for NAEs and AAEs throughout the training process. As illustrated in Fig. 3.18 (middle), the variation in prediction confidence for NAEs consistently exceeds zero, suggesting an effective adversarial perturbation that moves the training sample nearer to the decision boundaries. In contrast, we show that adversarial perturbations have a paradoxical effect on AAEs, resulting in abnormally negative prediction confidence.

Alongside the failure of AAEs to mislead DNNs, their logit distribution is also disrupted during the CO process. To examine the changes in logit distribution, we utilize the Euclidean distance (L2) as a metric to quantify the impact of adversarial perturbations. The calculation is formulated as follows:

$$\| f_\theta\left(\mathbf{x}+\eta+\delta\right)-f_\theta\left(\mathbf{x}+\eta\right)\|_2^2, \tag{3.22}$$

where f_θ is the DNN classifier parameterized by θ and $\|\cdot\|_2^2$ is the L2 distance.

The changes in logit distribution are depicted in Fig. 3.18 (right). During CO, the variation for AAEs significantly escalates, reaching a level 13 times greater than previously observed. Furthermore, the optimization of AAEs contributes to this disparity, with the alteration being more pronounced than prior to CO.

Abnormal Adversarial Examples Regularization To mitigate the generation of AAEs, we introduce a novel regularization approach, Abnormal Adversarial Examples Regularization (AAER). This method is designed to prevent the distortions in DNNs by explicitly hindering the number and outputs variation of AAEs. Part (*i*) categorizes training samples into NAEs and AAEs according to the criteria in

Eq. (3.20). The output variation of AAEs is simultaneously restricted by prediction confidence in part (*ii*) and logit distribution in part (*iii*).

Specifically, part (*i*) differentiates training samples into NAEs and AAEs based on the criteria in Eq. (3.20), subsequently imposing a penalty on the number of AAEs. Concurrently, the output variation of AAEs is constrained through part (*ii*) prediction confidence and part (*iii*) logit distribution. For part (*ii*), we impose penalties on any negative variations in AAEs that occur during the inner maximization process, which is defined as follows:

$$\mathrm{AAE_{CE}} = \frac{1}{n}\sum_{i=1}^{n}\left(\ell\left(\mathbf{x}_i^{\mathrm{AAE}} + \eta, y_i; \theta\right) - \ell\left(\mathbf{x}_i^{\mathrm{AAE}} + \eta + \delta, y_i; \theta\right)\right), \tag{3.23}$$

where n is the number of abnormal adversarial examples.

For part (*iii*), we calculate the differences in the logit distribution for AAEs and NAEs separately, as explained in Eq. (3.24) and (3.25):

$$\mathrm{AAE}_{L_2} = \frac{1}{n}\sum_{i=1}^{n}\left(\| f_\theta\left(\mathbf{x}_i^{\mathrm{AAE}} + \eta + \delta\right) - f_\theta\left(\mathbf{x}_i^{\mathrm{AAE}} + \eta\right) \|_2^2\right); \tag{3.24}$$

$$\mathrm{NAE}_{L_2} = \frac{1}{m-n}\sum_{j=1}^{m-n}\left(\| f_\theta\left(\mathbf{x}_j^{\mathrm{NAE}} + \eta + \delta\right) - f_\theta\left(\mathbf{x}_j^{\mathrm{NAE}} + \eta\right) \|_2^2\right), \tag{3.25}$$

where m is the number of training samples.

In order to control the variation in AAEs, we then use the variation in the logit distribution of NAEs as a reference. It should be noted that our optimization goal is to bring the variation in the logit distribution of AAEs closer to that of NAEs. In order to do this, we use the max function to establish a minimum value, which is as follows:

$$\text{Constrained\_Variation} = \max\left(\mathrm{AAE}_{L_2}\ (3.24) - \mathrm{NAE}_{L_2}\ (3.25), 0\right), \tag{3.26}$$

where $\max(,)$ is the max function.

Based on the aforementioned optimization objectives, we propose a novel regularization term, AAER. This term aims to mitigate AAEs by controlling their number, the variability in prediction confidence, and the distribution of logits, as elucidated in the formula below:

$$\mathrm{AAER} = (\lambda_1 \cdot \frac{n}{m}) \cdot (\lambda_2 \cdot \mathrm{AAE_{CE}}\ (3.23) + \lambda_3 \cdot \text{Constrained\_Variation}\ (3.26)), \tag{3.27}$$

where λ_1, λ_2, and λ_3 are the hyperparameters to control the strength of the regularization term.

Table 3.10 Hyperparameter settings for Cifar-10/100. Values are shown as CIFAR-10/CIFAR-100, where λ_2 represents the first row and λ_3 represents the second row. The value of λ_1 is fixed at 1.0 for all settings

Method	R-AAER				N-AAER			
ϵ	8/255	12/255	16/255	32/355	8/255	12/255	16/255	32/355
λ_2	2.5/3.5	5.0/3.5	7.0/6.0	5.75/5.0	1.5/1.5	5.0/3.5	8.5/6.0	2.75/3.5
λ_3	1.5/1.5	2.75/2.75	3.25/2.25	1.5/0.75	0.15/0.15	0.55/0.3	1.5/0.5	0.75/0.5

Experiments: Setup for AAER We employ the SGD optimizer with a momentum of 0.9, weight decay of 5×10^{-4}, and use the cyclical learning rate of 0.2 with 30 training epochs [338]. We directly integrate our method to the wildly-used baseline, R-FGSM [401] and N-FGSM [96], with their standard setting. Detailed hyperparameter configurations for Cifar-10/100 are outlined in Table 3.10. We also provide the results of PGD-2 and PGD-10 [258] to provide a reference for multistep AT.

Performance Evaluation: In Table 3.11, we can observe that when CO is not present in the baseline methods, our proposed AAER method can effectively enhance the robust accuracy of the original baseline. More importantly, when the CO occurs on the baseline methods, AAER exhibits the reliable ability to consistently prevent the occurrence of CO across all test scenarios, even facing the extremely 32/255 noise magnitudes.

Optimization Objectives We analyze the changes in the number of AAEs, prediction confidence, and logit distribution over the training period to demonstrate the effectiveness of AAER. Figure 3.20 demonstrates that AAER can efficiently regulate three optimization objectives, thus preventing distortion of the decision boundary and the emergence of CO.

Ablation Study We examine the specific influence of each component within AAER, as demonstrated in Table 3.12. Our findings suggest that neither the number of AAEs nor predictive confidence is adequate to entirely prevent CO. Although the logit distribution can inhibit the occurrence of CO, it fails to offer a comprehensive assessment of decision boundary distortion, resulting in suboptimal performance. Therefore, the integration of all components of the regularization term is crucial for the stable and effective elimination of CO.

Time-Cost Analysis One of the primary advantages of single-step AT over multistep AT is its efficiency. Thus, computational overhead is a crucial factor in the overall performance of single-step AT. As presented in Table 3.13, AAER introduces only a slight increase in training time, which is a negligible increase over baseline. In comparison, multistep AT takes much more time than our method.

Table 3.11 Performance comparison on CIFAR10/100 using PreActResNet-18 under L_∞ norm attacks. For each method and noise magnitude, the top number shows natural accuracy (%) and the bottom number shows PGD-50-10 accuracy (%). Results are averaged over three random seeds with standard deviation

Dataset	CIFAR10				CIFAR100			
Noise magnitude	8/255	12/255	16/255	32/255	8/255	12/255	16/255	32/255
R-FGSM	83.91±0.21	66.46±22.80	66.54±12.25	36.43±7.86	60.29±1.51	18.19±8.51	11.03±5.24	11.40±8.60
	46.01±0.18	0.00±0.00	0.00±0.00	0.00±0.00	10.58±13.1	0.00±0.00	0.00±0.00	0.00±0.00
R-AAER	83.83±0.27	74.40±0.79	64.56±1.45	31.58±1.13	57.71±0.29	44.06±0.93	33.10±0.05	18.50±1.68
	46.14±0.02	**32.17**±0.16	**23.87**±0.36	**10.62**±0.51	**25.31**±0.01	**16.41**±0.13	**11.80**±0.17	**4.90**±0.50
N-FGSM	80.48±0.21	71.30±0.12	62.96±0.74	29.79±3.87	54.92±0.28	46.16±0.13	37.93±0.22	18.18±4.55
	47.91±0.29	36.23±0.10	27.14±1.44	8.30±7.85	26.29±0.41	18.75±0.19	14.05±0.07	0.00±0.00
N-AAER	80.56±0.35	71.15±0.18	61.84±0.43	27.08±0.02	54.47±0.45	45.98±0.13	36.80±0.14	16.95±0.44
	48.31±0.23	**36.52**±0.10	**28.20**±0.71	**12.97**±0.57	**26.81**±0.13	**19.03**±0.04	**14.31**±0.05	**5.45**±0.14
PGD-2	85.07±0.12	78.97±0.23	72.31±0.40	48.45±0.71	60.09±0.20	53.46±0.27	47.50±0.28	31.89±0.69
	45.27±0.07	32.99±0.46	24.32±0.64	11.24±0.40	24.58±0.12	17.16±0.21	12.69±0.06	4.51±0.21
PGD-10	80.55±0.37	72.37±0.31	67.20±0.69	34.70±0.67	55.05±0.25	47.42±0.29	42.39±0.17	21.68±0.18
	50.67±0.40	**38.60**±0.39	**29.34**±0.18	**16.10**±0.20	**27.87**±0.12	**20.29**±0.18	**15.01**±0.21	**7.39**±0.38

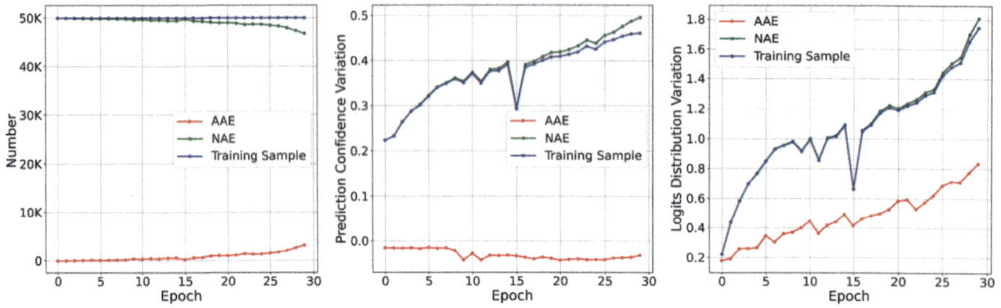

Fig. 3.20 The number, the variation of prediction confidence, and logit distribution (from left to right) for NAEs, AAEs, and training samples in R-AAER with 16/255 noise magnitude

Table 3.12 The impact of regularization terms on natural and robust accuracy

Term (i)	Term (ii)	Term (iii)	Natural Acc (%)	Robust Acc (%)
✓			75.14	0.00
	✓		77.22	0.00
		✓	13.61	9.77
✓	✓		76.19	0.00
✓		✓	56.65	23.29
	✓	✓	16.68	12.56
✓	✓	✓	64.56	23.87

Table 3.13 CIFAR10 training time (in seconds) on a single NVIDIA RTX 4090 GPU using PreactResNet-18, averaged over 30 epochs

Method	R/N-FGSM	R/N-AAER	PGD-2	PGD-10
Training time	11.0	11.2	16.4	59.1

3.2.2.3 Pseudo-Robust Shortcut Dependency in Catastrophic Overfitting

The above section explains the underlying process of *catastrophic overfitting* (CO), where optimizing abnormal adversarial examples causes rapid and extreme distortion in *deep neural networks (DNNs)*. However, the underlying factors that contribute to the slight distortion of the decision boundary, which in turn triggers the generation of the initial abnormal adversarial examples, remain unclear.

In this work, we demonstrate that the latter layers of DNNs exhibit resilience, whereas the former layers are more vulnerable, exhibiting earlier distortion transformations during the CO process [235]. The creation of *pseudo-robust shortcuts*, which can circumvent genuine-robustness learning and result in distorted decision boundaries, is the cause of this increased sensitivity. The robustness of CO-affected DNNs can be partially restored by eliminating these shortcuts and reducing inappropriate reliance on them. The analysis demonstrates how the model relies on these shortcuts to make decisions that result in CO.

By avoiding pseudo-robust shortcuts, we propose **L**ayer-**A**ware Adversarial Weight **P**erturbation (LAP) method, which introduces adaptive weight perturbations across layers to break stable shortcuts. This method reduces computational overhead by producing adversarial perturbations for both inputs and weights.

Method: Layers Transformation During CO Previous studies have identified distorted decision boundaries in CO-affected DNNs, but these analyses primarily focus on the DNNs' final output. However, the final output of L-layer DNN with parameters $\{\mathbf{W}_l\}_{l=1}^{L}$ emerges from a series of forward propagations $f_\theta(\mathbf{x}) = \mathbf{W}_L(\mathbf{W}_{L-1}\ldots(\mathbf{W}_1\mathbf{x}))$ for $l = 1, \ldots, L$. As a result, previous works often overlook the specific contributions of individual layers to the distorted decision boundaries.

In this section, we use the R-FGSM method to train a PreActResNet-18 network on the CIFAR-10 dataset with a noise magnitude of 16/255, and we analyze the single-step adversarial training (AT) by looking at transformations in each layer. To analyze the loss landscape, we introduce random perturbations to the input, denoted as $\mathbf{x} + \delta$, and quantify the corresponding loss variation as Δ. Additionally, we examine the effect of perturbing the weights of selected layers—specifically layers $l = 1, 5, 9, 13, 17$ represented as $\mathbf{W}_l + \delta$, to observe how the loss responds to weight-level perturbations.

Both the overall model and individual layers display a flattened loss landscape prior to the CO, as seen in Fig. 3.21 (upper row). Former layers show greater stability, as evidenced by smaller loss variations following weight perturbations.

As shown in Fig. 3.21 (middle row), the model's loss landscape noticeably distorts during CO, resulting in a reduction in robustness. According to the layer-aware analysis, later layers undergo negligible changes, whereas earlier layers exhibit greater sensitivity first, as evidenced by a sharper loss landscape.

Figure 3.21 shows how the DNN's decision boundaries are greatly warped after CO, making it more vulnerable to multistep adversarial attacks (lower row). A layer-by-layer examination reveals that while later layers are mostly unaffected, earlier layers experience the most extreme distortions, which are indicated by a sharp surface. Overall, during the CO process, the earlier layers in DNNs undergo significant changes, moving from a stable state to increased sensitivity.

Pseudo-Robust Shortcut Dependency By examining weight changes during the CO process, we delve into the factors that lead to sensitivity transformation in DNNs. Figure 3.22 illustrates the evaluation of singular values of convolutional kernels. Prior to CO, singular values have equal distribution throughout the layers; however, after CO, variance significantly increases, producing sharper model outputs. An increasing reliance on particular weights for decision-making is indicated by the rise in large singular values in earlier layers.

As shown in Fig. 3.23 (left column), we eliminated weights from the first five layers of a CO-affected model in order to investigate this dependency. A performance paradox was revealed when accuracy against single-step attacks (FGSM) declined as weight removal increased, while performance against multistep attacks (PGD) marginally improved. This implies that earlier layers' heightened sensitivity results

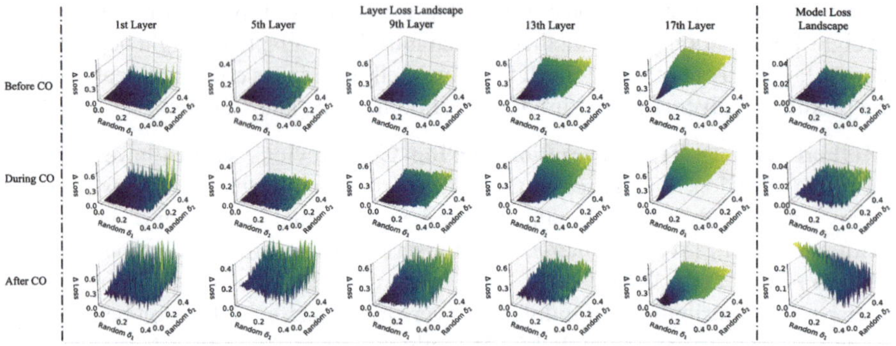

Fig. 3.21 The visualization presents the loss landscape across individual layers (from the 1st to 5th columns) and the entire model (6th column). The top, middle, and bottom rows correspond to the stages preceding, during, and following CO, respectively

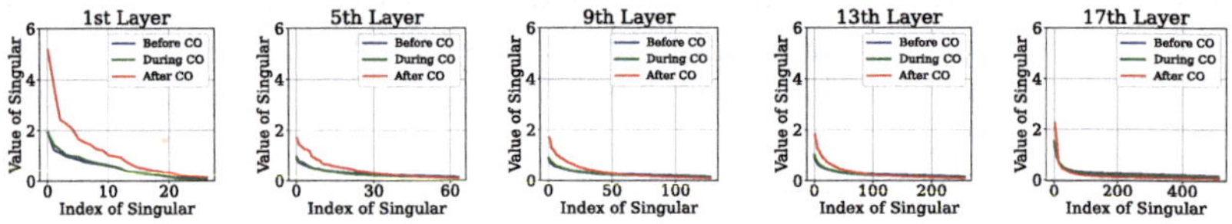

Fig. 3.22 Singular value of weights (convolution kernel) at different DNN layers. The blue, green, and red lines represent the model state before, during, and after CO, respectively

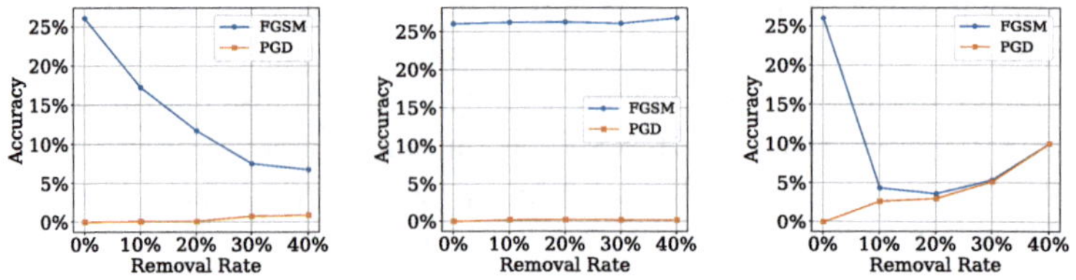

Fig. 3.23 Random weights, small-magnitude weights, and large-magnitude weights are removed from the first five layers, corresponding to the left, middle, and right columns, respectively

from their reliance on pseudo-robust shortcuts, which work well against single-step attacks but compromise true-robust learning.

We also examined the weights associated with these pseudo-robust shortcuts. Figure 3.23 (middle column) demonstrates that removing small weights has little effect on performance against both attack types. On the other hand, as Fig. 3.23 (right column) illustrates, eliminating larger weights breaks the pseudo-robust shortcuts, drastically decreasing the accuracy of single-step attacks while regaining robustness against multistep attacks.

Finally, to explain the emergence of CO, we adopt the perspective of pseudo-robust shortcut dependency. Although such shortcuts provide apparent robustness against single-step adversarial attacks, they hinder the development of truly robust

Table 3.14 Hyperparameter β settings for CIFAR-10

Method	8/255	12/255	16/255	32/255
V-LAP	0.030	0.058	0.070	0.480
R-LAP	0.002	0.030	0.050	0.300
N-LAP	0.001	0.002	0.005	0.075

representations. This leads to distorted decision boundaries and contributes to the performance paradox observed during CO.

Proposed Method Building on our insights, we propose **L**ayer-**A**ware Adversarial Weight **P**erturbation (LAP) approach. This method strategically applies adaptive weight perturbations across various layers to inhibit the generation of pseudo-robust shortcuts:

$$\min_{\theta} \frac{1}{n} \sum_{i=1}^{n} \max_{\delta_i, v_l} \ell \left(f_{\theta+v_l}(\mathbf{x}_i + \delta_i), y_i) \right), \tag{3.28}$$

where δ is the input perturbation, v_l is the layer-aware weight perturbation calculated by $\nabla_\theta \cdot \lambda_l$, f is the classifier parameterized by θ, and ℓ is the cross-entropy loss.

To meet our goals, we make three modifications to the objective function in Eq. (3.28). (i) It generates adversarial perturbations for both inputs and weights at the same time; (ii) it accumulates weight perturbations to consistently disrupt persistent shortcuts; (iii) the strategy applies a gradually decreasing scale of weight perturbations from earlier to later layers, minimizing unnecessary perturbations, as mentioned below:

$$\lambda_l = \beta \cdot \left(1 - \left(\frac{\ln(l)}{\ln(L+1)} \right)^{\gamma} \right), \quad \text{for } l = 1, \ldots, L \tag{3.29}$$

where λ_l is the layer-aware strength, β is the step size, and γ controls the different layers strength.

Experiments: Setup for LAP We utilize the SGD optimizer with a momentum of 0.9. We apply the L_∞ norm attack for input perturbations and the L_2 norm attack for weight perturbations. LAP is integrated into three widely used baseline methods, V-FGSM [124], R-FGSM [401], and N-FGSM [96]. For our hyperparameters, we set the γ to 0.3, and the detailed setting for β is provided in Table 3.14.

Performance Evaluation According to Table 3.15, LAP consistently outperforms other methods across all evaluation scenarios. Specifically, in scenarios where CO does not affect, our method shows a reliable improvement in robustness. Importantly, in scenarios where CO occurs, LAP not only successfully prevents CO but also significantly enhances performance.

Loss Landscape As demonstrated by the loss landscape of the individual layers and the entire model in Fig. 3.24, LAP is effective in preventing the formation

Table 3.15 Comparison of CIFAR-10 test accuracy (%) for various methods under different noise magnitudes. Results are averaged over three random seeds and reported with standard deviation

Noise magnitude	8/255		12/255		16/255		32/255	
Method	Natural	AutoAttack [76]	Natural	AutoAttack [76]	Natural	AutoAttack [76]	Natural	AutoAttack [76]
V-FGSM	84.26±4.18	0.00±0.00	79.92±1.82	0.00±0.00	72.72±3.04	0.00±0.00	65.52±2.15	0.00±0.00
V-LAP	79.09±0.78	41.24±0.51	66.20±0.42	24.07±0.34	56.02±0.07	15.17±0.31	17.76±3.11	7.12±0.64
R-FGSM	84.12±0.29	42.88±0.09	79.49±4.57	0.00±0.00	73.67±6.86	0.00±0.00	33.31±8.31	0.00±0.00
R-LAP	83.81±0.24	43.14±0.45	74.10±0.31	26.04±1.04	64.83±0.29	15.69±0.28	27.49±0.48	8.04±0.63
N-FGSM	80.40±0.16	44.21±0.47	71.44±0.16	30.25±0.06	62.91±1.03	18.58±2.28	27.66±3.57	0.00±0.00
N-LAP	80.76±0.15	**44.97**±0.24	71.91±0.19	**30.60**±0.27	63.73±0.27	**19.55**±0.18	29.19±1.00	**8.85**±1.48
PGD-2	84.72±0.08	42.92±0.60	79.13±0.25	28.30±0.35	72.50±0.51	17.89±0.16	48.99±0.19	3.76±0.02
PGD-10	80.91±0.52	**46.37**±0.76	72.03±0.30	**33.13**±0.28	67.61±0.83	**21.98**±0.30	35.28±0.78	**10.88**±0.41

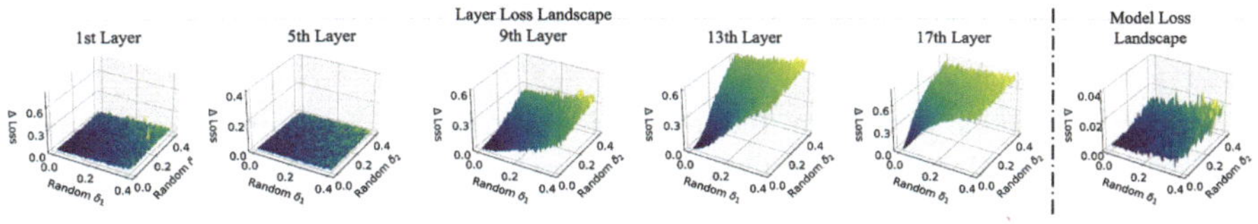

Fig. 3.24 Visualization of the loss landscape for individual layers (1st to 5th columns) and for the whole model (6th column)

Table 3.16 Comparison of test accuracy (%) for LAP with various optimization objectives. Results are averaged over three random seeds and reported with standard deviation

Method	Natural	AutoAttack [76]
LAP	64.83±0.29	15.69±0.28
Original AWP	88.47±0.75	0.00±0.00
Modified AWP	30.00±0.25	12.53±0.98
LAP-R	53.87±0.14	11.22±0.10
LAP-L_∞	20.38±0.38	13.67±0.35

Table 3.17 Comparison of training time (in seconds) averaged over 30 epochs. Results obtained on a single NVIDIA RTX 4090 GPU

Method	V/R/N-FGSM	V/R/N-LAP	PGD-2	PGD-10
Training time	**11.0**	11.8	16.4	59.1

of pseudo-robust shortcuts, which are typically associated with sharp decision boundaries. The results show a more flattened loss landscape for both the individual layers and the entire model.

Optimization Objectives We also investigate LAP in conjunction with a number of other optimization strategies, such as Original AWP [404], a Modified AWP that preserves accumulated weight perturbation, LAP-R, which incorporates Random weight perturbation, and LAP-L_∞, which employs a L_∞ norm attack for weight perturbation. According to Table 3.16, the Original AWP fails to prevent CO because it does not break persistent shortcuts, while Modified AWP improves CO prevention but produces unsatisfactory natural and robust acurracy. On the other hand, the LAP-family methods successfully prevent the formation of pseudo-robust shortcuts, thus preventing CO.

Time-Cost Analysis We provide a comparative analysis of the training time cost for different methods in Table 3.17. The LAP method's training cost is comparable to the baseline FGSM method, while Grad Align and PGD-10 take significantly longer than ours.

3.3 Diffusion Models for Adversarial Examples

Adversarial attacks transform the data distribution by injecting imperceptible perturbations, challenging the reliability of deep learning systems. Recent advances in diffusion models, which iteratively refine data through stochastic denoising processes, offer a novel paradigm to counteract such threats. By leveraging their inherent ability to model complex data manifolds and reconstruct clean samples from noisy inputs, diffusion-based approaches aim to purify adversarial examples by reversing perturbation effects or detect anomalies by quantifying deviations from natural data distributions. Research in this domain explores two directions: (1) adversarial purification through diffusion-guided generation, where corrupted inputs are projected back to the clean data manifold, and (2) adversarial detection by analyzing the alignment between adversarial samples and the diffusion model's generative trajectory. These methods bridge generative modeling with adversarial robustness, providing a probabilistic defense framework against adversarial attack strategies.

3.3.1 Diffusion-Based Adversarial Purification

Adversarial samples threaten DNNs' reliability in safety-critical applications [124]. While adversarial training [258] enhances robustness, it incurs high computational costs and exhibits attack-specific limitations. Adversarial purification methods [279, 333] leverage generative models but struggle with incomplete noise removal and inconsistent noise handling. We reveal that samples from generative models align with clean data manifolds while distancing from adversarial ones. This motivates our **Consistency Model-based Adversarial Purification (CMAP)** approach. By optimizing latent vectors in a pretrained consistency model's space [341], CMAP restores samples to clean manifolds through perceptual consistency restoration and latent distribution constraints.

Problem 1 (Adversarial Purification) *Let $\mathcal{X} \subset \mathbb{R}^d$ be a separable metric space and $\mathcal{D}$ be a probability distribution on $\mathcal{X}$. Giving IID samples $D = \{\mathbf{x}^{(i)}\}_{i=1}^n$ from the distribution $\mathcal{D}$ and a ground-truth labeling mapping $f : \mathbb{R}^d \to C$ with $C = \{1, \ldots, C\}$ being a label set. Assuming that the attacker has access to some well-trained classifier f_θ and generates samples $D' = \{\tilde{\mathbf{x}}^{(i)}\}_{i=1}^m$ that mislead the classifier, we wish to restore samples $\tilde{\mathbf{x}}_i$ in D' back to the corresponding $\mathbf{x}_i$ in D.*

Method Existing diffusion-based purification methods [279] inject Gaussian noise followed by denoising but face two limitations: (1) adversarial noise often differs from diffusion noise, leading to incomplete purification; (2) fixed diffusion timesteps struggle with varying attack intensities. To address this, we propose **Consistency Model-based Adversarial Purification (CMAP)**, which shifts defense operations to the latent space of a pretrained consistency model [341]. Our key

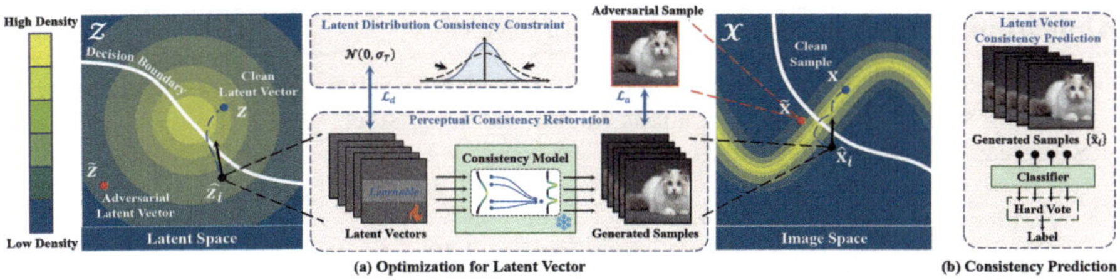

Fig. 3.25 Overview of the proposed CMAP. **(a)** For an input sample $\tilde{\mathbf{x}}$ (potentially adversarial), we optimize latent vectors $\{\hat{\mathbf{z}}_i\}$ in a pretrained consistency model's space $\mathcal{Z}$ to generate purified samples $\{\hat{\mathbf{x}}_i\}$. The optimization uses *perceptual consistency restoration*: MAE + SSIM losses ($\mathcal{L}_a$) to align generated samples with the clean manifold; *latent distribution constraint*: Gaussian constraint ($\mathcal{L}_d$) to keep $\{\hat{\mathbf{z}}_i\}$ within the valid manifold. **(b)** Final predictions aggregate outputs from $\{\hat{\mathbf{x}}_i\}$ via label voting, ensuring robustness against adversarial perturbations

insight lies in aligning adversarial samples with the clean data manifold through latent vector optimization, as in Fig. 3.25 and Alg. 2. The framework comprises three components:

Perceptual Consistency Restoration Traditional purification directly modifies input space, but adversarial perturbations often persist due to manifold misalignment. Instead, CMAP optimizes latent vectors $\{\hat{\mathbf{z}}_i\}_{i=1}^K$ to generate samples that simultaneously match the test input's pixel features and structural patterns. We minimize:

$$\mathcal{L}_a(\hat{\mathbf{z}}, \tilde{\mathbf{x}}) = \frac{1}{K}\sum_{i=1}^{K} \left\| h_\omega(\hat{\mathbf{z}}_i) - \tilde{\mathbf{x}} \right\|_1 - \alpha \cdot \mathrm{SSIM}(h_\omega(\hat{\mathbf{z}}_i), \tilde{\mathbf{x}}). \tag{3.30}$$

where h_ω is the pretrained consistency model with parameters ω, MAE enforces pixel fidelity and SSIM preserves perceptual consistency. Unlike GANs limited by low-dimensional latent spaces, the consistency model enables high-fidelity reconstruction through one-step generation.

Latent Distribution Consistency Constraint Although minimizing Eq. (3.30) can restore adversarial samples, a critical challenge arises: over-optimization may cause latent vectors to deviate from the clean manifold. This risk is revealed by:

Theorem 3.1 *Assuming that the distribution of natural data* $p(\mathbf{x}) = \mathcal{N}(\boldsymbol{\mu}_{\mathbf{x}}, \sigma_{\mathbf{x}}^2 \mathbf{I})$, *where* $\mathbf{I}$ *is an identity matrix, given a PF ODE sampling* $d\mathbf{x} = -t\nabla_{\mathbf{x}} \log p_t(\mathbf{x})$ *with* $\mathbf{f}(\mathbf{x}, t) = \mathbf{0}$ *and* $g(t) = \sqrt{2t}$, *then for* $\forall\, \mathbf{x} \in p(\mathbf{x})$ *and its adversarial sample* $\tilde{\mathbf{x}} = \mathbf{x} + \delta$, *we have*

$$\mathbf{x}_T - \tilde{\mathbf{x}}_T \sim \mathcal{N}(\mathbf{0}, 2\sigma_{cl}^2 \mathbf{I}) + \boldsymbol{\mu}_\delta, \tag{3.31}$$

where $\boldsymbol{\mu}_\delta = \left(\mathbb{E}_t \frac{t}{\sigma_{\mathbf{x}}^2 + t^2} - 1\right)\delta$ *and* $\sigma_{cl}^2 = \mathbb{E}_t \frac{t^2}{\sigma_{\mathbf{x}}^2 + t^2}$.

Algorithm 2 Adversarial purification with CMAP

Input: test sample $\tilde{\mathbf{x}}$, consistency model h_ω, latent distribution $\mathcal{N}(\mathbf{0}, \sigma_T \mathbf{I})$, the number of sampling K, total optimization iterations T, step size η, classifier f_θ
Sample $\hat{\mathbf{z}}^0 = \{\hat{\mathbf{z}}_i\}_{i=1}^K \sim \mathcal{N}(\mathbf{0}, \sigma_T \mathbf{I})$
for $j = 1, \ldots, T$ **do**
// Perceptual consistency restoration
Obtain the restoration loss $\mathcal{L}_a(\hat{\mathbf{z}}^{j-1}, \tilde{\mathbf{x}})$ using Eq. (3.30)
// Latent distribution consistency constraint
Obtain the latent distribution constrain loss $\mathcal{L}_d(\hat{\mathbf{z}}^{j-1})$ using Eq. (3.32)
Update the latent vectors by gradient descent:
$\hat{\mathbf{z}}^j \leftarrow \hat{\mathbf{z}}^{j-1} - \eta \cdot \frac{\partial[\mathcal{L}_a(\hat{\mathbf{z}}^{j-1}, \tilde{\mathbf{x}}) + \beta \mathcal{L}_d(\hat{\mathbf{z}}^{j-1})]}{\partial \mathbf{z}_{j-1}}$.
end for
// Latent vector consistency prediction
Obtain the generated samples $\hat{\mathbf{x}} \leftarrow h_\omega(\hat{\mathbf{z}}^{T_{\text{def}}})$
Obtain the prediction $\tilde{y}$ via the label voting using Eq. (3.33)
Output: $\tilde{y}$

To counteract this, we impose a moment-matching constraint:

$$\mathcal{L}_d(\hat{\mathbf{z}}) = \left\| \boldsymbol{\mu}(\{\hat{\mathbf{z}}_i\}) - \boldsymbol{\mu}_{\mathbf{z}} \right\|_2^2 + \left\| \sigma(\{\hat{\mathbf{z}}_i\}) - \sigma_{\mathbf{z}} \right\|_2^2, \tag{3.32}$$

forcing optimized vectors to align with the Gaussian prior ($\boldsymbol{\mu}_{\mathbf{z}} = \mathbf{0}, \sigma_{\mathbf{z}}^2 = \sigma_T \mathbf{I}$). This ensures that generated samples remain on the clean manifold.

Latent Vector Consistency Prediction Even with constraints, stochasticity in latent optimization may cause prediction variance. To address this, CMAP aggregates result from K samples via majority voting:

$$\hat{y} = \arg\max_y \sum_{i=1}^{K} \mathbb{I}\left[f_\theta(h_\omega(\tilde{\mathbf{z}}_i)) = y \right]. \tag{3.33}$$

This exploits ensemble diversity while suppressing outliers, which is a strategy proven effective against randomized defenses [417].

Experiments We evaluate CMAP against strong ℓ_∞ and ℓ_2 white-box attacks (PGD [258], AutoAttack [76]) under rigorous threat models, where gradients are approximated through the denoising process to enhance attack efficacy [214].

Results on CIFAR-10 From Tables 3.18 and 3.19 on CIFAR-10 with ℓ_∞ ($\epsilon = 8/255$) and ℓ_2 ($\epsilon = 0.5$) attacks, CMAP achieves state-of-the-art robust accuracy across WideResNet architectures. Against PGD+EOT attacks, it surpasses prior purification methods by 18.73% in robust accuracy under ℓ_∞ attacks, while maintaining clean accuracy parity. For AutoAttack, CMAP improves robust accuracy by 8.27% to 9.66% on WideResNet-28-10 and WideResNet-70-16, demonstrating consistent defense capability without attack-specific tuning.

Table 3.18 Standard and robust accuracy against PGD+EOT and AutoAttack with ℓ_∞-norm ($\epsilon = 8/255$) on CIFAR-10. Adversarial training (AT) and adversarial purification (AP) strategies are evaluated. (* Model is trained with extra data)

Strategy	Method	Standard	PGD+EOT	AutoAttack
WRN-28-10				
AT	Gowal et al. [127]	87.51	66.01	63.38
	Gowal et al. [126]*	88.54	65.93	62.76
	Pang et al. [285]	88.62	64.95	61.04
AP	ADP [249]	$85.66_{\pm 0.51}$	$33.48_{\pm 0.86}$	$59.53_{\pm 0.87}$
	DiffPure [279]	$90.27_{\pm 0.81}$	$48.27_{\pm 1.86}$	$64.93_{\pm 2.14}$
	GNSP [214]	$\mathbf{90.40}_{\pm \mathbf{1.40}}$	$55.87_{\pm 0.50}$	$70.40_{\pm 1.80}$
	CMAP (Ours)	$88.73_{\pm 0.50}$	$\mathbf{74.60}_{\pm \mathbf{1.59}}$	$\mathbf{78.67}_{\pm \mathbf{1.90}}$
WRN-70-16				
AT	Rebuffi et al. [308]*	**92.22**	69.97	66.56
	Gowal et al. [127]	88.75	69.03	66.10
	Gowal et al. [126] *	91.10	68.66	65.87
AP	ADP [249]	$86.76_{\pm 1.15}$	$37.11_{\pm 1.35}$	$37.11_{\pm 1.35}$
	DiffPure [279]	$90.00_{\pm 0.87}$	$50.93_{\pm 1.30}$	$63.87_{\pm 1.86}$
	GNSP [214]	$90.53_{\pm 0.58}$	$56.07_{\pm 0.83}$	$71.67_{\pm 0.64}$
	CMAP (Ours)	$88.27_{\pm 0.50}$	$\mathbf{74.80}_{\pm \mathbf{1.56}}$	$\mathbf{81.33}_{\pm \mathbf{0.31}}$

Results on ImageNet-100 From Table 3.20 for ℓ_∞ ($\epsilon = 4/255$) and ℓ_2 ($\epsilon = 0.5$) attacks on ImageNet-100, CMAP outperforms existing methods in both robust and clean accuracy. It achieves 6.47% to 8.79% absolute gains against PGD+EOT attacks and 0.62% to 3.74% improvements for PGD-ℓ_2+EOT while enhancing standard accuracy by 1.15% to 2.01%. Notably, CMAP employs a unified defense operation (e.g., fixed optimization iterations T) across all attacks and architectures, eliminating the need for attack-specific hyperparameters required by prior diffusion-based purification methods [214, 249, 279].

3.3.2 Diffusion-Based Adversarial Detection

Adversarial samples pose significant threats to machine learning systems [217], driving the need for robust defenses. While adversarial training enhances robustness by incorporating adversarial data, it suffers from high computational costs and performance degradation [208, 401]. Adversarial purification mitigates adversarial effects using generative models but compromises both natural and adversarial accuracy [279, 438]. In contrast, adversarial detection identifies and rejects adversarial examples with minimal impact on natural accuracy, making it compatible with existing systems and aiding in the identification of compromised inputs [1]. However, current detection methods are typically tailored to specific attacks or classifiers, limiting their generalization to unseen or transferable attacks [89, 105, 215, 257].

Table 3.19 Standard and robust accuracy against PGD+EOT and AutoAttack with ℓ_2-norm ($\epsilon = 0.5$) on CIFAR-10. Adversarial training (AT) and adversarial purification (AP) strategies are evaluated. (* Model is trained with extra data)

Strategy	Method	Standard	PGD-ℓ_2+EOT	AutoAttack-ℓ_2
WRN-28-10				
AT	Rebuffi et al. [308]*	91.79	85.05	78.80
	Augustin et al. [13]	**93.96**	**86.14**	78.79
	Sehwag et al. [329]	90.93	83.75	77.24
AP	ADP [249]	$85.66_{\pm 0.51}$	$73.32_{\pm 0.76}$	$79.57_{\pm 0.38}$
	DiffPure [279]	$90.27_{\pm 0.81}$	$82.00_{\pm 0.72}$	$83.40_{\pm 0.20}$
	GNSP [214]	$90.40_{\pm 1.40}$	$84.80_{\pm 1.59}$	$\mathbf{87.47_{\pm 0.70}}$
	CMPA (Ours)	$88.73_{\pm 0.50}$	$84.00_{\pm 1.71}$	$83.40_{\pm 1.44}$
WRN-70-16				
AT	Rebuffi et al. [308] *	**95.74**	**89.62**	82.32
	Gowal et al. [126]*	94.74	88.18	80.53
	Rebuffi et al. [308]	92.41	86.24	80.42
AP	ADP [249]	$86.76_{\pm 1.15}$	$75.66_{\pm 1.29}$	$80.43_{\pm 0.42}$
	DiffPure [279]	$90.00_{\pm 0.87}$	$82.07_{\pm 1.79}$	$82.67_{\pm 1.94}$
	GNSP [214]	$90.53_{\pm 0.58}$	$85.40_{\pm 0.60}$	$\mathbf{87.73_{\pm 1.01}}$
	CMPA (Ours)	$88.27_{\pm 0.50}$	$82.93_{\pm 1.45}$	$83.47_{\pm 0.46}$

Table 3.20 Standard and robust accuracy against PGD+EOT attack with ℓ_∞-norm ($\epsilon = 4/255$) and ℓ_2-norm ($\epsilon = 0.5$) on ImageNet-100

Classifier	Method	Standard	PGD+EOT	PGD-ℓ_2+EOT
ResNet50	DiffPure [279]	$58.40_{\pm 1.10}$	$28.57_{\pm 1.82}$	$46.96_{\pm 1.44}$
	GNSP [214]	$59.92_{\pm 1.24}$	$33.40_{\pm 1.44}$	$51.06_{\pm 1.74}$
	CMAP (Ours)	$\mathbf{61.93_{\pm 1.86}}$	$\mathbf{39.87}_{\pm 2.58}$	$\mathbf{54.67}_{\pm 1.40}$
ResNet101	DiffPure [279]	$58.93_{\pm 2.15}$	$28.97_{\pm 1.77}$	$50.26_{\pm 2.98}$
	GNSP [214]	$60.38_{\pm 2.89}$	$32.61_{\pm 1.39}$	$51.98_{\pm 2.15}$
	CMAP (Ours)	$\mathbf{61.53}_{\pm 1.45}$	$\mathbf{39.27}_{\pm 1.79}$	$\mathbf{52.60}_{\pm 3.47}$
WRN-50-2	DiffPure [279]	$61.44_{\pm 1.02}$	$28.57_{\pm 2.41}$	$51.46_{\pm 1.35}$
	GNSP [214]	$62.10_{\pm 1.21}$	$31.68_{\pm 0.12}$	$50.99_{\pm 2.10}$
	CMAP (Ours)	$\mathbf{63.33}_{\pm 2.32}$	$\mathbf{40.47}_{\pm 3.59}$	$\mathbf{54.73}_{\pm 3.18}$

Therefore, estimating and comparing data distributions in high-dimensional spaces remains a key challenge. One alternative approach is to estimate the *gradient* of log probability density with respect to the sample, i.e., score. This statistic has emerged as a powerful means for adversarial defense [279, 438] and diffusion models [340], but their potential for adversarial detection is underexplored. This work introduces a new statistic, called **expected perturbation score** (**EPS**), and a universal adversarial detection approach **EPS-AD**.

Method Considering a well-trained classification model f_θ trained on dataset $D=\{(\mathbf{x}_i, \mathbf{y}_i)\}_{i=1}^n$, where $\mathbf{x}_i \in \mathcal{X}$ denotes an input sample and $\mathbf{y}_i$ represents its

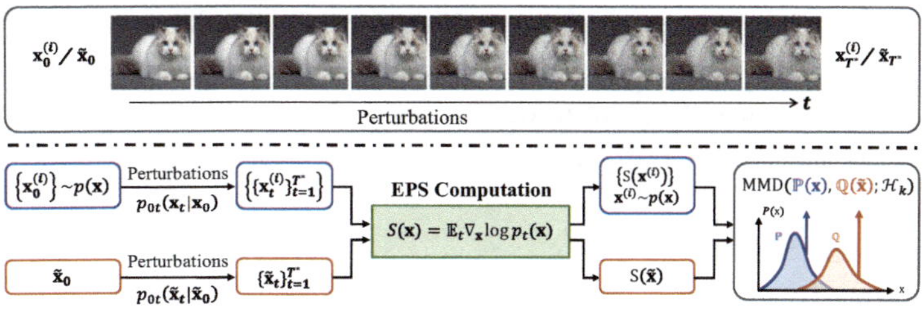

Fig. 3.26 Illustration of the proposed EPS-AD. EPS refers to the expected score derived from multiple perturbations of a sample utilizing a pretrained score model. Specifically, we concurrently introduce perturbations to a collection of natural images $\{\mathbf{x}_0^{(i)}\}$ and a test image $\tilde{\mathbf{x}}_0$ in accordance with the diffusion process at a timestep T^* to generate perturbed images. From these, we derive their EPSs $S(\mathbf{x})$ through the score model and compute the MMD between the EPS of the test sample and the EPSs of the natural samples

ground-truth label in $C = \{1, \ldots, C\}$, an adversarial sample $\tilde{\mathbf{x}}$ for $\mathbf{x}$ is constructed with perturbation ϵ through the optimization problem:

$$\tilde{\mathbf{x}} = \underset{\tilde{\mathbf{x}} \in \mathcal{B}(\mathbf{x}, \epsilon)}{\arg\max}\, \ell\left(f_\theta(\tilde{\mathbf{x}}), \mathbf{y}\right), \tag{3.34}$$

where $\mathcal{B}(\mathbf{x}, \epsilon) = \left\{\mathbf{x}' \in \mathcal{X} \mid d\left(\mathbf{x}, \mathbf{x}'\right) \leq \epsilon\right\}$, d is some distance (*e.g.*, ℓ_2 or ℓ_∞ distance), and ℓ is some loss function. For simplicity, we denote $\tilde{\mathbf{x}}$=$\mathbf{x} + \delta$ as the adversarial data.

Problem 2 (Adversarial Detection) *Let $\mathcal{X} \subset \mathbb{R}^d$ be a separable metric space equipped with a Borel-measurable probability measure, and IID samples $D_X = \{\mathbf{x}^{(i)}\}_{i=1}^n$ from $\mathcal{D}$ and a ground-truth labeling mapping f: $\mathbb{R}^d \rightarrow C$, where C=$\{1, \ldots, C\}$ is a label set. Considering the adversarial scenario where the attacker has access to a well-trained classifier f_θ on D_X and IID samples D'_X from $\mathcal{D}$, our goal is to tell whether each sample in D_Y=$\left\{\mathbf{x}'^{(i)}\right\}_{i=1}^m$ crafted with D'_X is following $\mathcal{D}$.*

Existing approaches, such as those in [438], depend on evaluating a "score" metric—defined as the log probability density gradient relative to a sample—to identify distributional differences between clean and adversarial data. However, relying solely on single-sample scores often results in unreliable detection due to limited data scope and unstable discrepancy measurements. To address this, we introduce a new statistic called **expected perturbation score** (EPS), which leverages multi-view perturbations of a sample to provide a more comprehensive measure of distributional discrepancy. Building on EPS, we introduce the **EPS-based adversarial detection framework (EPS-AD)**, shown in Fig. 3.26. Specifically, EPS-AD first computes EPS values for test samples using a pretrained diffusion model and then performs detection by comparing the test sample against a reference set of natural samples using maximum mean discrepancy (MMD) derived from EPS features. Next, we provide the definition of EPS.

Definition 3.1 (Expected Perturbation Score) Let $\mathcal{X} \subset \mathbb{R}^d$ be a separable metric space and $\mathcal{D}$ be a probability distribution on $\mathcal{X}$. Given a perturbation process transition distribution $p_{0t}(\mathbf{x}_t \mid \mathbf{x}_0)$ from $\mathbf{x}_0$ to $\mathbf{x}_t$, the expected perturbation score (EPS) of a sample $\mathbf{x} \sim \mathcal{D}$ is given by

$$S(\mathbf{x}) = \mathbb{E}_{t\sim U(0,T)} \nabla_{\mathbf{x}} \log p_t(\mathbf{x}), \tag{3.35}$$

where $p_t(\mathbf{x})$ is the marginal probability distribution of $\mathbf{x}_t$ with $p_0(\mathbf{x}) := p(\mathbf{x})$ and T is the last perturbation timestep.

Remark 3.1 The perturbation transition distribution $p_{0t}(\mathbf{x}_t \mid \mathbf{x}_0)$ admits arbitrary specifications, including but not limited to Gaussian distribution or uniform distribution. Through multistep noise injection, this mechanism generates augmented sample variants that enable better exploitation of its information. EPS can be estimated using a neural network trained via score matching as demonstrated in [340].

Based on Definition 2, we establish a theorem to further analyze EPS $S(\mathbf{x})$ for natural and adversarial data under Gaussian assumptions for both the perturbation transition $p_{0t}(\mathbf{x}_t \mid \mathbf{x}_0)$ and data distribution $\mathcal{D}$.

Theorem 3.2 *Assuming that the distribution of natural data $p(\mathbf{x})=\mathbb{N}(\boldsymbol{\mu}_{\mathbf{x}}, \sigma_{\mathbf{x}}^2\mathbf{I})$, where $\mathbf{I}$ is an identity matrix, given a perturbation transition kernel $p_{0t}(\mathbf{x}_t \mid \mathbf{x}_0) = \mathcal{N}(\gamma_t\mathbf{x}_0, \sigma_t^2\mathbf{I})$ with γ_t and σ_t being the time-dependent noise schedule, then the following three conclusions hold:*

(1) For $\forall\, \mathbf{x} \sim p(\mathbf{x})$, $S(\mathbf{x}) \sim \mathbb{N}(\mathbf{0}, \sigma_S^2\mathbf{I})$
(2) For $\forall\, \mathbf{x}' \sim p(\mathbf{x})$ and adversarial sample $\tilde{\mathbf{x}}'=\mathbf{x}'+\boldsymbol{\varepsilon}$, $S(\hat{\mathbf{y}}) \sim \mathbb{N}(-\boldsymbol{\mu}_S, \sigma_S^2\mathbf{I})$
(3) For $\forall\, \mathbf{x}, \mathbf{x}' \sim p(\mathbf{x})$ and adversarial sample $\tilde{\mathbf{x}}'=\mathbf{x}'+\boldsymbol{\varepsilon}$

$$S(\mathbf{x}) - S(\mathbf{x}') \sim \mathbb{N}(\mathbf{0}, 2\sigma_S^2\mathbf{I}); \tag{3.36}$$

$$S(\mathbf{x}) - S(\tilde{\mathbf{x}}') \sim \mathbb{N}(\boldsymbol{\mu}_S, 2\sigma_S^2\mathbf{I}), \tag{3.37}$$

where $\boldsymbol{\mu}_S=\mathbb{E}_{t\sim U(0,T)}\boldsymbol{\mu}_t$ with $\boldsymbol{\mu}_t=\frac{\delta}{\gamma_t^2\sigma_{\mathbf{x}}^2+\sigma_t^2}$ and $\sigma_S^2=\mathbb{E}_{t\sim U(0,T)}\zeta_t^2$ with $\zeta_t^2=\frac{1}{\gamma_t^2\sigma_{\mathbf{x}}^2+\sigma_t^2}$.

Theorem 3.2 reveals the following: (1) the first two findings indicate that, because of the additional term $\boldsymbol{\mu}_S$, the mean of the expected perturbation scores (EPS) for adversarial samples is different from that of natural samples. (2) The third finding shows that, compared with adversarial samples, the EPS of a natural sample is closer to that of other natural samples. Moreover, this difference becomes more significant when the perturbation transitions γ_t and σ_t are small. Consequently, we are inspired to use $S(\mathbf{x})$ for adversarial detection.

EPS for Adversarial Detection Due to the demonstrated effectiveness of MMD in evaluating the similarity between distributions [249], we employ it to characterize EPS. Let $D_X = \{\mathbf{x}^{(i)}\}_{i=1}^n$ represent a set of natural samples and $D_Y = \{\tilde{\mathbf{x}}\}$ denote a test sample. The MMD distance between these sets is computed as

$$\begin{aligned}\widehat{\mathrm{MMD}}_b^2 [D_X, D_Y; \mathbb{H}_k] = &\frac{1}{n^2}\sum_{i,j=1}^{n} k\left(S(\mathbf{x}^{(i)}), S(\mathbf{x}^{(j)})\right) \\ &- \frac{2}{n}\sum_{i=1}^{n} k\left(S(\mathbf{x}^{(i)}), S(\mathbf{x}')\right) + k\left(S(\mathbf{x}'), S(\mathbf{x}')\right),\end{aligned} \tag{3.38}$$

where k is the kernel of a reproducing kernel Hilbert space $\mathbb{H}_k$ such as the Gaussian kernel $k(\mathbf{a}, \mathbf{b}) = \exp\left(-\|\mathbf{a}-\mathbf{b}\|^2 / \left(2\sigma^2\right)\right)$. Following the approach in [244], we utilize deep kernel MMD in our framework. This allows us to measure the discrepancy between the test sample and natural samples, enabling effective adversarial detection.

Note that after the same perturbation process, the first and third terms in Eq. (3.38) are generally identical for each test sample (e.g., when k is a Gaussian kernel). Thus, we focus on the cross-term for the test sample.

Corollary 3.1 *Considering the Gaussian kernel* $k(\mathbf{a}, \mathbf{b}) = \exp\left(-\|\mathbf{a}-\mathbf{b}\|^2 / \left(2\sigma^2\right)\right)$ *and the assumption in Theorem 3.2, for* $\forall 0 < \eta < 1$*, the probability of* $P\{k\left(S(\mathbf{x}), S(\tilde{\mathbf{x}}')\right) > \eta\}$ *is given by*

$$P\{k\left(S(\mathbf{x}), S(\tilde{\mathbf{x}}')\right) > \eta\} = \int_0^C \chi_d^2(z)dz, \tag{3.39}$$

where $z = \|\boldsymbol{\mu}_S\|^2$ *with* $\boldsymbol{\mu}_S$ *being the mean of* $S(\mathbf{x}) - S(\tilde{\mathbf{x}}')$.

Corollary 3.1 demonstrates that the cross-term J increases when $\|\boldsymbol{\mu}_S\|^2$ approaches zero for a given η. By integrating the results from Eqs. (3.36) and (3.37), we infer that natural data exhibit a higher value of J compared to adversarial data with greater probability due to the additional term $\mathbb{E}_t \frac{\delta}{\gamma_t^2 \sigma_{\mathbf{x}}^2 + \sigma_t^2}$, which implies that the MMD between the EPSs of natural samples is smaller than the MMD between natural and adversarial samples.

Experiments The results in Tables 3.21 and 3.22 demonstrate that EPS-AD and EPS-N consistently outperform baselines on CIFAR-10 and ImageNet in AUROC, with EPS-AD achieving the highest detection performance across various attacks. Additionally, as shown in Tables 3.23 and 3.24, EPS-AD excels in detecting unseen and transferable attacks, where baseline methods, particularly non-diffusion-based ones (e.g., KD, LID, MD, and LiBRe), experience significant performance drops. By modeling distributional differences without relying on specific attack methods, EPS-AD maintains reliable detection capabilities, demonstrating its robustness in diverse attack scenarios.

Table 3.21 Comparison of different adversarial detection methods on CIFAR-10 w.r.t. AUROC under $\epsilon = 4/255$ attack

AUROC	FGSM	PGD	BIM	CW	FGSM-ℓ_2	BIM-ℓ_2	AA
KD	0.9213	0.9007	0.9037	0.8335	0.9121	0.9107	0.9135
LID	0.9236	0.9159	0.9217	0.9014	0.9169	0.9320	0.9419
MD	0.9990	0.9860	0.9758	0.9829	**0.9995**	0.9586	0.9830
S-N	**1.0000**	0.9998	0.9961	0.9998	0.9885	0.9674	0.9995
EPS-N (Ours)	**1.0000**	**1.0000**	0.9996	**1.0000**	0.9916	0.9883	**1.0000**
EPS-AD (Ours)	**1.0000**	**1.0000**	**0.9998**	**1.0000**	**0.9995**	**0.9991**	**1.0000**

Table 3.22 Comparison of different adversarial detection methods on ImageNet w.r.t. AUROC under $\epsilon = 4/255$ attack

AUROC	FGSM	PGD	BIM	CW	FGSM-ℓ_2	BIM-ℓ_2	AA
KD	0.7099	0.9720	0.9797	0.9413	0.7004	0.9775	0.9319
LID	0.8912	0.9750	0.9808	0.9528	0.8932	0.9816	0.9582
MD	0.8786	0.9773	0.9835	0.9609	0.8715	0.9830	0.9518
LiBRe	0.9889	0.9530	0.9269	0.9039	0.8708	0.9211	0.8724
S-N	0.9828	0.8974	0.7208	0.8969	0.8830	0.6762	0.9151
EPS-N (Ours)	0.9987	0.9978	0.9215	0.9978	0.8191	0.7172	0.9943
EPS-AD (Ours)	**1.0000**	**1.0000**	**1.0000**	**1.0000**	**1.0000**	**1.0000**	**1.0000**

Table 3.23 Comparison of different adversarial detection methods for detecting unseen attacks on CIFAR-10 w.r.t. AUROC, where "FGSM(seen)"denotes seen attack for KD, LiD, and MD methods

AUROC	FGSM(seen)	PGD	BIM	CW	FGSM-ℓ_2 (seen)	BIM-ℓ_2	AA
KD	0.9213	0.9007	0.9082	0.8339	0.9146	0.9146	0.9135
LID	0.9236	0.8964	0.9028	0.8828	0.9160	0.8984	0.9253
MD	0.9990	0.9855	0.9742	0.9835	0.9992	0.9503	0.9820
S-N	**1.0000**	0.9998	0.9961	0.9998	0.9885	0.9674	0.9995
EPS-N (Ours)	**1.0000**	**1.0000**	0.9996	**1.0000**	0.9916	0.9883	**1.0000**
EPS-AD (Ours)	**1.0000**	**1.0000**	**0.9998**	**1.0000**	**0.9995**	**0.9991**	**1.0000**

Table 3.24 Comparison of different adversarial detection methods for detecting transferable attacks on ImageNet w.r.t. AUROC, where KD, LID, MD, and LiBRe are trained with adversarial samples over ResNet-50 but detect adversarial ones over ResNet-101

AUROC	FGSM	PGD	BIM	CW	FGSM-ℓ_2	BIM-ℓ_2	AA
KD	0.7754	0.5999	0.5847	0.7632	0.7906	0.7756	0.7698
LID	0.8467	0.7627	0.7663	0.7704	0.8520	0.7925	0.7967
MD	0.8467	0.7698	0.7684	0.7665	0.8067	0.7759	0.7880
*L*iBRe	0.9849	0.8414	0.7161	0.8286	0.8489	0.7250	0.8485
S-N	0.9816	0.8965	0.7166	0.8963	0.8764	0.6705	0.9106
EPS-N (Ours)	0.9983	0.9975	0.9178	0.9979	0.8235	0.7215	0.9930
EPS-AD (Ours)	**1.0000**	**1.0000**	**1.0000**	**1.0000**	**1.0000**	**0.9998**	**1.0000**

Chapter 4
Trustworthy Machine Learning with Out-of-Distribution Data

It is well known that machine learning models are susceptible to out-of-distribution (OOD) data, which occurs when the distribution of data at test time differs significantly from that at training time. This inherent shift in distribution signifies that we cannot guarantee that the resulting models can perform well when deployed in real-world scenarios. This issue has spurred numerous studies on OOD learning, aiming to mitigate the negative impacts raised by such distribution shifts.

In practice, there are primarily two kinds of OOD data, associated with semantic shifts and covariate shifts, respectively. Semantic shifts occur when the semantics changes between training and test, typically echoing the cases where the ground-truth labels of test data may not be in the considered label space. In this case, conventionally trained models are infeasible to make correct predictions, raising the research in OOD detection. On the other side, covariate shifts consider the cases where the covariate distribution changes between training and test, for example, for the change in environment or style of data. Machine learning models typically do not perform well in such OOD cases, motivating the need for understanding OOD generalization. In this chapter, we focus on the key questions as well as solutions within the field of OOD learning, providing a comprehensive review for the current advances. By exploring these aspects, we aim to shed light on enhancing model reliability and performance under distributional changes.

4.1 Trustworthy Learning for Out-of-Distribution Detection

4.1.1 Introduction and Notation

OOD detection is the identification of inputs for which reliable predictions are lacking for which classification models. Usually, the data include ground-truth labels outside the considered label space. Machine learning models, especially deep

B. Han, T. Liu, *Trustworthy Machine Learning under Imperfect Data*,
https://doi.org/10.1007/978-981-96-9396-2_4

learning architectures, depend on OOD detection as, when exposed to OOD inputs, they may provide either inaccurate or too confident predictions. Medical analysis, autonomous driving, and financial uses all make great use of OOD detection.

Let $\mathcal{X} \subset \mathbb{R}^d$, where d represents the dimension, denotes the feature space and $\mathcal{Y}$ denote the label space for in-distribution (ID) data. Let $X_{\text{id}} \in \mathcal{X}$ and $X_{\text{ood}} \in \mathcal{X}$ correspondingly to represent the random variables connected with ID and OOD data, respectively. For ID data, random variables are represented as $Y_{\text{id}} \in \mathcal{Y}$ and for OOD data as $Y_{\text{ood}} \notin \mathcal{Y}$. Whereas $P(X_{\text{ood}}, Y_{\text{ood}})$ represents the joint distribution for OOD data, the joint distribution for ID data is shown as $P(X_{\text{id}}, Y_{\text{id}})$. For ID data, the marginal distributions are $P(X_{\text{id}})$ and for OOD data $P(X_{\text{ood}})$.

OOD Score Functions Let $D_{\text{id}}^{\text{Train}} = \{(\mathbf{x}_i, y_i)\}_{i=1}^n$ represent the training dataset of size n comprising ID samples obtained from the joint distribution $P(X_{\text{id}}, Y_{\text{id}})$, and C is the number of ID classes. According to the framework outlined in Fang et al. [104], the objective of OOD detection is to train a detector G utilizing the dataset $D_{\text{id}}^{\text{Train}}$, ensuring that for any test sample $\mathbf{x}$: (1) If $\mathbf{x}$ is sampled from $P(X_{\text{id}})$, then G classifies $\mathbf{x}$ into the correct ID classes accurately; and (2) if $\mathbf{x}$ is drawn from $P(X_{\text{id}})$, then G identifies $\mathbf{x}$ as OOD. With a threshold τ, a pretrained and a score function S, a sample $\mathbf{x}$ is categorized as ID if and only if $S(\mathbf{x}; f_\theta) \geq \tau$:

$$\begin{cases} G_\tau(\mathbf{x}) = \text{ID} & \text{, if } S(\mathbf{x}; f_\theta) \geq \tau \\ G_\tau(\mathbf{x}) = \text{OOD} & \text{, otherwise} \end{cases}. \tag{4.1}$$

The design of the score function S and the model f_θ must ensure that the scores allotted to OOD samples are frequently lower than those assigned to ID samples, evaluating the success of OOD detection strategies primarily relies on these aspects.

Outlier Exposure A fine-tuning approach termed outlier exposure (OE) [153] has been introduced to enhance OOD detection performance. OE utilizes surrogate OOD data $D_{\text{ood}} = \{\mathbf{x}_j^s\}_{j=1}^m$ of size m and implements a fine-tuning strategy grounded in empirical risk minimization, expressed as

$$\arg\min_\theta \frac{1-\alpha}{n} \sum_{i=1}^n \ell(f_\theta(\mathbf{x}_i), y_i) + \frac{\alpha}{m} \sum_{j=1}^m \ell_{\text{OE}}(f_\theta(\mathbf{x}_j^s)), \tag{4.2}$$

where α is a hyperparameter, ℓ signifies the loss function, and ℓ_{OE} denotes the surrogate OOD loss. Utilizing surrogate OOD data enables the model to position specific OOD samples in latent embeddings that are remote from all ID classes, thereby typically improving the efficacy of OOD detection.

4.1.2 Post Hoc OOD Detection

4.1.2.1 Post Hoc OOD Detection Under In-Distribution Class Imbalance

A common difficulty in real-world applications where certain classes are much underrepresented in the training data is a long-tail problem. OOD detection combined with the long-tail issue presents difficult problems for practical use. Underrepresented classes in training data aggravate the long-tail problem, therefore affecting OOD detection. This imbalance could aggravate the difficulties in OOD identification as models might find it difficult to generalize across both the tail classes and the unknown classes. Long-tailed distributions challenge OOD detection by increasing the likelihood of misclassifying tail classes as unknown classes, therefore resulting in possibly significant repercussions. In medical diagnosis, for example, a model could misidentify rare diseases (tail classes) as unknown diseases, therefore impacting the therapeutic decisions. This junction of OOD detection and long-tailed distributions raises more complicated problems needing for sophisticated solutions to guarantee continuous model performance. Therefore, researchers are driven to create techniques that not only solve the long-tail issue but also improve OOD detection abilities, thereby assuring that models can correctly recognize and classify both ID and OOD classes in many real-world settings [229].

However, several well-known OOD detection methods, such as those in [168, 229], are based on an unsubstantiated supposition. The following assumption is put forward: regarding an OOD data point $\mathbf{x}_{\text{ood}}$, the likelihood of it being assigned to each ID class i remains constant. Specifically, for any $\mathbf{x}_{\text{ood}}$, we have

$$[P(\text{label of } \mathbf{x}_{\text{ood}} \text{ is } 1), \ldots, P(\text{label of } \mathbf{x}_{\text{ood}} \text{ is } C)] = [1/C, \ldots, 1/C]_{1\times C} := \mathbf{U}, \tag{4.3}$$

where $\mathbf{U}$ denotes a uniform distribution and C represents the number of ID classes. Taking GradNorm [168], a cutting-edge OOD detection method, as an example, let $f_\theta(\mathbf{x})$ denote the output of the ID model for a data point $\mathbf{x}$, and the score function of GradNorm, $S_{\text{GradNorm}}(\mathbf{x})$, is defined as

$$\left\| \frac{\partial D_{\text{KL}}(\mathbf{U} \| \operatorname{softmax}(f_\theta(\mathbf{x})))}{\partial w} \right\|_{L^1}, \tag{4.4}$$

where $\operatorname{softmax}(f_\theta(\mathbf{x}))$ is a vector of predicted probabilities indicating the likelihood that $\mathbf{x}$ belongs to the ID classes, θ represents the parameters of the model f_θ, w is a component of the network parameter θ, and $D_{\text{KL}}(\cdot \parallel \cdot)$ is the Kullback-Leibler divergence function. GradNorm distinctly makes use of $\mathbf{U}$ in the role of a reference distribution for the purpose of differentiating between ID and OOD data. A less difference between $\operatorname{softmax}(f_\theta(\mathbf{x}))$ and $\mathbf{U}$ suggests that $\mathbf{x}$ is more likely to be an OOD data point. However, we are unable to ascertain the validity of this assumption since it remains unverified. In the event that this assumption turns out to be false, the score functions relying on $\mathbf{U}$ (e.g., the one presented in Eq. (4.4)) will lose their well-defined nature. This is because these functions are then unable to ensure that the lowest score is associated with the data that is most OOD.

Method For methods (e.g., MSP [150] and GradNorm) whose score functions are linked to the uniform distribution $\mathbf{U}$, a straightforward approach to modify them is to substitute the uniform distribution $\mathbf{U}$ with the ID-class-prior distribution $P_{Y_{\text{id}}}$. We therefore modify the score function of MSP, denoted as $S_{\text{RP+MSP}}(\mathbf{x})$ to be

$$\max_{i\in\{1,\ldots,C\}} (\text{softmax}(f_\theta(\mathbf{x}))_i - P_{Y_{\text{id}}}(i)), \tag{4.5}$$

and the score function of GradNorm is modified to

$$S_{\text{RP+GradNorm}}(\mathbf{x}) = \left\| \frac{\partial D_{\text{KL}}(P_{Y_{\text{id}}} \| \text{softmax}(f_\theta(\mathbf{x}))}{\partial w} \right\|_{L^1}. \tag{4.6}$$

For techniques devoid of obvious relation to the uniform distribution, we suggest the RW strategy as a complementary to the RP strategy (e.g., ODIN [229] and Energy [246]). The RW approach modifies the scores by considering the resemblance between the prior distribution of the ID classes, denoted as $P_{Y_{\text{id}}}$, and the result of applying the softmax function to $f_\theta(\mathbf{x})$. We expect that the weights employed in this process will not bring about significant changes to the scores of the OOD data.

In more detail, the RW method focuses on the relationship between how the ID classes are initially distributed, which is represented by $P_{Y_{\text{id}}}$, and the output of the softmax operation on $f_\theta(\mathbf{x})$. This relationship is used as a basis for adjusting the scores. Regarding the OOD scores, we foresee that the weights used in the adjustment mechanism will have a relatively minor impact on them, meaning that the overall values of these scores for OOD data will remain largely unchanged.

In this study, we utilize the cosine function as the weight function. This function is widely applied as a measure of distance and similarity in contrastive learning, as demonstrated in works such as [59, 130, 146]. The cosine function is selected primarily because it is a bounded function. After normalization, this property makes it suitable for use as a weighting factor. Specifically

$$S_{\text{RW+Method}}(\mathbf{x}) = S_{\text{Method}}(\mathbf{x}) \cdot W(f_\theta, P_{Y_{\text{id}}}), \tag{4.7}$$

where $W(f_\theta, P_{Y_{\text{id}}})$ is the weight function defined as

$$\begin{aligned} W(f_\theta, P_{Y_{\text{id}}}) =& (1 - \cos(\text{softmax}(f_\theta(\mathbf{x})), P_{Y_{\text{id}}})) \\ =& (1 - \frac{\text{softmax}(f_\theta(\mathbf{x})) \cdot P_{Y_{\text{id}}}^\top}{\| \text{softmax}(f_\theta(\mathbf{x}))\| \cdot \|P_{Y_{\text{id}}}\|}), \end{aligned} \tag{4.8}$$

where the cosine function with a constant 1 to guarantee the weight function stays nonnegative. Here, $S_{\text{Method}}(\mathbf{x})$ refers to a score function proposed in prior studies, such as ODIN and Energy.

Next, we will detail the reweighted versions of ODIN and Energy.

In comparison to MSP, the primary enhancement of ODIN is the implementation of a temperature scaling strategy. We can modify ODIN as follows: for a temperature $T > 0$, the reweighted score function $S_{\text{RW+ODIN}}(\mathbf{x})$ is given by

$$\max_{i\in\{1,\ldots,C\}} \frac{\exp\left(f_i(\mathbf{x})/T\right)}{\sum_{j=1}^{C}\exp\left(f_j(\mathbf{x})/T\right)} \cdot W(f_\theta, P_{Y_{\text{id}}}). \tag{4.9}$$

Energy [246] suggests replacing the softmax function with the energy function for OOD detection. The energy function has a feature directly connected to the distribution: systems with a more concentrated probability distribution show lower energy, while those with a more scattered probability distribution (more closer to the uniform distribution) show greater energy [212]. Consequently, the energy of ID data is lower than that of OOD data. Based on Eq. (4.7), we modify the Energy method as follows:

$$-T \cdot \log \sum_{i=1}^{C} e^{f_i(\mathbf{x})/T} \cdot W(f_\theta, P_{Y_{\text{id}}}). \tag{4.10}$$

Experiments In the experiment, for the training dataset, inspired by [242], a series of imbalanced datasets are constructed by sampling the ImageNet-1k [91] dataset using the Pareto distribution, and the ImageNet-LT-a8 dataset is the most imbalanced. For the inference phase, the large-scale ImageNet-1k benchmark [163] is utilized, and the OOD datasets consist of subsets from iNaturalist, SUN, Places, and Textures. It is emphasized that there are no overlapping classes between the ID and OOD datasets.

The strategies presented are relevant to various score functions employed in out-of-distribution detection methods. Comparatively to baseline techniques, our methods provide comprehensive data shown in Table 4.1. The changes applied to previous approaches using our techniques produced significant performance improvements, therefore proving their effectiveness. Among all the studied techniques, RP+GradNorm showed the lowest FPR95 at 70.12%, whereas RW+ODIN achieved the greatest AUROC of 77.73%. Moreover, our strategies consistently surpassed the performance of established baseline methods across all evaluation tasks. RW+ODIN enhanced the AUROC from 60.69% to 77.73%, while RP+GradNorm reduced the FPR95 from 80.45% to 70.12%. The results of our experiments show that our strategies are much better than the baselines when there are imbalanced ID classes.

Given the uneven distribution of samples in the training dataset, we leverage the classification strategy detailed in [242] to partition all ID classes into three distinct groups. The ID Head group consists of classes that have an ample number

Table 4.1 OOD detection performance comparison with other competitive score-based OOD detection methods. All methods are based on ResNet101 trained on ImageNet-LT-a8. ↑ indicates larger values are better and ↓ indicates smaller values are better. All values are percentages. The bold indicates the best performance while the underline indicates the second

Method	iNaturalist		SUN		Places		Textures		Average	
	AUROC ↑	FPR95 ↓	AUROC ↑	FPR95 ↓	AUROC ↑	FPR95 ↓	AUROC ↑	FPR95 ↓	AUROC ↑	FPR95 ↓
MSP [229]	63.95	97.72	66.60	93.13	66.84	92.11	42.74	98.79	60.03	95.44
ODIN [229]	60.14	98.70	70.63	93.13	70.14	91.96	41.83	98.30	60.69	95.52
Mahalanobis [215]	60.72	95.87	56.79	94.50	55.27	93.78	49.43	86.99	55.55	92.78
Energy [246]	55.99	98.74	71.12	93.11	70.24	91.30	42.38	98.07	59.93	95.30
GradNorm [168]	82.51	72.19	74.57	78.10	70.67	86.58	57.31	84.95	71.26	80.45
Maxlogit [151]	60.14	98.70	70.64	93.13	70.15	91.96	41.83	98.30	60.69	95.52
Dice [348]	85.80	58.96	73.17	76.90	67.83	87.89	58.43	80.59	71.31	76.11
RP+MSP(Ours)	64.95	96.44	67.39	91.79	67.46	91.16	43.05	98.51	60.71	94.48
RW+ODIN(Ours)	90.24	47.52	**81.33**	75.44	**77.90**	**79.90**	61.43	88.03	**77.73**	72.72
RW+Energy(Ours)	83.12	65.67	80.13	77.51	77.12	80.49	51.70	91.31	73.02	78.75
RP+GradNorm(Ours)	**91.23**	**43.87**	77.36	**73.53**	72.67	83.29	**62.94**	**79.80**	76.05	**70.12**
RW+Maxlogit(Ours)	78.16	77.68	77.26	81.91	75.15	83.28	48.34	94.06	69.73	84.23

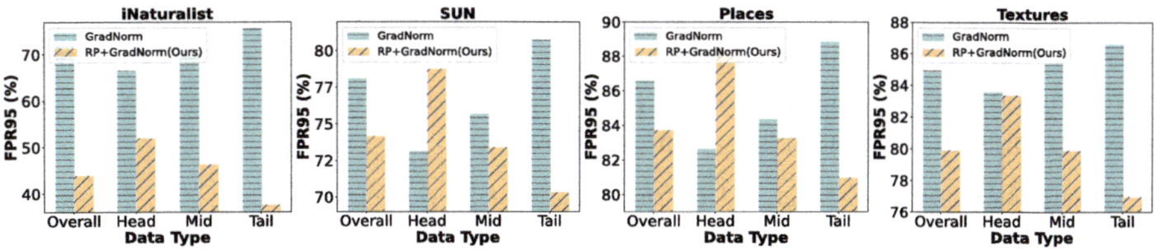

Fig. 4.1 Performance comparison of different data type. The figures show the OOD detection performance of GradNorm and RP+GradNorm in four OOD datasets

of samples, specifically more than 100, within the training data. Classes with a modest sample size ranging between 20 and 100 come within the ID Mid category. Conversely, the ID Tail group consists of classes with fewer than 20 in the training set, so a quite limited sample count. We then performed an in-depth analysis of three distinct datasets' OOD detection capability. These datasets were created by aggregating OOD data with each of the ID class groupings (ID Head, ID Mid, and ID Tail), therefore producing ID Head + OOD, ID Mid + OOD, and ID Tail + OOD datasets. If a technique performs better on the ID Tail + OOD dataset than the ID Head + OOD dataset, then one may deduce its superiority in handling OOD data and tail classes.

The experimental results shown in Fig. 4.1 amply show how much the RP + GradNorm method improves performance across all three of these datasets. A thorough investigation of these performance improvements indicates that the main causes of the general increase in performance of the RP + GradNorm technique are the significant increases attained on the Tail + OOD dataset. This suggests that conventional approaches, including GradNorm, most likely had difficulties precisely separating OOD data from ID tail classes. Their efficiency in OOD identification was therefore hampered. Nevertheless, the results of our trials show that the novel approaches we used have effectively overcome this difficulty and shown their improved capacity to manage such challenging situations.

Although the RW can achieve notable improvements in OOD performance under the standard OOD detection setup, it still suffers a common challenge under the post hoc setup: adapting the score functions are hard for specific detection task of interest, arising since we assume intact model parameters that limit our flexibility for adaptation. Therefore, in the following, we introduce a general strategy [389] that can adapt the post hoc OOD detection effectively.

4.1.2.2 Post Hoc OOD Detection with Watermarking

Classification-based OOD detection has achieved many attentions in the literature, which relies on the so-called score functions in discerning between ID and OOD data, without requiring the model parameters to be further tuned. The score functions can extract meaningful information from fixed models and can be defined by either logit outputs, gradients, or embedding features. Although significant progress has been achieved, previous studies have shown that there is no single score function that consistently emerges as the optimal choice. It motivates us to pose an important research question as follows:

How to adapt the score function for a specific task of our interest?

It is not an easy question as we assume the predefined score function, and we want to keep the proficiency of the classification-based methods that keep the model intact. In this work, the authors explore the reprogramming property of deep models, allowing the adaptation of a model trained on task to perform another task by altering the input patterns, without the requirements for retraining or modifying the model parameters. Accordingly, the authors proposed "watermark," a static pattern that matches the size of the original data $\mathbf{x}$ and can be added to the original inputs. We need to find a watermark $\boldsymbol{\omega}$, such that, after watermarking following $\mathbf{x} + \boldsymbol{\omega}$, the capability of our original model in OOD detection should be largely improved. Extensive experiments validate the universality and the effectiveness of watermarking in enhancing OOD detection.

Method The key concept in the algorithmic design is the watermark, denoted by $\boldsymbol{\omega} \in \mathbb{R}^d$, a unified pattern that is in the same size as original inputs. Since it is in the same size as inputs, it can be added to test data directly, and we name $\mathbf{x} + \boldsymbol{\omega}$ as the watermarked input with respect to the original data $\mathbf{x}$. Then, for the specific score function of our interest, we need to ensure that the model will improve its performance in OOD detection given the watermarked data, without further need to adjust model parameters.

To learn a proper watermark, we need to consider ID and OOD cases jointly, with the associated loss functions ℓ and ℓ_{OE}, respectively. Regarding ID, we typically have the original ID data, where we need high scores for their watermarked results. On the other hand, regarding OOD, we typically lack proper knowledge about them while we still hope our model, after watermarking, can make low scores for them. To tackle this issue, the authors suggest to further perturb the watermark during training. Specifically, the authors use the Gaussian noise and have the perturbed watermark following $\boldsymbol{\epsilon} + \boldsymbol{\omega}$ with $\boldsymbol{\epsilon} \sim \mathcal{N}(\mathbf{0}, \sigma_1 \boldsymbol{I}_d)$ the i.i.d. Gaussian noise. Then, we have the final results for the proper watermark as

$$\mathcal{L}_D(\boldsymbol{w}) = \sum_n \ell(f_\theta(\mathbf{x}_i + \boldsymbol{\omega}), y_i) + \beta \sum_n \ell_{\mathrm{OE}}(f_\theta(\boldsymbol{\epsilon}_j + \boldsymbol{\omega})), \tag{4.11}$$

with $\beta \geq 0$ the trade-off parameter and n the sample size. The first-order gradient updating rule can be adopted to minimize $\mathcal{L}_D(\boldsymbol{\omega})$. However, it is also well known that feature-level optimization within deep learning is typically hard, as the optimization dynamics can easily get stuck at some local solutions. Therefore, the gradient signums, rather than its original values, are adopted for updating $\boldsymbol{w}$, which is given by

$$\boldsymbol{\omega} \leftarrow \boldsymbol{\omega} - \alpha \texttt{sign}(\nabla_{\boldsymbol{\omega}} \mathcal{L}_D(\boldsymbol{\omega})), \tag{4.12}$$

where $\texttt{sign}(\cdot)$ denotes the signum function and $\alpha > 0$ is the step size.

Moreover, it is also well known that we typically prefer those solutions that lie in smooth loss landscapes, which can facilitate generality and robustness. To this end, the authors further refine the optimization rule for watermarking by adopting the sharpness-aware minimization, which can guarantee that the solutions have smooth landscapes with overall small singular values of Hessian. Overall, with sharpness aware minimization, the updating rule of $\boldsymbol{\omega}$ is given by

$$\boldsymbol{\omega} \leftarrow \boldsymbol{\omega} - \alpha \texttt{sign}(\nabla_{\boldsymbol{\omega}} \mathcal{L}_D(\boldsymbol{\omega} + \boldsymbol{\kappa})), \tag{4.13}$$

where $\boldsymbol{\kappa}$ is defined by

$$\boldsymbol{\kappa} = \rho \texttt{sign}(\nabla_{\boldsymbol{\omega}} \mathcal{L}_D(\boldsymbol{\omega})) \frac{|\nabla_{\boldsymbol{\omega}} \mathcal{L}_D(\boldsymbol{\omega})|^{q-1}}{\left(||\nabla_{\boldsymbol{\omega}} \mathcal{L}_D(\boldsymbol{\omega})||_q^q\right)^{1/p}}, \tag{4.14}$$

p, q are hyperparameters that satisfy $1/p + 1/q = 1$ and ρ the perturbation strength.

Watermarking is general in combining with different score functions—when we change the score function, we just need to change the corresponding definition of ℓ and ℓ_{OE}. Here, we take the softmax scoring and the free-energy scoring as two examples. When the softmax scoring is considered, of which the score function is

$$S(\mathbf{x}; f_\theta) = \max_c \ \texttt{softmax}_c \ f_\theta(\mathbf{x}), \tag{4.15}$$

we set $\ell^{\text{CE}}(\cdot)$ to be the cross-entropy loss and $\ell_{\text{OE}}^{\text{CE}}(\cdot)$ to be the cross-entropy regarding the uniform distribution, namely,

$$\begin{aligned} \ell^{\text{CE}}(f_\theta(\mathbf{x}), y) = -\log \texttt{softmax}_y f_\theta(\mathbf{x}) \ \text{ and } \ \ell_{\text{OE}}^{\text{CE}}(f_\theta(\mathbf{x})) \\ = -\sum_c \frac{1}{C} \log \texttt{softmax}_c f_\theta(\mathbf{x}). \end{aligned} \tag{4.16}$$

Instead, when the free-energy scoring is considered, which is given by

$$S(\mathbf{x}; f_\theta) = \log \sum_c \exp f_{\theta,c}(\mathbf{x})/T, \tag{4.17}$$

with $T > 0$ the temperature parameter, then we can specify

$$\ell^{\text{FE}}(f_\theta(\mathbf{x})) = \sum_c \exp -f_{\theta,c}(\mathbf{x}) \ \text{ and } \ \ell_{\text{OE}}^{\text{FE}}(f_\theta(\mathbf{x})) = \sum_c \exp f_{\theta,c}(\mathbf{x}), \tag{4.18}$$

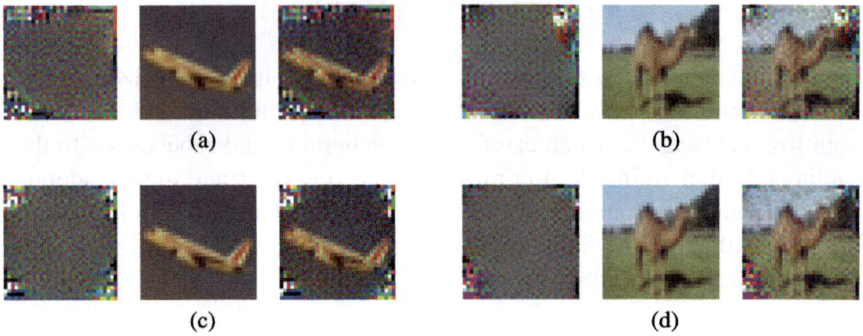

Fig. 4.2 The learned watermarks (left) and the example images with (middle) and without (right) the watermarks. All the pictures are clamped between 0 and 255 for the purpose of illustration. (**a**) Softmax scoring on CIFAR-10. (**b**) Softmax scoring on CIFAR-100. (**c**) Free-energy scoring on CIFAR-10. (**d**) Free-energy scoring on CIFAR-100

Table 4.2 Softmax scoring with/without watermarking on CIFAR benchmarks. ↓ (↑) indicates smaller (larger) values are preferred

	FPR95 ↓	AUROC ↑	AUPR ↑
	w/ (w/o) watermark		
CIFAR-10			
iSUN	**43.60** (55.55)	**93.53** (90.14)	**98.67** (97.84)
Places365	**60.75** (62.50)	**87.85** (87.41)	**96.98** (96.94)
Texture	**42.00** (59.30)	**92.83** (88.37)	**98.43** (97.14)
SVHN	**27.25** (49.10)	**96.00** (91.69)	**99.17** (96.54)
LSUN	**40.70** (52.05)	**94.36** (91.50)	**98.86** (98.16)
Average	**42.86** (55.70)	**92.91** (89.82)	**98.42** (97.32)
CIFAR-100			
iSUN	**77.85** (83.35)	**79.91** (75.28)	**95.35** (94.00)
Places365	83.25 (**82.20**)	74.28 (**74.40**)	**93.47** (93.44)
Texture	**79.10** (83.80)	**77.14** (72.83)	**94.26** (92.81)
SVHN	**82.95** (85.05)	**76.92** (70.64)	**94.72** (92.61)
LSUN	**76.75** (80.45)	**79.60** (76.25)	**95.27** (94.32)
Average	**79.98** (82.97)	**77.57** (73.88)	**94.61** (93.43)

Experiments Figure 4.2 shows the learned watermarks for softmax and free-energy scoring on the CIFAR benchmarks. As observed, the middle parts of these learned watermarks can largely maintain the original patterns of inputs. However, the surrounding areas of these learned watermarks change a lot, which integrate additional patterns that are critical to enhance OOD detection in a post hoc manner. Furthermore, Tables 4.2 and 4.3 show the results on CIFAR benchmarks, comparing the results before and after watermarking. We consider three evaluation metrics, (1) the false OOD rate with the true ID rate is at 95% (FPR95), (2) the area under the receiver operating characteristic curve (AUROC), and (3) the area under the precision-recall curve (AUPR). As shown, the results not only verify that watermarking is efficient and effective but also demonstrate its applicability across different scenarios.

Table 4.3 Free-energy scoring with/without watermarking on CIFAR benchmarks. ↓ (↑) indicates smaller (larger) values are preferred

	FPR95 ↓	AUROC ↑	AUPR ↑
	w/ (w/o) watermark		
CIFAR-10			
iSUN	**16.30** (32.10)	**96.97** (92.84)	**99.39** (98.33)
Places365	**36.25** (41.45)	**91.87** (89.65)	**97.94** (97.21)
Texture	**32.60** (52.05)	**93.14** (85.43)	**98.08** (95.52)
SVHN	**16.45** (35.25)	**97.11** (90.91)	**99.39** (97.68)
LSUN	**16.85** (27.50)	**96.97** (93.98)	**99.38** (98.59)
Average	**23.69** (37.67)	**95.21** (90.56)	**98.83** (97.46)
CIFAR-100			
iSUN	**75.05** (81.80)	**83.07** (79.04)	**96.15** (94.98)
Places365	**80.45** (80.50)	**77.78** (74.99)	**94.45** (93.37)
Texture	**75.15** (80.20)	**79.55** (76.00)	**94.79** (93.53)
SVHN	**82.85** (85.10)	**75.26** (74.20)	**94.18** (93.70)
LSUN	**71.85** (80.45)	**84.01** (78.29)	**96.33** (94.69)
Average	**77.07** (81.61)	**79.93** (76.50)	**95.18** (94.05)

Besides classical deep models, it is also an interesting question about how to conduct OOD detection for more advanced foundation models. In the following, as a case study, we explore how to realize post hoc OOD detection for pretrained vision-language models, such as CLIPs.

4.1.2.3 Post Hoc OOD Detection for Pretrained Vision-Language Models

Traditional visual OOD detection techniques mostly concentrate on picture data [160, 168, 352], therefore neglecting the rich linguistic information included in labels. Vision-language models (VLMs) could improve OOD detection by use of multimodal information. Recently, a number of new methods have come into existence with the goal of developing customized OOD detectors specifically for VLMs. For instance, ZOC [102] has introduced a novel zero-shot OOD detection task. In this task, a trainable captioner is utilized to generate suitable OOD labels for OOD images. However, when dealing with large-scale datasets like ImageNet-1k, which contains numerous ID classes, the captioner may encounter difficulties in providing appropriate OOD labels. This limitation can result in subpar performance.

Another approach, MCM [267], makes use of the maximum logit of a scaled softmax function to identify OOD images. Despite its merits, MCM fails to fully exploit the text interpretation capabilities that are intrinsically present in VLMs. It merely draws on data from the ID label segment. Thus, we believe that there is still a great deal of unexploited potential in using VLMs to improve OOD detection.

In this paper, we put forward a strategy to better utilize the information stored within VLMs for the purpose of OOD detection. In our approach, we incorporate a large number of negative labels. This inclusion serves to empower the model, allowing it to make more accurate and in-depth distinctions between OOD data. Our

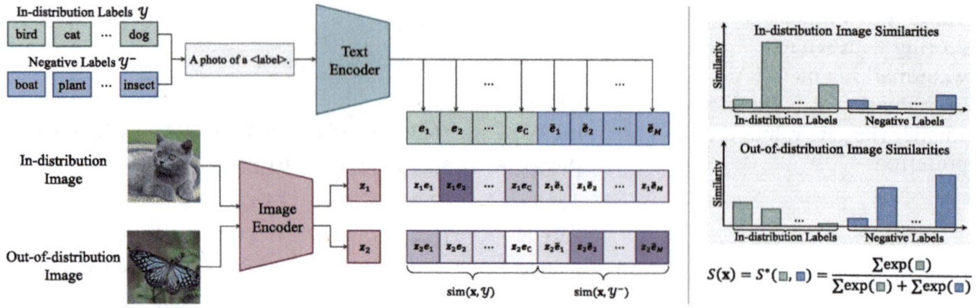

Fig. 4.3 Overview of NegLabel. The input images are processed by the image encoder, which then extracts them into image embeddings $\mathbf{z}$. Meanwhile, the text encoder takes ID and negative labels and extracts them into text embeddings $\boldsymbol{e}$. The negative labels are carefully chosen by NegMining from a vast large-scale corpus. The similarities between images and text are measured and quantified using the operation $\mathbf{z} \cdot \boldsymbol{e}$. These similarities are graphically represented by purple blocks, where darker shades are used to denote higher levels of similarity. The right-hand side of the relevant illustration shows that ID images generally result in lower similarity values when compared with negative labels in contrast to OOD images. Our NegLabel score combines the similarities of image-ID labels (depicted in green) and image-negative labels (depicted in blue). This NegLabel approach provides a more detailed and refined perspective for zero-shot OOD detection

newly put forward method, named NegLabel, is designed to spot OOD samples. It accomplishes this objective by meticulously scrutinizing the associations that exist among ID labels and negative labels. As can be observed in Fig. 4.3, because of the differences in affinity, ID samples tend to have higher OOD scores compared to OOD samples. The model, with the help of the added negative labels, gains a better understanding of the data landscape, enabling it to more precisely pick out which data points fall outside the normal distribution.

We introduce the NegMining approach. Its main objective is to sift through large corpus sets and identify high-quality negative labels. This technique evaluates the suitability of negative labels through the lens of semantic divergence. Specifically, it takes the distance between negative labels and ID labels as a crucial metric. This method makes use of negative labels that exhibit significant semantic disparities from ID labels. By doing so, it reinforces the distinction between ID and OOD data. Additionally, we have developed a novel OOD scoring system. Within this system, the affinities of photos and ID labels are positively correlated, whereas they are negatively correlated with negative labels. This new scoring system provides a more nuanced way of assessing whether a sample is OOD or not. This score effectively maximizes the utilization of information from both the negative label space and the ID label space. As a result, it enhances the text comprehension capabilities of VLMs.

Definition of the Problem For zero-shot image classification, a pretrained model similar to CLIP is utilized. This model is responsible for assigning images to the designated label names, which we also refer to as ID classes. As [306] suggests, this process is carried out without any fine-tuning on these particular classes. In the case of the zero-shot classification task, the current work integrated a pretrained CLIP-like model with a preestablished set of class labels. We name these particular categories as ID classes. Apart from the classes utilized during the model's pretraining, these ID classes are clearly meant for the zero-shot classification task. In a similar vein, the concept of OOD is based on these ID categories instead of the pretraining set data distribution. As a result, in our study, the OOD detection method that depends on VLM primarily attempts to identify samples that deviate from any of the known ID classes as OOD. Simultaneously, it maintains a high degree of accuracy in the classification of ID classes.

Method The semantic labels of ID samples are confined to $\mathcal{Y}$, whereas OOD labels come from other domains. Different from classical image classification, zero-shot classifiers are trained on large-scale datasets [181, 306] to enhance comprehension. The restricted ID label space $\mathcal{Y}$ might lead to a decrease in the model's ability to detect OOD samples. This is because it may not fully encapsulate the differences between ID and OOD samples, potentially hampering the model's sensitivity in identifying OOD data. CLIP-like models may scale label space with any language prompts and no computational cost. NegLabel expands the label space to improve the model's zero-shot generalization, using VLMs' text comprehension for zero-shot OOD detection.

Selection of Negative Labels Negative labels, which are denoted as $y^- \in \mathcal{Y}^-$, are characterized as labels y^- that do not belong to $\mathcal{Y}$. These labels demonstrate lower affinities with ID images compared to their affinities with OOD images. This characteristic property of negative labels plays a crucial role in enabling more effective OOD detection. The aim is to optimize the difference between $\mathcal{Y}$ and $\mathcal{Y}^-$.

The extensive corpus database encompasses nearly all prevalent concepts, enabling CLIP-like models to associate input samples with more comprehensive labels. During the process of negative label selection, we eliminate words that bear a strong resemblance to $\mathcal{Y}$. The purpose of this action is to guarantee that the negative labels we ultimately choose display lower affinities with ID samples in comparison to their affinities with OOD samples. To find appropriate negative labels, we put forward the NegMining algorithm.

We make use of a lexical database such as WordNet [107] to build a candidate label space. This space is denoted as $\mathcal{Y}^{\mathrm{K}} = \{\tilde{y}_1, \tilde{y}_2, \ldots, \tilde{y}_K\}$, where the variable K represents the overall count of candidate words. For every $\tilde{y}_i$ $(i = 1, 2, \ldots, K)$, we calculate negative cosine similarities. This computation enables us to assess the separation between the candidate label and the ID label space.

We use the following formula to measure the distance between the negative label and ID label space:

$$d_i = \mathrm{per}_\eta(\{-\cos(\tilde{\boldsymbol{e}}_i, \boldsymbol{e}_c)\}_{c=1}^{C}). \tag{4.19}$$

In this formula, $\tilde{\boldsymbol{e}}_i = f_\theta^{\text{text}}(\text{prompt}(\tilde{y}_i))$ represents the text embedding of the candidate word, and f_θ represents the ID model with the parameter θ. Here, C

indicates the quantity of ID classes, and per_η denotes the value corresponding to the 100η-th percentile within the dataset. We typically choose $\eta = 0.05$ for robustness against outlier ID labels.

The negative label selection criteria are defined as

$$\mathcal{Y}^- = \text{topk}(\{d_1, d_2, \ldots, d_C\}, \mathcal{Y}^{\text{K}}, M). \tag{4.20}$$

In this context, the function $\text{topk}(A, B, M)$ is designed to pinpoint the indices of the M largest elements within set A. Subsequently, it picks out the elements from set B that correspond to those identified indices. For achieving consistent and reliable performance across various benchmarks, we have set $M = 10000$ as the standard number of negative labels.

NegLabel Score Taking into account the selected set of negative labels $\mathcal{Y}^-$, we merge it with the ID labels $\mathcal{Y}$ to construct an extended label space $\mathcal{Y}^{\text{Ext}} = \mathcal{Y} \cup \mathcal{Y}^-$. The extended labels are then processed by the CLIP text encoder to produce text embeddings. The proposed OOD score is defined as follows:

$$S(\mathbf{x}) = S^*(\text{sim}(\mathbf{x}, \mathcal{Y}), \text{sim}(\mathbf{x}, \mathcal{Y}^-)), \tag{4.21}$$

where $\text{sim}(\cdot, \cdot)$ is the similarity function and the function $S^*(\cdot, \cdot)$ combines the similarity of a sample with both ID labels and negative labels. Because the selected negative labels have little resemblance to $\mathcal{Y}$, ID samples display a lower degree of similarity when contrasted with OOD samples. Therefore, S^* has been crafted such that it presents an inverse connection with the similarity within the negative label space and a positive connection with the similarity in the ID label space.

In the context of the extended label space $\mathcal{Y}^{\text{Ext}}$, the task of OOD detection is considered as the model's level of confidence regarding an image's belonging to $\mathcal{Y}$. We introduce a Neg Label score, which is devised through a sum-softmax approach:

$$S(\mathbf{x}) = \sum_{i=1}^{C} e^{\cos(\mathbf{z}, \boldsymbol{e}_i)/T} \Big/ \Big(\sum_{i=1}^{C} e^{\cos(\mathbf{z}, \boldsymbol{e}_i)/T} + \sum_{j=1}^{M} e^{\cos(\mathbf{z}, \tilde{\boldsymbol{e}}_j)/T} \Big), \tag{4.22}$$

where C represents the number of ID labels, T denotes the temperature coefficient, and M indicates the number of negative labels. The reduced affinities of ID images to negative labels, in contrast to OOD images, lead to elevated OOD scores for ID samples. The score formulation addresses two aspects relevant to CLIP-like models: the cosine function constrains similarity scores to the range $[-1, 1]$, potentially failing to capture the variance in confidence adequately. Consequently, a temperature coefficient T is defined to adjust the similarity. Zero-shot models that are not fine-tuned on task-specific ID data exhibit a greater susceptibility to super-class confusion. The numerator of Eq. (4.22) consolidates the sample's similarity to all ID labels, thereby improving robustness to confusion and domain shifts.

Grouping Strategy The NegMining algorithm identifies negative labels that are distanced from ID embeddings to broaden the label space and reduce false positives. False positives may still occur due to model limitations and semantic overlaps, particularly with more negative labels, which increases the variance of out-of-distribution scores. We propose a grouping strategy to balance the advantages of negative labels with the potential risk of false positives. The M negative labels are partitioned into n_g groups, with each group containing $\lfloor M/n_g \rfloor$ negative labels, while any remainder is discarded. A separate NegLabel score is calculated for each group, and the scores are averaged to derive the final NegLabel score. This approach lowers noise by averaging and lowers the false positive risk within every group.

Experiments We evaluate NegLabel in Table 4.4 against other currently used OOD detection techniques. We evaluated conventional techniques reused utilizing a fine-tuned CLIP on ImageNet-1k [90] as well as MCM [267], an approach especially meant for zero-shot OOD detection. NegLabel performs better than all the baselines on all the datasets, so demonstrates its exceptional quality. Moreover, with calibrated CLIP, our approach can even outperform conventional techniques as it shows the great OOD detecting power of CLIP in zero-shot situations. Due to pretraining on a large-scale image-text dataset, CLIP can finely grainedly interpret pictures. Moreover, the negative labels we select provide supplementary semantic details. As a result, this enhances the distinguishability between OOD and ID samples.

In order to assess the domain generalization ability, we utilize multiple diverse renditions of ImageNet, each of which features unique domain shifts, as ID samples. The findings presented in Table 4.5 demonstrate that the zero-shot OOD detection performance of MCM experiences a substantial decline across various domain shifts. This situation reveals the challenges associated with zero-shot OOD detection under these circumstances. Conversely, our NegLabel method manages to obtain outstanding outcomes across a wide spectrum of ID datasets. This convincingly shows its extraordinary robustness in the face of various domain shifts.

The MCM method depends heavily on the model's classification capabilities within the ID dataset. So, when a domain shift occurs and causes a drop in the ID classification accuracy, the OOD detection performance of MCM will decline accordingly. In contrast, our NegLabel method perceives the ID labels $\mathcal{Y}$ as an integral entity. This means it is less influenced by interclass confusion, thus being more resilient to domain shifts.

4.1.3 Outlier Exposure

4.1.3.1 Diversified Outlier Exposure for OOD Detection

One effective approach for OOD detection [431], known as OE [153, 246], involves incorporating examples of auxiliary OOD data during the training process. By exposing the model to these "outliers," the goal is to teach it to better differentiate

Table 4.4 OOD detection performance comparison with baselines. All methods are based on CLIP-B/16. The ID data are ImageNet-1k. All values are percentages. ↑ indicates larger values are better and ↓ indicates smaller values are better. The shadow part represents our method

	OOD dataset									
	iNaturalist		SUN		Places		Textures		Average	
Methods	AUROC↑	FPR95↓	AUROC↑	FPR95↓	AUROC↑	FPR95↓	AUROC↑	FPR95↓	AUROC↑	FPR95↓
	Requires training (or w. fine-tuning)									
MSP [150]	87.44	58.36	79.73	73.72	79.67	74.41	79.69	71.93	81.63	69.61
ODIN [229]	94.65	30.22	87.17	54.04	85.54	55.06	87.85	51.67	88.80	47.75
Energy [246]	95.33	26.12	92.66	35.97	91.41	39.87	86.76	57.61	91.54	39.89
GradNorm [168]	72.56	81.50	72.86	82.00	73.70	80.41	70.26	79.36	72.35	80.82
ViM [388]	93.16	32.19	87.19	54.01	83.75	60.67	87.18	53.94	87.82	50.20
KNN [353]	94.52	29.17	92.67	35.62	91.02	39.61	85.67	64.35	90.97	42.19
VOS [98]	94.62	28.99	92.57	36.88	91.23	38.39	86.33	61.02	91.19	41.32
NPOS [363]	96.19	16.58	90.44	43.77	89.44	45.27	88.80	46.12	91.22	37.93
	Zero-shot (no training required)									
MCM [267]	94.59	32.20	92.25	38.80	90.31	46.20	86.12	58.50	90.82	43.93
NegLabel	**99.49**	**1.91**	**95.49**	**20.53**	**91.64**	**35.59**	**90.22**	**43.56**	**94.21**	**25.40**

Table 4.5 Zero-shot OOD detection performance robustness to domain shift. All methods are based on CLIP-B/16. The ID data are ImageNet-Sketch. All values are percentages. ↑ indicates larger values are better and ↓ indicates smaller values are better

		OOD dataset									
		iNaturalist		SUN		Places		Textures		Average	
ID dataset	Method	AUROC↑	FPR95↓	AUROC↑	FPR95↓	AUROC↑	FPR95↓	AUROC↑	FPR95↓	AUROC↑	FPR95↓
ImageNet-Sketch	MCM	87.74	63.06	85.35	67.24	81.19	70.64	74.77	79.59	82.26	70.13
	NegLabel	**99.34**	**2.24**	**94.93**	**22.73**	**90.78**	**38.62**	**89.29**	**46.10**	**93.59**	**27.42**
ImageNet-A	MCM	79.50	76.85	76.19	79.78	70.95	80.51	61.98	86.37	72.16	80.88
	NegLabel	**98.80**	**4.09**	**89.83**	**44.38**	**82.88**	**60.10**	**80.25**	**64.34**	**87.94**	**43.23**
ImageNet-R	MCM	83.22	71.51	80.31	74.98	75.53	76.67	67.66	83.72	76.68	76.72
	NegLabel	**99.58**	**1.60**	**96.03**	**15.77**	**91.97**	**29.48**	**90.60**	**35.67**	**94.54**	**20.63**
ImageNetV2	MCM	91.79	45.90	89.88	50.73	86.52	56.25	81.51	69.57	87.43	55.61
	NegLabel	**99.40**	**2.47**	**94.46**	**25.69**	**90.00**	**42.03**	**88.46**	**48.90**	**93.08**	**29.77**

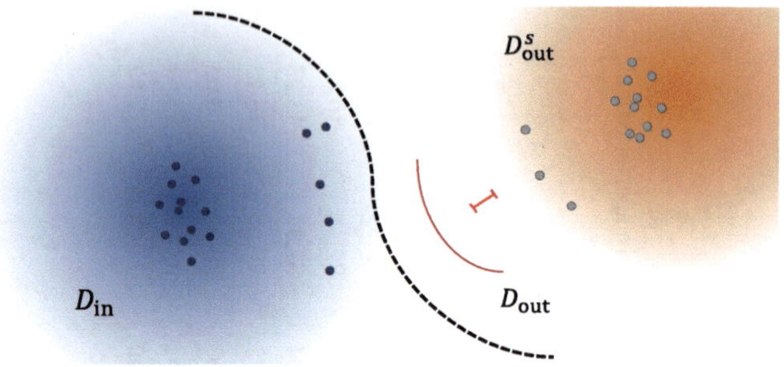

Fig. 4.4 Generally, the surrogate OOD data D^s_{out} is assumed to be informative in representing the the unseen OOD data D_{out}. However, we cannot always hold the assumption as predefining and collecting such boundary outliers (particularly those close to D_{in}) properly is not only impracticable but also impossible

between ID and OOD inputs. However, a key challenge (as illustrated in Fig. 4.4) arises: the OOD data available during training may not fully represent the OOD data the model will encounter in the real world. This discrepancy can limit the model's ability to generalize effectively to unseen OOD inputs. This raises a fundamental question: how can we use the limited auxiliary outliers to achieve robust OOD detection?

The core difficulty lies in the fact that the distribution of OOD data in real-world scenarios is often complex and unpredictable [104, 431]. It's difficult to know and collect all the types of OOD inputs a model might encounter. Consequently, the effectiveness of OE methods is often tied to the quality and representativeness of the available outlier data. If the training outliers are not sufficiently informative, the model's ability to detect novel OOD inputs will be hampered. A promising strategy is to expand the effective range of the available outlier data, creating a more comprehensive representation of the OOD space.

Method Diversified outlier exposure (DivOE) [494] is introduced as a general framework for outlier exposure when using limited or uninformative auxiliary data to address this issue. It is designed to produce new, diverse outliers that expand the representation coverage of the original OOD data, as demonstrated in Fig. 4.5. Specifically, DivOE introduces a new objective that aims to extrapolate the original outlier distribution through an adversarial optimization target in order to produce a more diverse OOD sample. This expansion of the surrogate OOD distribution allows the model to learn a more robust decision boundary between ID and OOD data. To implement this idea, it considers an automatically updated version of the surrogate OOD distribution during the optimization within the general framework:

$$\mathcal{L}^*_{\text{OE}} = \mathbb{E}_{D_{\text{in}}}\left[\ell_{\text{CE}}(f_\theta(\mathbf{x}), y)\right] + \lambda \mathbb{E}_{D^e_{\text{out}}}\left[\ell_{\text{OE}}(f_\theta(\tilde{\mathbf{x}}))\right], \tag{4.23}$$

where ℓ_{CE} is the cross-entropy loss, λ is a hyperparameter to control the extrapolated strength, D^e_{out} indicates the extrapolated surrogate OOD data sampled from the

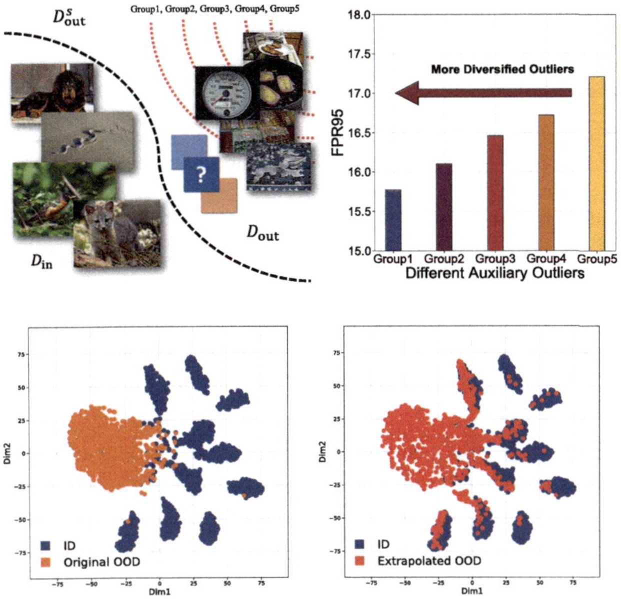

Fig. 4.5 In the upper panels, we provide an illustration of the auxiliary outliers being divided into various informative levels based on the extent to which they can encompass the unknown decision area. This concept is synonymous with diversity. In the lower panels, we have plotted the tSNE results of the original auxiliary outliers against our extroplated outliers. This demonstrates that the diversification can enhance the detection performance of OE.

surrogate OOD distribution $\mathcal{D}^{\mathrm{s}}_{\mathrm{out}}$, and $\tilde{x}$ indicates the perturbed samples for diverse outlier synthesis, distinct from those in the standard OE usage. The first term addresses the standard classification task, while the second term introduces our novel informative extrapolation. To diversify $\mathcal{D}^{\mathrm{s}}_{\mathrm{out}}$, the objective can be reformulated as

$$\begin{aligned}\mathbb{E}_{\mathcal{D}^{\mathrm{e}}_{\mathrm{out}}}\left[\ell_{\mathrm{OE}}(f(\tilde{x}))\right] =&(1-\beta)\ell_{\mathrm{OE}}(f_\theta;\mathcal{D}^{s}_{\mathrm{out}}) \\ &+\underbrace{\beta\max_{\Delta}\left[\ell_{\mathrm{OE}}(f_\theta;\mathcal{D}^{s+\Delta}_{\mathrm{out}})-\ell_{\mathrm{OE}}(f_\theta;\mathcal{D}^{s}_{\mathrm{out}})\right]}_{\text{Informative Extrapolation}},\end{aligned} \tag{4.24}$$

where β is a balancing factor controlling the extrapolation ratio and $\mathcal{D}^{s+\Delta}_{\mathrm{out}}$ is a synthesized distribution obtained through manipulation Δ. This objective encourages the model to learn not only from the original OOD distribution but also to generalize to different OOD distributions via the maximization term. The fundamental idea is to use the diverse extrapolation to promote a wider OOD distribution. The distribution-level difference between the generated and original OOD distributions is maximized to get the informative extrapolation for the surrogate outliers. The specific objective can be formulated as the follows:

$$\mathbb{E}_{D_{\text{out}}^{\text{e}}}\left[\ell_{\text{OE}}(f_\theta(\tilde{\mathbf{x}}))\right] = \underbrace{\frac{1}{n-rn}\sum_{n-rn}\ell_{\text{OE}}(f_\theta(\mathbf{x}^s))}_{\text{Original Distribution}} + \underbrace{\frac{1}{rn}\sum_{rn}\max_{\delta}\ell_{\text{OE}}(f_\theta(\mathbf{x}^s+\delta))}_{\text{Informative Extrapolation}}, \tag{4.25}$$

where n indicates the number of auxiliary outliers and r indicates the extrapolation ratio, related to β in Eq. (4.24); $\tilde{\mathbf{x}}$ is the original surrogated outlier sample and δ is the instance-wise perturbation. The second term represents the informative extrapolation, realized by instance-level maximization:

$$\delta = \arg\max_{\delta} \ell_{\text{OE}}(f_\theta(\mathbf{x}^s + \delta)). \tag{4.26}$$

In order to achieve this goal, DivOE generates new outliers close to the decision boundary, increasing the coverage of auxiliary outliers and their informativeness in defining the decision space between ID and OOD data. The optimization process for δ can be achieved by projected gradient descent (PGD) [258].

Experiments Table 4.6 reports the performance comparison of DivOE against several established baseline methods [62, 150, 246, 268] for OOD detection. The experiments are conducted on standard benchmark datasets, CIFAR-10 and CIFAR-100, using a consistent evaluation protocol. The results validate the general effectiveness of DivOE compared with both post hoc score functions [150] and other different outlier sampling methods [62, 268].

In addition to the heuristic framework of DivOE to tackle the discrepancy between training and test OOD cases, it is also an interesting question in how to handle this issue in a more systematic and formal way. In the following, we suggest a theoretical-driven framework based on distributionally robust learning that can effectively handle OOD distribution discrepancy.

4.1.3.2 Augmented Outlier Exposure for OOD Detection

The authors formalize the phenomenon where the distribution gap exists between auxiliary and real OOD data, quantifying its adverse effect on real-world OOD detection. One step further, the authors also raise the research question about the enhanced methodologies to alleviate the impacts of the distribution discrepancy and propose a new learning framework named Distributional-Augmented OOD Learning (DAL). In particular, DAL introduces a set of distributions inside a Wasserstein ball to the original distribution. By learning from the worst OOD distribution in such an expanded distribution range, DAL guarantees that the models will perform well therein. The authors conducted a series of experiments to show their effectiveness in real-world OOD detection.

Problem Definition We consider the feature space as well as the label space defined by $\mathcal{X}$ and $\mathcal{Y} = \{1, \ldots, C\}$, respectively. We also have the ID and OOD

Table 4.6 Main results of comparison with different OOD detection baselines (%). The experiments of outlier exposure are conducted with different random seed and results are presented with value in the form of mean±std

D_{in}	Method	FPR95 ↓	AUROC ↑	AUPR ↑	ID-ACC ↑	w./w.o $\mathcal{D}_{aux}$	Operation for $\mathcal{D}_{aux}$
CIFAR-10	MSP	52.21	90.61	97.84	94.84		
	ODIN	32.84	90.67	97.29	94.84		
	Mahalanobis	36.27	92.69	98.31	94.84		
	Energy	32.82	91.99	97.86	94.84		
	ASH	33.01	92.11	97.92	94.84		
	OE	13.76±0.23	97.53±0.03	99.43±0.01	94.51±0.06	✓	Vanilla
	Energy	18.15±0.32	90.90±0.18	96.34±0.10	94.51±0.01	✓	Vanilla
	VOS	34.30±0.02	91.27±0.01	97.64±0.01	94.02±0.01	✓	Virtually synthesized
	NTOM	17.76±0.84	91.08±0.44	96.40±0.18	**95.04±0.03**	✓	Greedy sampled
	POEM	17.81±0.78	90.96±0.37	96.36±0.15	94.90±0.04	✓	Thompson sampled
	MixOE	13.57±0.25	97.47±0.07	99.39±0.02	94.98±0.02	✓	Mixup with ID data
	DivOE (Ours)	**11.66±0.04**	**97.82±0.04**	**99.48±0.01**	94.66±0.12	✓	**Extrapolated**
CIFAR-100	MSP	79.71	76.13	94.07	75.95		
	ODIN	63.29	82.44	95.49	75.95		
	Mahalanobis	52.90	83.86	95.73	75.95		
	Energy	71.93	80.27	95.00	75.95		
	ASH	61.82	83.42	95.84	75.87		
	OE	27.67±0.43	91.89±0.12	97.93±0.04	75.41±0.10	✓	Vanilla
	Energy	27.86±0.26	90.70±0.07	97.40±0.03	75.03±0.11	✓	Vanilla
	VOS	70.87±0.00	80.00±0.00	94.89±0.00	**76.25±0.00**	✓	Virtually synthesized
	NTOM	27.58±0.88	91.08±0.34	97.52±0.10	75.41±0.10	✓	Greedy sampled
	POEM	26.50±0.10	91.33±0.06	97.58±0.02	75.74±0.14	✓	Thompson sampled
	MixOE	35.26±0.80	90.31±0.16	97.58±0.04	75.74±0.04	✓	Mixup with ID data
	DivOE (Ours)	**24.80±0.51**	**92.91±0.11**	**98.22±0.03**	75.24±0.01	✓	**Extrapolated**

distributions $\mathcal{D}_{\text{id}}$ and $\mathcal{D}_{\text{ood}}^{\text{Train}}$, respectively. Then, for outlier exposure, given the sets of samples drawn from ID and OOD distributions, we have the optimization problem:

$$\min_{\theta} \left[\hat{\mathcal{L}}_{\text{I}}(\theta) + \alpha \hat{\mathcal{L}}_{\text{A}}(\theta)\right], \tag{4.27}$$

where $\hat{\mathcal{L}}_{\text{I}}$ and $\hat{\mathcal{L}}_{\text{A}}$ denote the empirical risks for ID and OOD samples. However, we usually use auxiliary OOD data from a distribution $\mathcal{D}_{\text{ood}}^{\text{Train}}$ that might have a distribution gap compared with $\mathcal{D}_{\text{ood}}^{\text{Real}}$. The authors quantify its impact on detection performance using the Wasserstein-1 distance. Under certain assumptions, it can be proved that with high probability, the following generalization bound holds:

$$\mathcal{L}_{\mathcal{D}}(\hat{\theta}) \leq \min_{\theta} (\mathcal{L}_{\text{I}}(\theta) + \alpha \mathcal{L}_{\text{A}}(\theta)) + \alpha L_c \mathrm{W}_c(\mathcal{D}_{\text{ood}}^{\text{Real}}, \mathcal{D}_{\text{ood}}^{\text{Train}}) + O(1/\sqrt{n}) + O(1/\sqrt{m}), \tag{4.28}$$

where $\hat{\theta}$ is the parameter learned by Eq. (4.27), i.e., $\hat{\theta} \in \arg\min_{\theta} \hat{\mathcal{L}}_{\text{I}}(\theta) + \alpha \hat{\mathcal{L}}_{\text{A}}(\theta)$, L_c is the Lipschitz constant of ℓ_{OE}, and n and m the sample sizes of ID and OOD data. As observed, $\mathrm{W}_c(\mathcal{D}_{\text{ood}}^{\text{Real}}, \mathcal{D}_{\text{ood}}^{\text{Train}})$ characterizes the distribution gap, with its larger values indicating weaker guarantees of their real-world OOD performance.

Method To combat the adverse impacts of the distribution gap, the authors introduce an enlarged OOD distribution coverage centered on the original OOD distribution, which is defined by the Wasserstein ball as follows:

$$\mathfrak{D} = \{\mathcal{D}_{X'} : \mathrm{W}_c(\mathcal{D}_{X'}, \mathcal{D}_{\text{ood}}^{\text{Train}}) \leq \rho\},$$

where W_c denotes the Wasserstein distance and ρ denotes the radius of the Wasserstein ball. Thereby, we can consider the following optimization problem

$$\min_{\theta} \hat{\mathcal{L}}_{\mathcal{D}}(\theta; \rho) = \min_{\theta} \left[\hat{\mathcal{L}}_{\text{I}}(\theta) + \alpha \hat{\mathcal{L}}_{\text{O}}(\theta; \rho)\right], \tag{4.29}$$

where

$$\hat{\mathcal{L}}_{\text{O}}(\theta; \rho) = \sup_{\mathrm{W}_c(\mathcal{D}_{X'}, \mathcal{D}_{\text{ood}}^{\text{Train}}) \leq \rho} \mathbb{E}_{\mathbf{x} \sim \mathcal{D}_{X'}} \ell_{\text{OE}}(f_{\theta}(\mathbf{x})). \tag{4.30}$$

However, Eq. (4.30) is not a tractable optimization problem, yet we can address its dual problem by the theorem below.

Theorem 4.1 *Let* $\phi_{\theta}(\mathbf{x}, \gamma) = \sup_{\mathbf{x}' \in \mathcal{X}} \{\ell(f_{\theta}(\mathbf{x}')) - \gamma c(\mathbf{x}', \mathbf{x})\}$. *For any* $\rho > 0$, *we have*

$$\hat{\mathcal{L}}_{D}(\theta; \rho) = \hat{\mathcal{L}}_{\text{I}}(\theta) + \alpha \inf_{\gamma \geq 0} \{\gamma \rho + \frac{1}{m} \sum_{i=1}^{m} \phi_{\theta}(\mathbf{x}_{\text{A}}^{i}; \gamma)\}. \tag{4.31}$$

We can verify that the following theorem holds given the optimization problem following Eq. (4.29), which shows a tighter bound over the original OE and can be controlled by the hyperparameter of ρ (or γ equivalently).

Theorem 4.2 (Risk Estimation) *We assume that* $0 \leq \ell(f_\theta(\mathbf{x}), y) \leq M_\ell$, $0 \leq \ell_{\mathrm{OE}}(f_\theta(\mathbf{x})) \leq M_{\ell_{\mathrm{OE}}}$, *and* $c(\cdot,\cdot)$ *is a continuous metric. Also, let* $\hat{\theta}$ *be the solution of Eq.* (4.29), *which is given by* $\hat{\theta} \in \arg\min_\theta \hat{\mathcal{L}}_{\mathcal{D}}(\theta;\rho)$. *If* $\ell_{\mathrm{OE}}(f_\theta(\mathbf{x}))$ *is* L_c*-Lipschitz w.r.t.* $c(\cdot,\cdot)$, *i.e.,* $|\ell_{\mathrm{OE}}(f_\theta(\mathbf{x})) - \ell_{\mathrm{OE}}(f_\theta(\mathbf{x}'))| \leq L_c c(\mathbf{x},\mathbf{x}')$, *then with the probability at least* $1 - 4e^{-t} > 0$

$$\mathcal{L}_D(\hat{\theta}) - \overbrace{\min_\theta \mathcal{L}_D(\theta;\rho)}^{\text{approximate risk}} \leq \underbrace{\alpha L_c \max\{\mathrm{W}_c(\mathcal{D}_{ood}^{\mathrm{Train}}, \mathcal{D}_{ood}^{\mathrm{Real}}) - \rho, 0\} + \epsilon(n,m;t)}_{\text{estimation error}},$$

for any $\rho > 0$, *where* $\epsilon(n,m;t)$ *is a constant that depends on the sample sizes* n *and* m, *as well as the confident level* t.

Experiments Table 4.7 shows the results on CIFAR benchmarks, comparing the results before and after watermarking. The authors consider three evaluation metrics, (1) the false OOD rate with the true ID rate is at 95% (FPR95), (2) the area under the receiver operating characteristic curve (AUROC), and (3) the area under the precision-recall curve (AUPR). As shown, the results show that DAL performs much better than the baselines in real-world OOD detection.

Similarly to post hoc OOD detection, it is still an interesting question about how to realize effective OOD detection for pretrained vision-language models based on outlier exposure. The key challenge therein is about how to handle hard OOD sample, which will be discussed in the following.

4.1.3.3 Envision Outlier Exposure via Large Language Models

Although zero-shot OOD detection [267] has achieved promising OOD detection capability by leveraging the pretrained VLMs such as CLIP [306], existing methods usually only utilize ID label names to construct textual classifiers, which largely limits VLM's inherent ability to identify samples from a large open-label space. We introduce Envisioning Outlier Exposure (EOE) [49], a novel approach that leverages large language models (LLMs) to generate potential outlier class labels without using actual OOD data. Since LLM contains expert knowledge of common categories and knows the visual features of each category, that is, LLM knows which categories are visually similar, and categories that are difficult to distinguish by OOD detector are often visually similar. Therefore, we prompt LLM output categories that are similar to ID, thereby building a more powerful text classifier. Experiments show that EOE can be easily expanded to large datasets such as ImageNet-1k.

Table 4.7 Benchmarking on OOD performance on the CIFAR benchmarks

Method	SVHN		LSUN		iSUN		Textures		Places365		Average	
	FPR95 ↓	AUROC ↑	FPR95 ↓	AUROC ↑	FPR95 ↓	AUROC ↑	FPR95 ↓	AUROC ↑	FPR95 ↓	AUROC ↑	FPR95 ↓	AUROC ↑
CIFAR-10												
Using ID data only												
MSP	48.89	91.97	25.53	96.49	56.44	89.86	59.68	88.42	60.19	88.36	50.15	91.02
Free Energy	35.21	91.24	4.42	99.06	33.84	92.56	52.46	85.35	40.11	90.02	33.21	91.64
ASH	33.98	91.79	4.76	98.98	34.38	92.64	50.90	86.07	40.89	89.79	32.98	91.85
Mahalanobis	12.21	97.70	57.25	89.58	79.74	77.87	15.20	95.40	68.81	82.39	46.64	88.59
KNN	26.56	95.93	27.52	95.43	33.55	93.15	37.62	93.07	41.67	91.21	33.38	93.76
KNN+	3.28	99.33	2.24	98.90	17.85	95.65	10.87	97.72	30.63	94.98	12.97	97.32
CSI	17.37	97.69	6.75	98.46	12.58	97.95	25.65	94.70	40.00	92.05	20.47	96.17
VOS	36.55	93.30	9.98	98.03	28.93	94.25	52.83	85.74	39.56	89.71	33.57	92.21
Using ID data and auxiliary OOD data												
OE	2.36	99.27	1.15	**99.68**	2.48	99.34	5.35	98.88	11.99	97.23	4.67	98.88
Energy-OE	0.97	99.54	1.00	99.15	2.32	99.27	3.42	**99.18**	9.57	97.44	3.46	98.91
ATOM	1.00	99.59	0.61	99.53	2.15	**99.40**	2.52	99.10	7.93	97.27	2.84	98.97
DOE	1.80	99.37	**0.25**	99.65	2.00	99.36	5.65	98.75	10.15	97.28	3.97	98.88
POEM	1.20	99.53	0.80	99.10	**1.47**	99.26	2.93	99.13	7.65	97.35	2.81	98.87
DAL	**0.80**	**99.65**	0.90	99.46	1.70	99.34	**2.30**	99.14	**7.65**	**97.45**	**2.68**	**99.01**
CIFAR-100												
Using ID data only												
MSP	84.39	71.18	60.36	85.59	82.63	75.69	83.32	73.59	82.37	73.69	78.61	75.95
Free Energy	85.24	73.71	23.05	95.89	81.11	79.02	79.63	76.35	80.18	75.65	69.84	80.12
ASH	70.09	83.56	13.20	97.71	69.87	82.56	63.69	83.59	79.70	74.87	59.31	84.46
Mahalanobis	51.00	88.70	91.60	69.69	38.48	91.86	47.07	89.09	82.70	74.18	72.37	82.70

KNN	52.10	88.83	68.82	79.00	42.17	90.59	42.79	89.07	92.21	61.08	59.62	81.71
KNN+	32.50	93.86	47.41	84.93	39.82	91.12	43.05	88.55	63.26	79.28	45.20	87.55
CSI	64.50	84.62	25.88	95.93	70.62	80.83	61.50	86.74	83.08	77.11	61.12	95.05
VOS	78.06	92.59	40.40	92.90	85.77	70.20	82.46	77.22	82.31	75.47	73.80	91.67
Using ID data and auxiliary OOD data												
OE	46.73	90.54	16.30	96.98	47.97	88.43	50.39	88.27	54.30	87.11	43.14	90.27
Energy-OE	35.34	94.74	16.27	97.25	33.21	93.25	46.13	90.62	50.45	90.04	36.28	93.18
ATOM	24.80	95.15	17.83	96.76	47.83	91.06	44.86	**91.80**	53.92	88.88	37.84	92.73
DOE	43.10	91.83	13.95	**97.56**	47.25	87.88	49.40	88.62	51.05	88.08	40.95	90.79
POEM	22.27	**96.28**	**13.66**	97.52	42.46	91.97	45.94	90.42	49.50	90.21	34.77	93.28
DAL	**19.35**	96.21	16.05	96.78	**26.05**	**94.23**	**37.60**	91.57	**49.35**	**90.81**	**29.68**	**93.92**

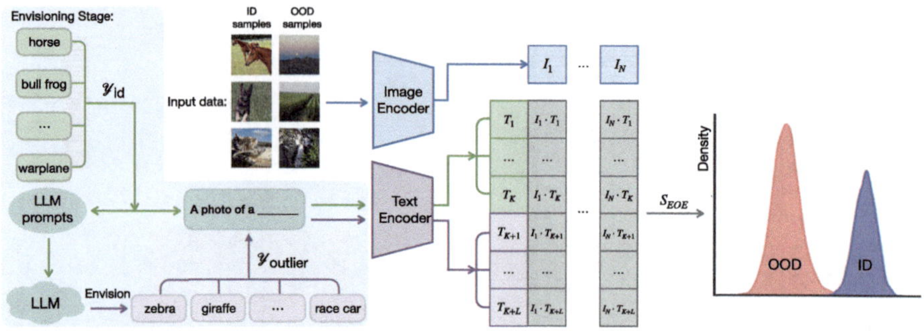

Fig. 4.6 EOE's framework. LLM generates outlier class label $\mathcal{Y}_{out}$ based on the given ID class label $\mathcal{Y}_{id}$. During testing, the image embedding is similarly calculated with the output of the text encoder, and the detection score function S calculates the OOD score

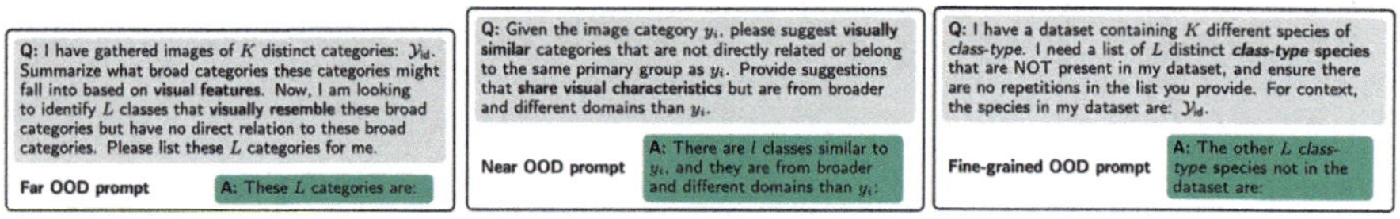

(a) Far OOD detection (b) Near OOD detection (c) Fine-grained OOD detection

Fig. 4.7 LLM prompts for OOD detection, including Question (Q) and Answer (A). (**a**) Far OOD detection, (**b**) near OOD detection, (**c**) fine-grained OOD detection

Method Using LLM to generate potential outlier class labels requires addressing two key challenges: how to guide LLM to generate useful outlier class labels and how to effectively use the outlier class labels. Given a set of ID class labels $\mathcal{Y}_{id}$, the proposed method operates in two stages. We found that OOD classes that are easily confused with IDs are usually visually similar to ID classes. Thus, we can ask LLM to generate OOD classes which are similar to ID classes. Our method's framework is shown in Fig. 4.6.

In far OOD detection, the model can already distinguish most of the OOD samples well, so we only need to provide some visually similar broader categories. This approach scales efficiently to large datasets like ImageNet-1k [91], as shown in Fig. 4.7a. The OOD classes in near OOD detection are difficult to distinguish. For instance, *husky* and *wolf* are examples of OODs that are easily identified as IDs. Therefore, we ask LLM to construct OOD class labels for each ID class label (Fig. 4.7b). In the fine-grained OOD detection, all categories are considered to be part of the same category. For instance, the "Golden Retriever" and "Husky" are both considered to be members of the "dog" category. We require LLMs to generate OOD class labels within the same broad category (Fig. 4.7c).

We propose an OOD score function S to help OOD detection with the generated OOD class labels. The similarity score $s_i(\mathbf{x})$ for image $\mathbf{x}$ and class labels is

$$s_i(\mathbf{x}) = \frac{\mathcal{I}(\mathbf{x}) \cdot \mathcal{T}(\mathbf{t}_i)}{\|\mathcal{I}(\mathbf{x})\| \cdot \|\mathcal{T}(\mathbf{t}_i)\|}, \quad \mathbf{t}_i \in \mathcal{Y}_{\text{id}} \cup \mathcal{Y}_{\text{out}}, \tag{4.32}$$

where $\mathcal{I}$ is the image encoder and $\mathcal{T}$ is the text encoder.

In order to reflect the relationship between the input samples and the generated OOD categories, S penalizes similar potential outliers:

$$S(\mathbf{x}) = \max_{i \in [1,C]} \frac{e^{s_i(\mathbf{x})}}{\sum_{j=1}^{C+L} e^{s_j(\mathbf{x})}} - \max_{c \in (C,C+L]} \frac{\beta \cdot e^{s_c(\mathbf{x})}}{\sum_{j=1}^{C+L} e^{s_j(\mathbf{x})}}, \tag{4.33}$$

where C and L are the number of ID and envisioned OOD class labels, respectively, and $\beta = 0.25$. The OOD detection decision is made based on a threshold τ to classify the sample as ID or OOD:

$$G_\tau(\mathbf{x}) = \begin{cases} \text{ID}, & S(\mathbf{x}) \geq \tau \\ \text{OOD}, & S(\mathbf{x}) < \tau \end{cases}, \tag{4.34}$$

where G is the OOD detector with a threshold τ.

Experiments Our evaluation of EOE's performance in three separate OOD detection tasks is carried out with the assistance of a variety of datasets. We select ImageNet-1k as the ID dataset and use the subset of Texture, SUN, iNaturalist, and Places as the OOD dataset from Huang and Li [163] in far OOD detection. And we use ImageNet-10 and ImageNet-20 created by MCM [267] in near OOD detection. In terms of the fine-grained OOD detection, we randomly divide the existing four datasets (CUB-200-2011 [381], STANFORD-CARS [202], Food-101 [41], Oxford-IIIT Pet [288]) into two, half of which is used as the ID dataset and the rest as the OOD dataset. As shown in Table 4.8, EOE outperforms existing methods on ImageNet-1k, with an FPR95 of 30.09% and an AUROC of 92.96%. In terms of near OOD detection, results in Table 4.9 demonstrate that EOE improves average FPR95 by 2.13% and AUROC by 0.30% over MCM while also outperforming Energy [246] and MaxLogit [151]. As for fine-grained OOD detection, Table 4.10 shows that EOE outperforms MCM with an average FPR95 improvement of 3.59%. On datasets such as Food-50 and Oxford-IIIT Pet, EOE also surpasses CLIPN [392] despite its reliance on auxiliary datasets, further demonstrating EOE's robustness.

4.1.4 Representation OOD Detection

4.1.4.1 Exploring the Intrinsic OOD Capability of Model Representations

Despite the promising progress in OOD detection [150, 215, 229, 246, 431], there has been limited focus on whether a well-trained model is the optimal starting point for OOD detection. Models are often trained for different original tasks [369]

Table 4.8 Zero-shot far OOD detection results. The **black bold** indicates the best performance

	OOD dataset									
	iNaturalist		SUN		Places		Texture		Average	
	FPR95 ↓	AUROC ↑	FPR95 ↓	AUROC ↑	FPR95 ↓	AUROC ↑	FPR95 ↓	AUROC ↑	FPR95 ↓	AUROC ↑
CLIPN	19.13	96.20	25.69	94.18	32.14	92.26	44.60	88.93	30.39	92.89
Energy	81.08	85.09	79.02	84.24	75.08	83.38	93.65	65.56	82.21	79.57
MaxLogit	61.66	89.31	64.39	87.43	63.67	85.95	86.61	71.68	69.08	83.59
MCM	30.92	94.61	37.59	92.57	44.71	89.77	57.85	86.11	42.77	90.77
EOE (Ours)	12.29	97.52	20.40	95.73	30.16	92.95	57.53	85.64	**30.09**	**92.96**

Table 4.9 Zero-shot near OOD detection results. The **bold** indicates the best performance on each dataset

	ID	ImageNet-10		ImageNet-20		Average	
	OOD	ImageNet-20		ImageNet-10			
Method		FPR95 ↓	AUROC ↑	FPR95 ↓	AUROC ↑	FPR95 ↓	AUROC ↑
CLIPN		7.80	98.07	13.67	97.47	10.74	97.77
Energy		10.30	97.94	16.40	97.37	13.35	97.66
MaxLogit		9.70	98.09	14.00	97.81	11.85	97.95
MCM		5.00	98.71	17.40	97.87	11.20	98.29
EOE (Ours)		4.20	99.09	13.93	98.10	**9.07**	**98.59**

from the OOD detection, such as multi-class classification. However, existing methods typically use these models directly for OOD detection. This raises the question: does a well-trained model possess the best OOD discriminative capability, and if not, how can a more suitable model be found? An interesting observation is made [492]: there is often an inconsistency between achieving better OOD discrimination and improving performance on ID data during training. Empirically, as shown in Fig. 4.8, there exists a training stage where the model exhibits higher OOD detection performance than the final well-trained model. In-depth analysis (as shown in Fig. 4.9) suggests a possible explanation: memorizing atypical samples, which are semantically different from other ID samples, harms generalization, and makes the model overly confident in unseen OOD inputs.

This analysis motivates a new method, called Unleashing Mask (UM) [492], to uncover the hidden OOD detection capability of a well-trained model by mitigating the memorization of atypical samples. The goal is to revert the model to a previous stage with better OOD discriminative capabilities. First, since atypical samples are more sensitive to parameter changes, a mask with a specific cutting rate is initialized to identify these samples based on parameter discrepancy. Second, using a loss reference estimated by the mask, a constrained gradient ascent is performed for model forgetting, encouraging the model to stabilize around an optimal stage. To prevent significant performance degradation on ID data, a variant called UM Adopts Pruning (UMAP) is proposed, which tunes the mask using a new objective.

Method The overall framework and learning objectives for the proposed UM and UMAP for OOD detection are presented as illustrated in Fig. 4.10. The framework consists of two critical components for uncovering intrinsic OOD detection capability: (1) an initialized mask with a specific masking rate for constructing output discrepancy with the original model and (2) subsequent adjustment to alleviate the memorization. The workflow starts by estimating the loss value of misclassified atypical samples and then tunes the model or masked output to forget them.

Based on the previous component, the forgetting objective is introduced as

$$\min \mathcal{L}_{\text{UM}} = \min_{m_\eta \in [0,1]^n} |\ell_{\text{CE}}(f) - \hat{\ell}_{\text{CE}}(m_\eta \odot f^c)| + \hat{\ell}_{\text{CE}}(m_\eta \odot f^c), \tag{4.35}$$

Table 4.10 Zero-shot fine-grained OOD detection results. The **bold** indicates the best performance on each dataset

Method	ID	CUB-100		Stanford-Cars-98		Food-50		Oxford-Pet-18		Average	
	OOD	CUB-100		Stanford-Cars-98		Food-51		Oxford-Pet-19			
		FPR95 ↓	AUROC ↑	FPR95 ↓	AUROC ↑	FPR95 ↓	AUROC ↑	FPR95 ↓	AUROC ↑	FPR95 ↓	AUROC ↑
CLIPN		73.54	74.65	53.33	82.25	43.33	88.89	53.90	86.92	56.05	83.18
Energy		76.13	72.11	73.78	73.82	44.95	89.97	68.51	88.34	65.84	81.06
MaxLogit		76.89	73.00	72.18	74.80	41.73	90.79	65.66	88.49	64.11	81.77
MCM		83.58	67.51	83.99	68.71	43.38	91.75	63.92	84.88	68.72	78.21
EOE (Ours)		74.74	73.41	76.83	71.60	37.95	91.96	52.55	90.33	**60.52**	**81.82**
Ground Truth		61.23	81.42	58.31	83.71	11.34	97.79	29.17	95.58	40.01	89.63

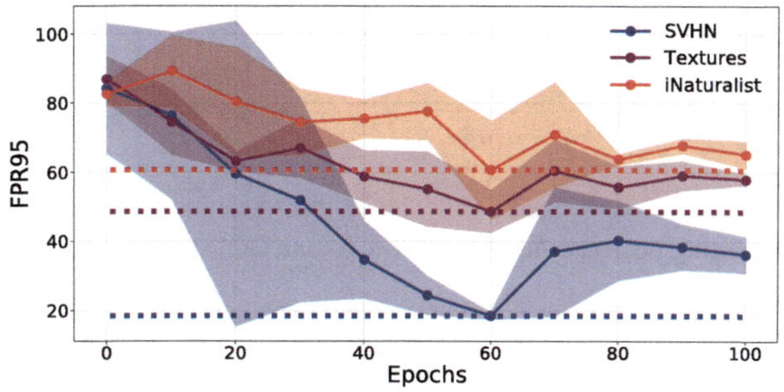

Fig. 4.8 Critical reveal of the intermediate stage with better OOD detection performance across various setups: the curves of FPR95 (false positive rate of OOD examples when the true positive rate of ID examples is at 95%) based on Energy score [246] across three different OOD datasets during the training on CIFAR-10 dataset

where m_η is the layer-wise mask with masking rate η and element size n, ℓ_{CE} is the cross-entropy (CE) loss, $\hat{\ell}_{\text{CE}}$ is the averaged CE loss over the ID training data, $|\cdot|$ denotes the absolute value computation, and $m_\eta \odot f^c$ represents the masked output of the fixed pretrained model, used to estimate the loss constraint for the forgetting objective. The value of $\hat{\ell}_{\text{CE}}(\cdot)$ remains constant during the fine-tuning process. Specifically, the well-trained model will optimize itself again if it memorizes atypical samples and achieves a near-zero loss value. A positive gradient signal is provided when the current loss value is lower than the estimated one and vice versa. The model is expected to stabilize around the stage that can forget atypical samples. Considering the potential negative effect on the original task performance when tuning for forgetting, a variant called UM Adopts Pruning (UMAP) is proposed. UMAP performs tuning based on the masked output (e.g., replacing $\ell_{\text{CE}}(f)$ with $\ell_{\text{CE}}(\hat{m}_p \odot f)$ in Eq. (4.35)) using a functionally different mask $\hat{m}_p$ with a pruning rate p, as follows:

$$\min \mathcal{L}_{\text{UMAP}} = \min_{\hat{m}_p \in [0,1]^n \; m_\eta \in [0,1]^n} |\ell_{\text{CE}}(\hat{m}_p \odot f) - \hat{\ell}_{\text{CE}}(m_\eta \odot f^c)| + \hat{\ell}_{\text{CE}}(m_\eta \odot f^c). \tag{4.36}$$

Unlike UM's objective (Eq. (4.23)), which minimizes the loss value over model parameters, UMAP's objective minimizes the loss over the mask $\hat{m}_p$ to forget atypical samples. UMAP provides an extra mask to restore detection capacity without affecting the model parameters for inference on original tasks, making it a more practical choice in real-world applications (as empirically verified in Table 4.11).

Experiments Table 4.11 presents the performance comparison of the proposed method, i.e., UM and UMAP in the OOD detection task. Specifically, they verify

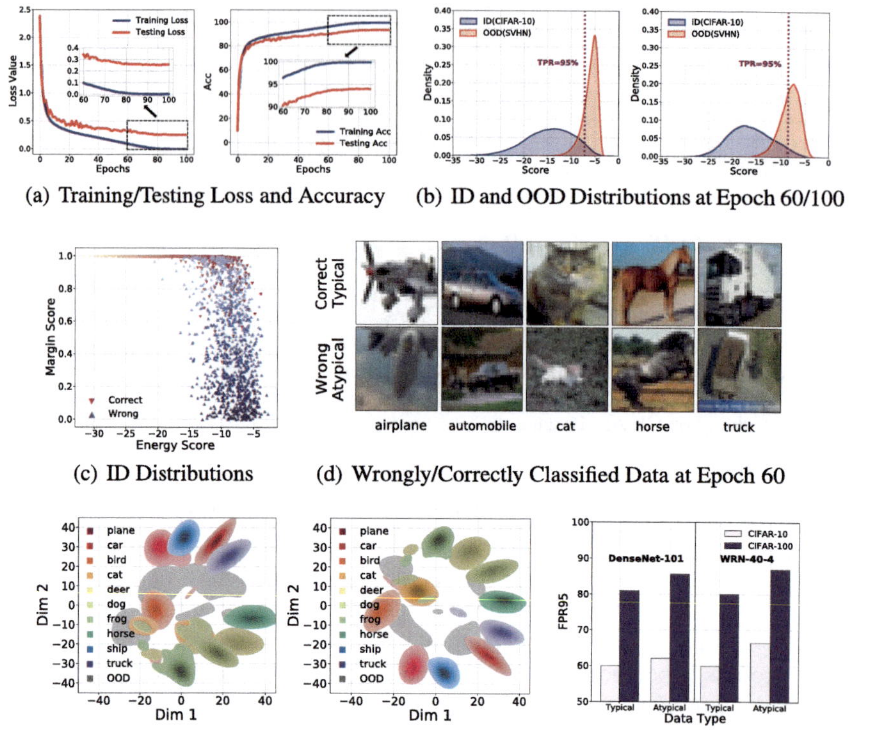

(a) Training/Testing Loss and Accuracy (b) ID and OOD Distributions at Epoch 60/100

(c) ID Distributions (d) Wrongly/Correctly Classified Data at Epoch 60

(e) TSNE Visualization at Epoch 60/100 (f) Diff. Effects

Fig. 4.9 Delve into the data-level attribution of the phenomenon with the original multi-classification on CIFAR-10: (**a**) training/testing loss and accuracy on ID data; (**b**) comparison of ID/OOD distributions based on Energy score at Epoch 60/100; (**c**) scatterplot of wrongly/correctly classified samples at Epoch 60 using Margin value and Energy score; (**d**) visualization of wrongly/correctly classified samples at Epoch 60; (**e**) TSNE visualization of the feature embedding on ID/OOD data at Epoch 60/100. (**f**) Effects on OOD detection of tuning with those identified typical/atypical samples. Through comparison from various perspectives, we find that achieving a reasonably small loss value (at round Epoch 60) on ID data is enough for OOD detection. However, continually optimizing on those atypical samples may impair OOD detection

the effectiveness of our UM and UMAP with two mainstreams of OOD detection approaches: (i) fundamental score function methods; (ii) outlier exposure methods involving auxiliary samples, in achieving better OOD detection efficacy (lower FPR95 and higher AUROC and AUPR) while maintaining the classification performance.

Fig. 4.10 Overview of unleashing mask: Given a well-trained model, it initializes a mask for mining the atypical samples that are sensitive to the changes in model parameters. Then, the framework fine-tune the original model or adopt pruning with the estimated forgetting threshold, i.e., the loss value estimated by the UM. The final model can serve as the base of those representative score functions to utilize the discriminative features and also as a new initialization of OE methods

4.2 Trustworthy Learning for Out-of-Distribution Generalization

4.2.1 Introduction and Notation

In a changing world, distribution shifts are inevitable when deploying models in real-world scenarios due to the inherent variability of the data curation process under different environments. For example, an autopilot system typically encounters diverse weather and road conditions that may not have been presented during training. To overcome the distribution shifts, the objective of out-of-distribution (OOD) generalization is to train a model $f^* : \mathcal{X} \rightarrow \mathcal{Y}$ making predictions $\mathcal{Y}$ from input samples $\mathcal{X}$ that performs robustly across all (unseen) environments $\mathcal{E}_a$:

$$f^* = \arg\min_{f_\theta} \max_{e \in \mathcal{E}_a} \mathcal{L}_e(f_\theta),$$

where $\mathcal{L}_S^e(f_\theta)$ denotes the empirical risk in environment e. Moreover, the data collected from each training environment, i.e., $D_e = \{\mathbf{x}_i^e, y_i^e\}_{i=1}^{n_e}$, is typically assumed to be independently and identically distributed (IID). The training set in OOD generalization contains data from several training environments, $D_{\text{tr}} = \{D_e\}_{e \in \mathcal{E}_{\text{tr}}}$. Furthermore, the model $f_\theta = w \circ \varphi$ can be decomposed into a featurizer $\varphi : \mathcal{X} \rightarrow \mathcal{Z}$ that extracts informative features from $\mathcal{Z}$ and a classifier $w : \mathcal{Z} \rightarrow \mathcal{Y}$ that makes predictions based on those features.

A common strategy for realizing the framework of out-of-distribution (OOD) generalization is invariant risk minimization (IRM) [8]. IRM seeks to achieve OOD generalization by learning an invariant representation φ so that a classifier operating on φ is simultaneously optimal across multiple environments, including $\mathcal{E}_{\text{tr}}$. This formulation gives rise to a challenging bi-level optimization problem, as

$$\min_{w,\varphi} \sum_{e \in \mathcal{E}_{\text{tr}}} \mathcal{L}_S^e(w \circ \varphi), \text{ s.t. } w \in \arg\min_{\bar{w}:\mathcal{Z}\rightarrow\mathcal{Y}} \mathcal{L}_S^e(\bar{w} \circ \varphi), \ \forall e \in \mathcal{E}_{\text{tr}}. \tag{4.37}$$

Table 4.11 Main results (%). Comparison with competitive OOD detection baselines, ↑ indicates the larger the better, while ↓ indicates the smaller the better. (averaged by multiple trials)

D_{in}	Method	AUROC ↑	AUPR ↑	FPR95 ↓	ID-ACC ↑	w./w.o auxiliary outliers
CIFAR-10	MSP	89.90 ± 0.30	91.48 ± 0.43	60.08 ± 0.76	**94.01±0.08**	
	ODIN	91.46 ± 0.56	91.67 ± 0.58	42.31 ± 1.38	**94.01±0.08**	
	Mahalanobis	75.10 ± 1.04	72.32 ± 1.92	61.35 ± 1.25	**94.01±0.08**	
	Energy	92.07 ± 0.22	92.72 ± 0.39	42.69 ± 1.31	**94.01±0.08**	
	Energy+UM (ours)	93.73 ± 0.36	94.27 ± 0.60	33.29 ± 1.70	92.80 ± 0.47	
	Energy+UMAP (ours)	**93.97±0.11**	**94.38±0.06**	**30.71±1.94**	**94.01±0.08**	
	OE	97.07 ± 0.01	97.31 ± 0.05	13.80 ± 0.28	92.59 ± 0.32	✓
	Energy (OE-style)	94.58 ± 0.64	94.69 ± 0.65	18.79 ± 2.31	80.91 ± 3.13	✓
	POEM	94.37 ± 0.07	94.51 ± 0.06	18.50 ± 0.33	77.24 ± 2.22	✓
	OE+UM (ours)	**97.60±0.03**	**97.87±0.02**	**11.22±0.16**	93.66 ± 0.12	✓
	OE+UMAP (ours)	97.48 ± 0.01	97.74 ± 0.00	12.21 ± 0.09	**94.01±0.08**	✓
CIFAR-100	MSP	74.06 ± 0.69	75.37 ± 0.73	83.14 ± 0.87	**74.86±0.21**	
	ODIN	76.18 ± 0.14	76.49 ± 0.20	78.93 ± 0.31	**74.86±0.21**	
	Mahalanobis	63.90 ± 1.91	64.31 ± 0.91	78.79 ± 0.50	**74.86±0.21**	
	Energy	**76.29±0.24**	**77.06±0.55**	78.46 ± 0.06	**74.86±0.21**	
	Energy+UM (ours)	76.22 ± 0.42	76.39 ± 1.03	74.05 ± 0.55	64.55 ± 0.24	
	Energy+UMAP (ours)	75.57 ± 0.59	75.66 ± 0.07	**72.21±1.46**	**74.86±0.21**	
	OE	90.55 ± 0.87	90.34 ± 0.94	34.73 ± 3.85	73.59 ± 0.30	✓
	Energy (OE-style)	88.92 ± 0.57	89.13 ± 0.56	37.90 ± 2.59	57.85 ± 2.65	✓
	POEM	88.95 ± 0.54	88.94 ± 0.31	38.10 ± 1.30	56.18 ± 1.92	✓
	OE+UM (ours)	91.04 ± 0.11	**91.13±0.24**	34.71 ± 0.81	**75.15±0.18**	✓
	OE+UMAP (ours)	**91.10±0.16**	90.99 ± 0.23	**33.62±0.26**	74.76 ± 0.11	✓

Given the training environments $\mathcal{E}_{\mathrm{tr}}$, and the function spaces $\mathcal{W}$ for w and Φ for φ, predictors of the form $f = w \circ \varphi$ that meet the constraint in Eq. (4.37) are referred to as invariant predictors and are denoted by $\mathcal{I}(\mathcal{E}_{\mathrm{tr}})$. In practice, finding invariant predictors is particularly difficult because characterizing $\mathcal{I}(\mathcal{E}_{\mathrm{tr}})$ is challenging. Consequently, it is common to limit $\mathcal{W}$ to the space of linear functions over $\mathcal{Z} = \mathbb{R}^d$ [176]. Furthermore, Arjovsky et al. [8] suggest that linear classifiers do not provide greater representational capacity than scalar classifiers, meaning that when $d = 1$ and $\mathcal{W} = \mathbb{R}^1$, this scalar constraint leads to a practical variant known as $\mathrm{IRM}_{\mathcal{S}}$.

$$\min_{\varphi} \sum_{e \in \mathcal{E}_{\mathrm{tr}}} \mathcal{L}_{\mathcal{S}}^{e}(\varphi), \text{ s.t. } \nabla_{w|w=1} \mathcal{L}_{\mathcal{S}}^{e}(w \cdot \varphi) = 0, \ \forall e \in \mathcal{E}_{\mathrm{tr}}. \tag{4.38}$$

Since Eq. (4.38) is still formulated as a constrained optimization problem, Arjovsky et al. [8] propose a relaxed variant called IRMv1, which is formulated as follows:

$$\min_{\varphi} \sum_{e \in \mathcal{E}_{\mathrm{tr}}} \mathcal{L}_{\mathcal{S}}^{e}(\varphi) + \lambda |\nabla_{w|w=1} \mathcal{L}_{\mathcal{S}}^{e}(w \cdot \varphi)|^2. \tag{4.39}$$

4.2.2 *Invariant Risk Minimization*

4.2.2.1 Pareto Invariant Risk Minimization

The interplay between optimization and generalization is essential to the success of deep learning [4, 5, 9, 176, 461]. Empirical risk minimization (ERM) [375] has shown to be effective in navigating the complicated non-convex loss landscapes of deep neural networks when optimized with simple algorithms [321]. Yet, under distribution shifts, optimization can often be biased by irrelevant or spurious correlations, leading to poor performance on out-of-distribution (OOD) data [28, 87, 119]. As a result, addressing the challenge of OOD generalization has garnered increasing attention in recent years. However, most existing approaches primarily concentrate on developing enhanced optimization objectives that aim to regularize ERM [2, 66, 74, 201, 205, 247, 286, 295, 307, 312, 334, 382]. Despite the progress and success, an important aspect has often been overlooked: the optimization with OOD objectives is inherently more challenging than optimizing with a simple ERM objective.

In practice, implementing OOD objectives typically requires compromises due to the complexity in the original formulation, e.g., bi-level optimization in the original IRM formulation [8]. Consequently, OOD objectives are frequently relaxed and reformulated as penalty terms when incorporated into the ERM objective [2, 8, 201, 205, 295, 307]. However, as Fig. 4.11a shows, the relaxed formulations often have a significant gap from the original objectives [8]. Moreover, because ERM and OOD objectives inherently conflict, achieving a balance between ERM and OOD

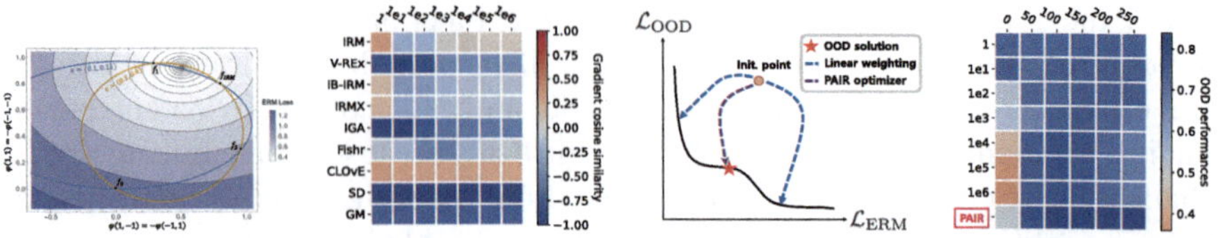

(a) Theoretical failure. (b) Gradient conflicts. (c) Unreliable opt. (d) Hp. tuning.

Fig. 4.11 The OOD optimization issue. (**a**) Relaxations in OOD objectives: For the ease of optimization, the practical OOD objectives are often relaxed several times from the original OOD objectives. Nevertheless, the practical formulation such as IRMv1 can behave very differently from the original formulation such as IRM. In the figure, the ellipsoids represent solutions that satisfy the invariance constraints of the practical IRM variant (IRMv1). When optimized with ERM, IRMv1 tends to prefer the solution f_1 rather than the desired IRM solution f_{IRM} (the predictor generated by IRM), which shows how the optimization process can diverge from the target objective. (**b**) Gradient conflicts: For various OOD objectives, it has widely observed the existence of gradient conflicts between the ERM objectives. Along with different penalty weights (x-axis), the conflicts will remain and vary. These conflicts complicate the optimization process, as balancing the two objectives is challenging. (**c**) Linear combination of ERM and OOD objectives: ERM and OOD objectives are often combined through linear summation, which is limited to reach desired solutions in the non-convex region of the Pareto front due to the use of a fixed weight. In contrast, the proposed method PAIR can find an adaptive descent direction to the desired solution despite the existence of gradient conflicts. (**d**) Sensitivity in hyperparamter tuning: The figure shows the performance of using IRMv1 on a modified ColoredMNIST dataset, with the y-axis representing penalty weights and the x-axis representing pretraining epochs. We can find that, in order to obtain optimal OOD performance, it usually requires exhaustive hyperparameter tuning

performance often involves trade-offs (see Fig. 4.11b). Studies such as Sagawa et al. [319] and Zhai et al. [454] indicate that improving OOD generalization typically necessitates sacrificing some degree of ERM performance. Additionally, careful tuning of OOD penalty hyperparameters is usually required [471] (see Fig. 4.11d), but this tuning can either diminish the efficacy of OOD objectives or cause them to dominate excessively, resulting in overly cautious models that fail to capture important patterns.

Method We adopt a multi-objective optimization (MOO) perspective to address OOD optimization. Taking IRM [8] as a representative example, we identify two primary issues that contribute to OOD optimization failures. The first issue is the weakened robustness of OOD objectives caused by the relaxations implemented in practical variants. In some cases, these relaxations can even remove the desired invariant solution from the Pareto front defined by the ERM objective and the OOD penalty (see Fig. 4.11a). As a consequence, simply optimizing the ERM objective along with the relaxed OOD penalty rarely leads to the targeted invariant solution. The second issue arises even when the Pareto front does include the desired solution. As illustrated in Fig. 4.11c, conventional linear weighting schemes—which combine the ERM and OOD objectives via a linear reweighting—cannot reach the solution if it resides in the non-convex region of the Pareto front [42]. Moreover, even if

Algorithm 3 The algorithm of PAIR-o

Input: Train set $D_{\text{tr}} = \{\mathbf{x}_i, y_i\}_{i=1}^N$ with partitions of environments $D_{\text{tr}} = \{D^e\}_{e \in \mathcal{E}_{\text{tr}}}$; learning rate η; batch size b; number of sampled environments d; OOD preference $\mathbf{p}_{\text{ood}}$ for ERM loss $\mathcal{L}_{\text{erm}}$ and m OOD losses $\mathbf{L}_{\text{ood}} = \{\mathcal{L}^i_{\text{ood}}\}_{i=1}^m$; pretraining epochs e_p; maximum training epochs for "balance" phase e_b; Trainable parameters at the "balance" phase θ
Randomly initialize the model $f = w \circ \varphi$
for $i = 1$ **to** e_p **do**
  Sample batches of data $\{\mathbf{x}_j, y_j\}_{j=1}^b$
  Make predictions with f: $\{\hat{y}_j\}_{j=1}^b = f(\{\mathbf{x}_j\}_{j=1}^b)$
  Calculate the empirical loss $\mathcal{L}_{\text{erm}}$ with $\{\hat{Y}_j\}_{j=1}^b$
  Update parameters of f with the empirical loss $\mathcal{L}_{\text{erm}}$ using learning rate η
end for
for $i = 1$ **to** e_b **do**
  for $D^e \in \texttt{permute}(\{D^e\}_e \in \mathcal{E}_{\text{tr}})$ **do**
    Sample a batch of data from D^e, $\{\mathbf{x}^e_j, y^e_j\}_{j=1}^b \sim D^e$
    Make predictions using f: $\{\hat{y}^e_j\}_{j=1}^b = f(\{\mathbf{x}^e_j\}_{j=1}^b)$
  end for
  Calculate empirical and OOD losses $\mathcal{L}_{\text{erm}}$ and $\mathcal{L}_{\text{ood}}$ and obtain the overall losses $\mathbf{L}$
  Obtain gradients $\mathbf{G} = \partial \mathbf{L} / \partial \theta$
  Calculate the loss weights β^*
  Update parameters via $\theta^{i+1} = \theta^i - \eta \mathbf{G} \beta^*$
end for

the desired OOD solution lies within the convex region, it still necessitates careful tuning of the OOD penalty weights, as shown in Fig. 4.11d.

To overcome these challenges, we introduce a novel optimization framework for OOD generalization called Pareto invariant risk minimization (PAIR), which consists of a new optimizer (PAIR-o) and a new model selection criterion (PAIR-s). By leveraging a multi-objective optimization (MOO) formulation, PAIR-o enables cooperative optimization alongside other OOD objectives, thereby enhancing the robustness of practical OOD formulations. Despite the substantial gap between IRMv1 and the ideal IRM, we demonstrate that integrating VREx [205] with IRMv1 can provably restore causal invariance [8] for certain problem instances. This integration leads to the derivation of the following IRMX objective:

$$\text{(IRMX)} \qquad \min_{\varphi} (\mathcal{L}_{\text{erm}}, \mathcal{L}_{\text{irm}}, \mathcal{L}_{\text{vrex}})^T. \tag{4.40}$$

When provided with robust OOD objectives, PAIR-o identifies a descent path using adaptive penalty weights. This process results in a Pareto optimal solution that appropriately balances ERM and OOD performance. The PAIR-o algorithm is detailed in Algorithm 3.

In addition, the MOO analysis inspires PAIR-s, a model selection strategy that takes into account the trade-offs between ERM and OOD objectives to facilitate better OOD model selection. We validate our approach through extensive experiments on challenging OOD benchmarks.

Experiments *Optimization with PAIR-o* Empirical results from applying PAIR-o on 6 datasets from Wilds [197] are summarized in Table 4.12. The findings show that PAIR-o not only substantially improves upon IRMv1 and IRMX but also achieves top-ranking OOD performance among all state-of-the-art methods across a variety of realistic distribution shifts. These results underscore the effectiveness and importance of resolving the optimization dilemma in OOD generalization. Furthermore, the performance gains of PAIR-o over IRMX confirm its ability to better trade-off between ERM and OOD objectives.

Model Selection with PAIR-s To evaluate the effectiveness of PAIR-s, we integrate it with several representative OOD methods and assess whether it can enhance model selection under rigorous hyperparameter tuning [133] on datasets such as ColoredMNIST [8], PACS [218], and TerraIncognita [28]. Intuitively, selecting models based solely on ERM performance often leads to a bias toward environments that closely match the distribution of the corresponding validation set. This bias can result in higher performance variance across different environments or lower worst-case performance.

To explore this issue in various scenarios [364], we use training-domain validation accuracy for ColoredMNIST and TerraIncognita and test-domain validation accuracy for PACS. As given in Table 4.13, the experimental results across three datasets from DomainBed [133] show that PAIR-s improves the performance of the selected OOD models by up to 10%, highlighting the importance of considering the trade-offs between ERM and OOD objectives during optimization and model selection.

Since PAIR primarily addresses the optimization of the objectives using the learned features, the next step is to explore how to acquire suitable features for effective OOD optimization.

4.2.2.2 Feature Augmented Training

It is critical to understand how features are learned by neural networks in order to understand their generalizability [3, 45, 48, 315, 331, 336]. Deep networks trained with ERM tend to acquire highly predictive features that perform remarkably well on IID data [125, 375]. However, ERM is also prone to learning *spurious* features or shortcuts—such as relying on image backgrounds—that correlate with labels in the training distribution but fail to hold in OOD scenarios [28, 87, 119, 474]. This reliance on non-robust features leads to significant performance degradation on OOD data [197]. As a result, a common consensus is that ERM's failure to reliably learn features that exhibit *invariant* correlations with labels across varying distributions is the main reason for OOD failures in deep neural networks [28]. On the contrary, Izmailov et al. [175], Kirichenko et al. [194], and Rosenfeld et al. [316] show that ERM-trained models *already learn sufficiently robust features* for OOD generalization. Furthermore, empirically, the success of OOD objectives, i.e., the penalty terms used to regularize ERM feature learning [2, 66, 201, 205, 286, 295, 307, 312, 334, 382, 488], is highly contingent on prior

Table 4.12 OOD generalization performances on Wilds benchmark

	CAMELYON17	CIVILCOMMENTS	FMOW	IWILDCAM	POVERTYMAP	RXRX1	
	Avg. acc. (%)	Worst acc. (%)	Worst acc. (%)	Macro F1	Worst Pearson r	Avg. acc. (%)	AVG. RANK(↓)†
ERM	70.3 (±*6.4*)	56.0 (±*3.6*)	32.3 (±*1.25*)	30.8 (±*1.3*)	0.45 (±*0.06*)	29.9 (±*0.4*)	4.50
CORAL	59.5 (±*7.7*)	65.6 (±*1.3*)	31.7 (±*1.24*)	**32.7** (±*0.2*)	0.44 (±*0.07*)	28.4 (±*0.3*)	5.50
GroupDRO	68.4 (±*7.3*)	70.0 (±*2.0*)	30.8 (±*0.81*)	23.8 (±*2.0*)	0.39 (±*0.06*)	23.0 (±*0.3*)	6.83
IRMv1	64.2 (±*8.1*)	66.3 (±*2.1*)	30.0 (±*1.37*)	15.1 (±*4.9*)	0.43 (±*0.07*)	8.2 (±*0.8*)	7.67
V-REx	71.5 (±*8.3*)	64.9 (±*1.2*)	27.2 (±*0.78*)	27.6 (±*0.7*)	0.40 (±*0.06*)	7.5 (±*0.8*)	7.00
Fish	74.3 (±*7.7*)	73.9 (±*0.2*)	34.6 (±*0.51*)	24.8 (±*0.7*)	0.43 (±*0.05*)	10.1 (±*1.5*)	4.33
LISA	**74.7** (±*6.1*)	70.8 (±*1.0*)	33.5 (±*0.70*)	24.0 (±*0.5*)	**0.48** (±*0.07*)	**31.9** (±*0.8*)	2.67
IRMX	67.0 (±*6.6*)	74.3 (±*0.8*)	33.7 (±*0.78*)	26.6 (±*0.9*)	0.45 (±*0.04*)	28.7 (±*0.2*)	4.00
PAIR-o	74.0 (±*7.0*)	**75.2** (±*0.7*)	**35.5** (±*1.13*)	27.9 (±*0.7*)	0.47 (±*0.06*)	28.8 (±*0.1*)	**2.17**

† Averaged rank is reported because of the dataset heterogeneity. A lower rank is better

Table 4.13 OOD generalization performances using DomainBed evaluation protocol

		ColoredMNIST†				PACS‡					TerraIncognita†				
	PAIR-s	**+90%**	**+80%**	**10%**	Δ wr.	A	C	P	S	Δ wr.	L100	L38	L43	L46	Δ wr.
ERM		71.0	**73.4**	10.0		87.2	79.5	95.5	76.9		46.7	**41.8**	57.4	39.7	
DANN		71.0	**73.4**	10.0		86.5	79.9	97.1	75.3		46.1	41.2	56.7	35.6	
DANN	✓	71.6	73.3	10.9	+0.9	87.0	81.4	96.8	77.5	+2.2	43.1	41.1	55.2	38.7	+3.1
GroupDRO		72.6	73.1	9.9		87.7	82.1	98.0	79.6		48.4	40.3	57.9	40.0	
GroupDRO	✓	**72.7**	73.2	13.0	+3.1	86.7	**83.2**	**97.8**	81.4	+1.8	48.4	40.3	57.9	40.0	+0.0
IRMv1		72.3	72.6	9.9		82.3	80.8	95.8	78.9		48.4	35.6	55.4	40.1	
IRMv1	✓	67.4	64.8	**24.2**	+14.3	85.3	81.7	97.4	79.7	+0.8	40.4	38.3	48.8	37.0	+1.4
Fishr		72.2	73.1	9.9		**88.4**	82.2	97.7	81.6		49.2	40.6	57.9	40.4	
Fishr	✓	69.1	70.9	22.6	+12.7	87.4	82.6	97.5	**82.2**	+0.6	**51.0**	40.7	**58.2**	**40.8**	+0.3

† Using the training domain validation accuracy. ‡ Using the test domain validation accuracy

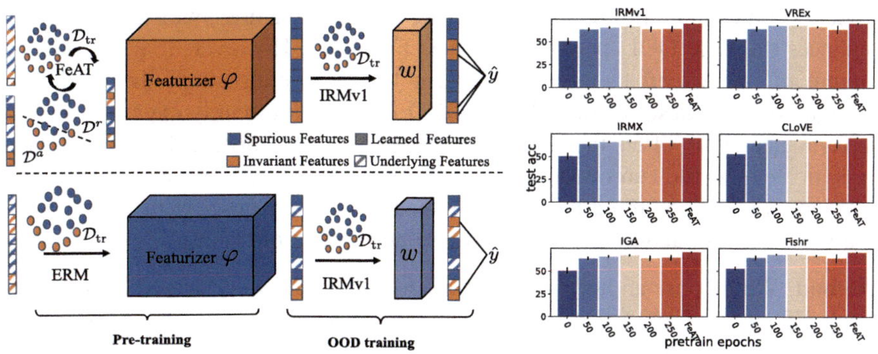

Fig. 4.12 (**a**) Comparison between FeAT (top row) and ERM (bottom row). Different types of features are denoted via different colors. Spurious features are illustrated using the color blue, and training the model solely with ERM tends to result in dominant learning of these spurious features while limiting the learning of invariant features (represented in orange). As a result, subsequent training with IRMv1 struggles due to insufficient invariant feature learning, ultimately hindering OOD generalization. In contrast, FeAT mitigates this issue by iteratively dividing the training set D_{tr} into an augmentation set D^a and a retention set D^r. The augmentation set contains features that the model has not yet learned, while the retention set includes features already captured by the current model in each iteration. At every round, FeAT improves the model by introducing new features from the expanding augmentation set while ensuring that previously learned features from the retention set are preserved. This process enables the model to acquire a richer feature representation for OOD training, thereby improving its OOD performance. (**b**) Impact of ERM pretraining epochs on OOD performance for ColoredMNIST-025. The effectiveness of various OOD objectives is heavily influenced by the quality of feature learning during ERM pretraining. When ERM fails to identify certain essential features, OOD performance becomes limited. In comparison, FeAT can effectively capture a broader set of features within only 2 rounds (or 300 epochs), improving OOD performance.

pretraining with ERM [67, 471]. As illustrated in Fig. 4.12b, the ultimate OOD performance is *highly sensitive* to the number of ERM pretraining epochs. This leads to an intriguing research question: *what features are captured by ERM and OOD objectives individually, and how do these features contribute to generalization on both in-distribution and OOD data?*

To investigate this question, we build on a theoretical framework involving a two-layer nonlinear, non-convex CNN model [3, 48] to analyze and understand the feature learning dynamics of ERM and a widely used OOD objective, IRMv1 [8]. In order to model invariant and spurious features, we consider two types of features with varying degrees of correlation to the labels [188].

In line with existing empirical findings, our first theoretical result reveals that ERM inherently learns *both* spurious and invariant features, with the speed of feature acquisition determined by the respective strength of their correlations with the labels. In addition, our second theoretical result shows that training the model solely with IRMv1 objective *fails to acquire new* features. As a result, the final OOD performance largely depends on the *quality* of features learned during ERM training. Specifically, as the number of ERM pretraining epochs increases, the model becomes more proficient at capturing invariant features, which subsequently enhances the final OOD performance (see Fig. 4.12). However, if ERM does not capture *all* the useful features needed for OOD generalization, the model will struggle to learn these features during OOD training, ultimately capping the OOD performance. It is often the case that ERM's feature learning biases [331], or limitations in model architectures [154] for a variety of reasons, such as low invariant correlation strength. Thus, the feature learning process in ERM can act as a *bottleneck* for achieving effective OOD generalization [309].

Method We propose an iterative feature learning framework, termed Feature Augmented Training (FeAT), designed to promote the acquisition of a more diverse and comprehensive set of features by the model. The pipeline of FeAT is illustrated in Fig. 4.12a. In each iteration, FeAT partitions the training set into two subsets based on whether the underlying features have already been learned or not. Specifically, samples for which the model makes correct predictions form the Retention set D^r, while those with incorrect predictions form the Augmentation set D^a. The FeAT objective at round k is then given by

$$\mathcal{L}_{\text{FeAT}} = \max_{D_i^a \in R^a} \mathcal{L}_{D_i^a}(w_k \circ \varphi) + \lambda \cdot \sum_{D_i^r \in R^r} \mathcal{L}_{D_i^r}(w_i \circ \varphi). \tag{4.41}$$

Here, $\mathcal{L}_{D_i}(w \circ \varphi)$ denotes the empirical risk of the model $f_\theta = w \circ \varphi$ evaluated on the subset D_i with w as the classifier and φ as the feature extractor, and $\{w_i \mid 1 \leq i \leq k-1\}$ represents the classifiers obtained from previous rounds. Intuitively, D^a and D^r are designed to capture distinct sets of features, with these features being systematically segregated across different rounds.

Next, FeAT applies distributionally robust optimization (DRO) [277, 471] across all subsets. This process *augments* the model by encouraging the learning of new features through minimizing the maximal ERM losses over all D^a while simultaneously *retaining* the features already captured by minimizing the ERM losses over all D^r. As the augmentation and retention sets expand, FeAT progressively equips the model with a richer set of features for OOD training, ultimately improving OOD performance. The FeAT process halts when no additional predictive features can be learned (see Algorithm 4).

Experiments We perform extensive experiments on ColoredMNIST [67] and Wilds [197] to demonstrate that FeAT can learn richer features than both ERM and the state-of-the-art algorithm Bonsai [471].

Proof-of-Concept Study on ColoredMNIST We begin with a proof-of-concept study on ColoredMNIST [67] to assess FeAT's feature learning performance under different conditions. We consider both the original ColoredMNIST, with $\mathcal{E}_{\text{tr}} = (0.25, 0.1), (0.25, 0.2)$ (denoted as ColoredMNIST-025), where spurious features exhibit stronger correlation with the labels, and a modified version (denoted as ColoredMNIST-01) with $\mathcal{E}_{\text{tr}} = (0.1, 0.2), (0.1, 0.25)$, where invariant features are more strongly correlated with the labels. We compare the OOD performance of

Algorithm 4 FeAT: feature augmented training

Input: Training data D_{tr}; the maximum augmentation rounds K; predictor $f := w \circ \varphi$; length of inner training epochs t; termination threshold p
Initialize groups $R^a \leftarrow D_{\text{tr}}$, $R^r \leftarrow \{\}$
for $k \in [1, \ldots, K]$ **do**
    Randomly initialize w_k
    for $j \in [1, \ldots, t]$ **do**
        Obtain $\mathcal{L}_{\text{FeAT}}$ with P via Eq. (4.41)
        Update w_k, φ with $\mathcal{L}_{\text{FeAT}}$
    end for
    // Early Stop if $f_k = w_k \circ \varphi$ fails to find new features
    if Training accuracy of f_k is smaller than p **then**
        Set $K = k - 1$ and terminate the loop
    end if
    Split D_{tr} into groups D_k^a, D_k^r according to whether f_k classifies the examples in D_{tr} correctly or not
    Update groups $R^a \leftarrow R^a \cup \{D_k^a\}$, $R^r \leftarrow R^r \cup \{D_k^r\}$
end for
Synthesize the final classifier $w \leftarrow \frac{1}{K} \sum_{i=1}^{K} w_i$
return $f_\theta = w \circ \varphi$;

features obtained via FeAT, ERM, and the state-of-the-art rich feature learning method Bonsai [471]. Based on the features learned by these methods, we leverage various state-of-the-art OOD objectives—including IRMv1 [8], VREx [205], IRMX [67], IB-IRM [2], CLOvE [382], IGA [201], and Fishr [307]—to assess the practical quality of the learned representations. To prevent distortion of the pretrained features, the feature representations are frozen once initialized for OOD training [207]. Additionally, we compare FeAT with the standard training approach that uses unfrozen ERM features, referred to as ERM-NF.

For Bonsai, we trained for 2 rounds following Zhang et al. [471], whereas FeAT automatically terminated at round 2 in ColoredMNIST-025 and at round 3 in ColoredMNIST-01. For ERM, we pretrained the model for the same total number of epochs as FeAT in ColoredMNIST-01 but applied early stopping (after one round's worth of epochs) in ColoredMNIST-025 to mitigate overfitting. All methods utilized the same backbone and followed the same protocol in prior works [67, 471].

The results reported in Table 4.14 reveal that ERM fails to learn sufficiently robust features under both scenarios of strong spurious correlations and strong invariant correlations. Moreover, while Bonsai succeeds in learning richer features in ColoredMNIST-025 and boosts OOD performance, it occasionally yields suboptimal outcomes in ColoredMNIST-01, likely due to instability in its feature learning. In contrast, FeAT consistently enhances the OOD performance across all evaluated OOD objectives and for both ColoredMNIST datasets, underscoring the advantages of directly controlling feature learning through FeAT over both Bonsai and ERM.

Experiments on Real-World Benchmarks We further compare FeAT with both ERM and Bonsai on six real-world OOD generalization datasets from Koh et al. [197], which involve complex features and distribution shifts. The features learned by

Table 4.14 OOD performance on ColoredMNIST datasets initialized with different representations

	ColoredMNIST-025				ColoredMNIST-01			
	ERM-nf	ERM	Bonsai	FeAT	ERM-nf	ERM	Bonsai	FeAT
ERM	17.14 (±*0.73*)	12.40 (±*0.32*)	11.21 (±*0.49*)	**17.27** (±*2.55*)	73.06 (±*0.71*)	73.75 (±*0.49*)	70.95 (±*0.93*)	**76.05** (±*1.45*)
IRMv1	67.29 (±*0.99*)	59.81 (±*4.46*)	70.28 (±*0.72*)	**70.57** (±*0.68*)	76.89 (±*3.25*)	73.84 (±*0.56*)	76.71 (±*4.10*)	**82.33** (±*1.77*)
V-Rex	68.62 (±*0.73*)	65.96 (±*1.29*)	70.31 (±*0.66*)	**70.82** (±*0.59*)	83.52 (±*2.52*)	81.20 (±*3.27*)	82.61 (±*1.76*)	**84.70** (±*0.69*)
IRMX	67.00 (±*1.95*)	64.05 (±*0.88*)	70.46 (±*0.42*)	**70.78** (±*0.61*)	81.61 (±*1.98*)	75.97 (±*0.88*)	80.28 (±*1.62*)	**84.34** (±*0.97*)
IB-IRM	56.09 (±*2.04*)	59.81 (±*4.46*)	70.28 (±*0.72*)	**70.57** (±*0.68*)	75.81 (±*0.63*)	73.84 (±*0.56*)	76.71 (±*4.10*)	**82.33** (±*1.77*)
CLOvE	58.67 (±*7.69*)	65.78 (±*0.00*)	65.57 (±*3.02*)	**65.78** (±*2.68*)	75.66 (±*10.6*)	74.73 (±*0.36*)	72.73 (±*1.18*)	**75.12** (±*1.08*)
IGA	51.22 (±*3.67*)	62.43 (±*3.06*)	**70.17** (±*0.89*)	67.11 (±*3.40*)	74.20 (±*2.45*)	73.74 (±*0.48*)	74.72 (±*3.60*)	**83.46** (±*2.17*)
Fishr	69.38 (±*0.39*)	67.74 (±*0.90*)	68.75 (±*1.10*)	**70.56** (±*0.97*)	77.29 (±*1.61*)	82.23 (±*1.35*)	84.19 (±*0.66*)	**84.26** (±*0.93*)
Oracle	71.97 (±*0.34*)				86.55 (±*0.27*)			

each method are subsequently evaluated using several representative state-of-the-art OOD objectives in Wilds, including GroupDro [319], IRMv1 [8], VREx [205], and IRMX [67]. By default, ERM, Bonsai, and FeAT are trained for an equal number of steps, and the number of rounds for Bonsai and FeAT are kept identical (although Bonsai still requires an additional round for feature synthesis). The only exception occurs in RxRx1, where both Bonsai and FeAT require more steps than ERM to achieve convergence. To ensure a fair comparison, we adhere to the standard evaluation protocols established in prior works [67, 197, 334, 471].

To further understand the essential feature learning beyond the OOD performance entailed by different OOD objectives, we assess the learned features using Deep Feature Reweighting (DFR) [194]. DFR conducts logistic regression based on the learned features on an extra OOD validation set, in which the *spurious correlation does not hold*. Intuitively, DFR provides a reliable measure of the quality of invariant features learned [175]. For datasets that lack a proper OOD validation set (e.g., Camelyon17), we adopt an alternative implementation that relies on a random split of the training and test data to evaluate the quality of invariant features [316].

We also report DFR-s in order to assess the extent of spurious feature learning across different methods, which run regression with the environment labels, when the environment labels are available.

Table 4.15 presents the results, which reveal that as the tasks become more challenging and neural architectures grow more complex, the quality of features learned by ERM can degrade. For example, in FMoW, ERM struggles to capture all useful features, while in CivilComments, it has a tendency to overfit to spurious correlations. Additionally, we observe that Bonsai often exhibits instability when attempting to learn richer features, sometimes even underperforming ERM in terms of both rich feature learning and OOD generalization across several datasets.

In contrast, FeAT consistently achieves superior performance in learning invariant features across various challenging and realistic datasets. Furthermore, compared to ERM and Bonsai, FeAT effectively reduces overfitting to spurious features caused by spurious correlations. As a result, when the features learned by FeAT are applied to different OOD objectives, they deliver consistent improvements in performance.

4.2.3 Contrastive-Based OOD Generalization

4.2.3.1 Causality-Inspired Contrastive Causal Representation Learning on Graphs

We now broaden our discussion to address OOD generalization in more complex data structures, such as graphs. Graph representation learning with graph neural networks (GNNs) has demonstrated remarkable success in tasks involving relational information [135, 193, 378, 425, 428]. However, these methods generally operate

Table 4.15 OOD generalization performances on Wilds benchmark

		CAMELYON17	CIVILCOMMENTS	FMOW	IWILDCAM	AMAZON	RXRX1
INIT.	METHOD	Avg. acc. (%)	Worst acc. (%)	Worst acc. (%)	Macro F1	10-th per. acc. (%)	Avg. acc. (%)
ERM	DFR†	95.14 (±*1.96*)	**77.34** (±*0.50*)	41.96 (±*1.90*)	23.15 (±*0.24*)	48.00 (±*0.00*)	–
ERM	DFR-s†	–	82.24 (±*0.13*)	56.17 (±*0.62*)	52.44 (±*0.34*)	–	–
Bonsai	DFR†	95.17 (±*0.18*)	77.07 (±*0.85*)	43.26 (±*0.82*)	21.36 (±*0.41*)	46.67 (±*0.00*)	–
Bonsai	DFR-s†	–	81.26 (±*1.86*)	58.58 (±*1.17*)	50.85 (±*0.18*)	–	–
FeAT	DFR†	**95.28** (±*0.19*)	**77.34** (±*0.59*)	**43.54** (±*1.26*)	**23.54** (±*0.52*)	**49.33** (±*0.00*)	–
FeAT	DFR-s†	–	79.56 (±*0.38*)	57.69 (±*0.78*)	52.31 (±*0.38*)	–	–
ERM	ERM	74.30 (±*5.96*)	55.53 (±*1.78*)	33.58 (±*1.02*)	28.22 (±*0.78*)	51.11 (±*0.63*)	30.21 (±*0.09*)
ERM	GroupDRO	76.09 (±*6.46*)	69.50 (±*0.15*)	33.03 (±*0.52*)	28.51 (±*0.58*)	52.00 (±*0.00*)	29.99 (±*0.13*)
ERM	IRMv1	75.68 (±*7.41*)	68.84 (±*0.95*)	33.45 (±*1.07*)	28.76 (±*0.45*)	52.00 (±*0.00*)	30.10 (±*0.05*)
ERM	V-REx	71.60 (±*7.88*)	69.03 (±*1.08*)	33.06 (±*0.46*)	28.82 (±*0.47*)	52.44 (±*0.63*)	29.88 (±*0.35*)
ERM	IRMX	73.49 (±*9.33*)	68.91 (±*1.19*)	33.13 (±*0.86*)	28.82 (±*0.47*)	52.00 (±*0.00*)	30.10 (±*0.05*)
Bonsai	ERM	73.98 (±*5.30*)	63.34 (±*3.49*)	31.91 (±*0.51*)	28.27 (±*1.05*)	48.58 (±*0.56*)	24.22 (±*0.44*)
Bonsai	GroupDRO	72.82 (±*5.37*)	70.23 (±*1.33*)	33.12 (±*1.20*)	27.16 (±*1.18*)	42.67 (±*1.09*)	22.95 (±*0.46*)
Bonsai	IRMv1	73.59 (±*6.16*)	68.39 (±*2.01*)	32.51 (±*1.23*)	27.60 (±*1.57*)	47.11 (±*0.63*)	23.35 (±*0.43*)
Bonsai	V-REx	76.39 (±*5.32*)	68.67 (±*1.29*)	33.17 (±*1.26*)	25.81 (±*0.42*)	48.00 (±*0.00*)	23.34 (±*0.42*)
Bonsai	IRMX	64.77 (±*10.1*)	69.56 (±*0.95*)	32.63 (±*0.75*)	27.62 (±*0.66*)	46.67 (±*0.00*)	23.34 (±*0.40*)
FeAT	ERM	77.80 (±*2.48*)	68.11 (±*2.27*)	33.13 (±*0.78*)	28.47 (±*0.67*)	**52.89** (±*0.63*)	**30.66** (±*0.42*)
FeAT	GroupDRO	**80.41** (±*3.30*)	**71.29** (±*0.46*)	33.55 (±*1.67*)	28.38 (±*1.32*)	52.58 (±*0.56*)	29.99 (±*0.11*)
FeAT	IRMv1	77.97 (±*3.09*)	70.33 (±*1.14*)	**34.04** (±*0.70*)	**29.66** (±*1.52*)	**52.89** (±*0.63*)	29.99 (±*0.19*)
FeAT	V-REx	75.12 (±*6.55*)	70.97 (±*1.06*)	34.00 (±*0.71*)	29.48 (±*1.94*)	**52.89** (±*0.63*)	30.57 (±*0.53*)
FeAT	IRMX	76.91 (±*6.76*)	71.18 (±*1.10*)	33.99 (±*0.73*)	29.04 (±*2.96*)	**52.89** (±*0.63*)	29.92 (±*0.16*)

† DFR/DFR-s use an additional OOD dataset to evaluate invariant and spurious feature learning, respectively

under the assumption that the training and test graphs are drawn from the same distribution—a condition that is often violated in real-world scenarios [162, 167, 197, 498].

The mismatch between training and test distributions, referred to as *distribution shifts*, arises due to environmental factors related to data collection or processing. These shifts can severely degrade the performance of deployed models [28, 87]. Such failures in out-of-distribution (OOD) generalization present a critical challenge for the practical application of graph representation learning [498]. At the same time, there has been growing interest in achieving OOD generalization for Euclidean data, with numerous solutions proposed in recent literature [2, 8, 31, 74, 201, 205, 319]. Many of these approaches are grounded in the invariance principle derived from causality [291, 293, 312]. This principle is based on the Independent Causal Mechanism (ICM) assumption [291, 294], which posits that predictions relying solely on the causal factors of the label will remain invariant across a broad range of distribution shifts [2, 8, 293].

Although the invariance principle has achieved notable success in Euclidean data, its application to graph data presents unique challenges due to the inherent complexity of graphs. First, distribution shifts in graphs are significantly more intricate, occurring at both the attribute and structural levels. These shifts can manifest in diverse forms, such as variations in graph sizes, subgraph densities, and levels of homophily [34, 406, 434]. Additionally, such shifts can lead to spurious correlations with labels in multiple ways [2, 8, 275], making it difficult to disentangle and identify invariant features.

Second, many OOD algorithms designed for Euclidean data rely on the availability of environment (or domain) labels to pinpoint the sources of distribution shifts [8]. However, obtaining such environment labels for graphs is often prohibitively expensive. This is because graphs are abstract and complex, necessitating significant expert knowledge for accurate labeling [162].

These challenges highlight an even more demanding question: *how can we extend the invariance principle to achieve OOD generalization on graphs?*

Method We introduce **C**ausality Inspired **I**nvariant **G**raph Le**A**rning (CIGA), a novel framework aimed at capturing invariance properties in graphs to ensure robust OOD generalization under diverse distribution shifts. Specifically, we construct three structural causal models (SCMs) [291] to characterize the potential distribution shifts in graphs:

- Graph generation SCM: models the process by which graphs are generated
- Invariant-spurious interaction SCMs: model the interactions between invariant and spurious features during graph generation, corresponding to two key scenarios, Fully Informative Invariant Feature (FIIF) and Partially Informative Invariant Feature (PIIF).

Using these models, we extend the invariance principle to graphs for OOD generalization: a GNN achieves invariance to distribution shifts if it focuses exclusively on an invariant and critical subgraph, denoted as $\mathcal{G}_c$. This subgraph

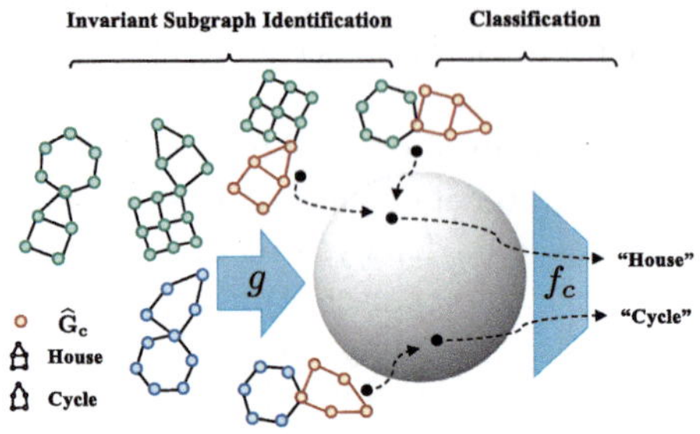

(a)

Algorithm	OOD Guarantee	Regime	E Known	SCM Support
IRM [8]	Yes	$\mathbb{R}$	Yes	PIIF
IB-IRM [2]	Yes	$\mathbb{R}$	Yes	PIIF&FIIF
EIIL [73]	Yes	$\mathbb{R}$	No	PIIF
DANN [115]	N/A	$\mathbb{R}$	Yes	N/A
MatchDG [260]	N/A	$\mathbb{R}$	Yes	FIIF
GroupDro [321]	N/A	$\mathbb{R}$	Yes	N/A
CNC [467]	N/A	$\mathbb{R}$	No	N/A
GIB [441]	Yes	$\mathcal{G}$	No	FIIF
DIR [409]	No	$\mathcal{G}$	No	FIIF
CIGA (Ours)	**Yes**	$\mathcal{G}$	**No**	**PIIF&FIIF**

(b)

Fig. 4.13 (**a**) CIGA is designed to address OOD generalization in graph classification tasks by focusing on invariant subgraphs. For instance, GNNs are tasked with classifying graphs based on specific motifs (e.g., "House" or "Cycle"). The framework operates as follows: The featurizer g extracts an (orange-colored) subgraph $\widehat{G}_c$ from each input, which is then used by the classifier f_c to predict the label. The training objective for g is designed using a contrastive approach, where the latent distribution of $\widehat{G}_c$ is optimized to maximize intra-class mutual information, ensuring that the predictions remain invariant to distribution shifts. (**b**) An overview of potential solutions

retains the majority of the information in the input graph $\mathcal{G}$ about the underlying causes of the label. Consequently, achieving OOD generalization on graphs can be broken down into two essential processes: (a) Identifying the invariant subgraph (b) Predicting the label based on the invariant subgraph

As illustrated in Fig. 4.13a, we propose a prototypical invariant graph learning algorithm that decomposes a GNN into two components: (a) Featurizer (g) (identifies the invariant subgraph $\mathcal{G}_c$ from the input graph $\mathcal{G}$) and (b) Classifier (f_c) (makes predictions solely based on $\mathcal{G}_c$).

To extract the desired subgraph $\mathcal{G}_c$, we derive an information-theoretic objective for the featurizer. This objective aims to identify subgraphs that maximally preserve invariant intra-class information across a set of diverse (and unknown) environments:

$$\max_{\mathcal{G}_c \subseteq \mathcal{G}} I(\mathcal{G}_c, Y) - \lambda I(\mathcal{G}_c, E),$$

where $I(\mathcal{G}_c, Y)$ represents the mutual information between the invariant subgraph $\mathcal{G}_c$ and the label Y; $I(\mathcal{G}_c, E)$ represents the mutual information between $\mathcal{G}_c$ and the environment E (to minimize spurious correlations); and λ is a trade-off parameter to balance these terms.

This objective encourages the featurizer to extract invariant subgraphs that are informative about the label while being independent of the environment, thereby enabling robust OOD generalization:

$$\text{(CIGAv1)} \qquad \max_{f_c, g} I(\widehat{\mathcal{G}}_c; Y), \text{ s.t. } \widehat{\mathcal{G}}_c \in \underset{\widehat{\mathcal{G}}_c = g(\mathcal{G}), |\widehat{\mathcal{G}}_c| \leq s_c}{\arg\max} I(\widehat{\mathcal{G}}_c; \widetilde{\mathcal{G}}_c | Y), \tag{4.42}$$

where $\widetilde{\mathcal{G}}_c = g(\widetilde{\mathcal{G}})$ and $\widetilde{\mathcal{G}} \sim \mathbb{P}(\mathcal{G}|Y)$, meaning that $\widetilde{\mathcal{G}}$ is sampled from the set of training graphs sharing the same label Y as $\mathcal{G}$. However, this approach still depends on assumptions about the size of the underlying invariant subgraph. To further address this limitation, we derive an advanced variant:

$$\begin{aligned} &\max_{f_c, g} \; I(\widehat{\mathcal{G}}_c; Y) + I(\widehat{\mathcal{G}}_s; Y), \;\; \text{s.t. } \widehat{\mathcal{G}}_c \in \arg\max_{\widehat{\mathcal{G}}_c = g(\mathcal{G})} I(\widehat{\mathcal{G}}_c; \widetilde{\mathcal{G}}_c | Y), \\ &(\text{CIGAv2}) \qquad\qquad I(\widehat{\mathcal{G}}_s; Y) \leq I(\widehat{\mathcal{G}}_c; Y), \; \widehat{\mathcal{G}}_s = \mathcal{G} - g(\mathcal{G}), \end{aligned} \tag{4.43}$$

We provide theoretical guarantees demonstrating that, under mild assumptions, our approach can reliably identify the underlying subgraph $\mathcal{G}_c$.

Experiment We conduct comprehensive experiments on both synthetic and real-world datasets exhibiting various distribution shifts, including a particularly challenging scenario from AI-aided drug discovery [498].

Benchmarks with Structure and Mixed Graph Distribution Shifts In Table 4.16, we report the test accuracies of different methods (excluding GIB due to convergence issues). The experiments explore varying degrees of distribution shifts under different bias settings. While most methods achieve training accuracies above 99%, their test accuracies degrade significantly as the bias increases and multiple distribution shifts are combined. Although interpretable GNNs and standard OOD objectives provide some improvement in OOD performance, they often exhibit high variance because they lack formal guarantees for OOD generalization. In contrast, both CIGAv1 and CIGAv2 consistently outperform all baselines by a significant margin—up to 10%—and exhibit lower variance. These results underscore the effectiveness and robust OOD generalization ability of CIGA.

Benchmarks with Structure and Mixed Graph Realistic Shifts Table 4.17 evaluates the performance of CIGA on real-world datasets that involve more complex and heterogeneous distribution shifts. For these tasks, we report both the averaged accuracy and the corresponding ranks due to the diverse nature of the datasets. In these more challenging scenarios, interpretable GNNs and standard OOD objectives often perform comparably to or even worse than ERM baselines, consistent with prior observations in nonlinear benchmarks [133, 498]. However, both CIGAv1 and CIGAv2 consistently outperform all other methods, including specialized approaches such as Γ GNNs [34], which are designed specifically to handle graph size shifts. These results demonstrate the generality and superiority of CIGA across diverse graph-based tasks. Importantly, they highlight the ability of CIGA to adapt to a wide range of distribution shifts, including those arising from graph structure, size, and heterogeneity.

4.2.3.2 Enhancing OOD Methods for Evolving Domain Generalization

Traditional approaches to domain generalization primarily emphasize learning features that remain constant across multiple source domains [8, 219, 319, 349].

Table 4.16 OOD generalization performance on structure and mixed shifts for synthetic graphs

	SPMotif-Struc†			SPMotif-Mixed†			
	bias=0.33	bias=0.60	bias=0.90	bias=0.33	bias=0.60	bias=0.90	Avg
ERM	59.49 (3.50)	55.48 (4.84)	49.64 (4.63)	58.18 (4.30)	49.29 (8.17)	41.36 (3.29)	52.24
ASAP	64.87 (13.8)	64.85 (10.6)	**57.29 (14.5)**	66.88 (15.0)	59.78 (6.78)	**50.45 (4.90)**	60.69
DIR	58.73 (11.9)	48.72 (14.8)	41.90 (9.39)	67.28 (4.06)	51.66 (14.1)	38.58 (5.88)	51.14
IRM	57.15 (3.98)	61.74 (1.32)	45.68 (4.88)	58.20 (1.97)	49.29 (3.67)	40.73 (1.93)	52.13
V-Rex	54.64 (3.05)	53.60 (3.74)	48.86 (9.69)	57.82 (5.93)	48.25 (2.79)	43.27 (1.32)	51.07
EIIL	56.48 (2.56)	60.07 (4.47)	55.79 (6.54)	53.91 (3.15)	48.41 (5.53)	41.75 (4.97)	52.73
IB-IRM	58.30 (6.37)	54.37 (7.35)	45.14 (4.07)	57.70 (2.11)	50.83 (1.51)	40.27 (3.68)	51.10
CNC	70.44 (2.55)	**66.79 (9.42)**	50.25 (10.7)	65.75 (4.35)	59.27 (5.29)	41.58 (1.90)	59.01
CIGA **v1**	**71.07 (3.60)**	63.23 (9.61)	51.78 (7.29)	**74.35 (1.85)**	**64.54 (8.19)**	49.01 (9.92)	**62.33**
CIGA**v2**	**77.33 (9.13)**	**69.29 (3.06)**	**63.41 (7.38)**	**72.42 (4.80)**	**70.83 (7.54)**	**54.25 (5.38)**	**67.92**
Oracle (IID)		88.70 (0.17)			88.73 (0.25)		

† Higher accuracy and lower variance indicate better OOD generalization ability

Table 4.17 OOD generalization performance on complex distribution shifts for real-world graphs

Datasets	Drug-Assay	Drug-Sca	Drug-Size	CMNIST-sp	Graph-SST5	Twitter	Avg (Rank)†
ERM	71.79 (0.27)	68.85 (0.62)	66.70 (1.08)	13.96 (5.48)	43.89 (1.73)	60.81 (2.05)	54.33 (6.00)
ASAP	70.51 (1.93)	66.19 (0.94)	64.12 (0.67)	10.23 (0.51)	44.16 (1.36)	60.68 (2.10)	52.65 (8.33)
GIB	63.01 (1.16)	62.01 (1.41)	55.50 (1.42)	15.40 (3.91)	38.64 (4.52)	48.08 (2.27)	47.11 (10.0)
DIR	68.25 (1.40)	63.91 (1.36)	60.40 (1.42)	15.50 (8.65)	41.12 (1.96)	59.85 (2.98)	51.51 (9.33)
IRM	72.12 (0.49)	68.69 (0.65)	66.54 (0.42)	31.58 (9.52)	43.69 (1.26)	63.50 (1.23)	57.69 (4.50)
V-Rex	72.05 (1.25)	68.92 (0.98)	66.33 (0.74)	10.29 (0.46)	43.28 (0.52)	63.21 (1.57)	54.01 (6.17)
EIIL	72.60 (0.47)	68.45 (0.53)	66.38 (0.66)	30.04 (10.9)	42.98 (1.03)	62.76 (1.72)	57.20 (5.33)
IB-IRM	72.50 (0.49)	68.50 (0.40)	66.64 (0.28)	**39.86 (10.5)**	40.85 (2.08)	61.26 (1.20)	58.27 (5.33)
CNC	72.40 (0.46)	67.24 (0.90)	65.79 (0.80)	12.21 (3.85)	42.78 (1.53)	61.03 (2.49)	53.56 (7.50)
CIGAv1	**72.71 (0.52)**	**69.04 (0.86)**	**67.24 (0.88)**	19.77 (17.1)	**44.71 (1.14)**	**63.66 (0.84)**	**56.19 (2.50)**
CIGAv2	**73.17 (0.39)**	**69.70 (0.27)**	**67.78 (0.76)**	**44.91 (4.31)**	**45.25 (1.27)**	**64.45 (1.99)**	**60.88 (1.00)**
Oracle (IID)	85.56 (1.44)	84.71 (1.60)	85.83 (1.31)	62.13 (0.43)	48.18 (1.00)	64.21 (1.77)	

† Averaged rank is also reported in the blankets because of dataset heterogeneity. Lower rank is better

However, such methods may exhibit limited effectiveness, necessitating more domain-specific solutions [419, 421]. Especially, domains could also be nonstationary and evolve along with certain structures [114, 302, 390, 433]. The task of generalizing under such shifts, known as *evolving domain generalization* (EDG), involves training models on examples from sequential source domains to generalize well to evolving unseen target domains [302, 390].

To tackle EDG, a common belief is to inherit the spirit of domain invariant learning and to learn evolving domain *invariant features* for generalizing to new domains [114]. In contrast, plentiful empirical studies challenged the position and showed that it is also needed to additionally learn the dynamic information which is useful for predicting evolving domain patterns [19, 390]. We consider the problem from the feature learning perspective. Specifically, we build structural causal models to characterize the distribution shifts caused by the underlying invariant and dynamic factors in EDG. Our causal analysis shows that it is essential to learn both dynamic and invariant features for better EDG.

To this end, we propose a principled EDG method called Mutual Information-Based Sequential Autoencoders (MISTS) [420]. Specifically, MISTS adopts a variational inference strategy to identify the underlying invariant and dynamic features. To encourage more complete separation between the invariant and dynamic latent representations, MISTS employs a novel information-theoretic objective that minimizes the mutual information between them. We theoretically show the new objective of MISTS is a valid evidence lower bound (ELBO) of the data log-likelihood with respect to our causal models.

Method Firstly, let us denote the input samples by $\mathbf{x}$ and their corresponding labels by y. Then, we can define the following probabilistic generative model for the input data of all source domains as

$$\begin{aligned} &P(\mathbf{x}_{1:T}, y_{1:T}, z_c, z_{1:T}, \theta_{1:T}) \\ &\quad = P(\mathbf{x}_{1:T}, z_{1:T}, z_c) P(y_{1:T}, \theta_{1:T} | z_{1:T}, z_c), \end{aligned} \tag{4.44}$$

where x in Eq. (4.44) separates the generation of $(\mathbf{x}_{1:T}, y_{1:T})$ into two parts via the decomposition of $z_{1:T}$ into the invariant feature z_c and the dynamic feature z_t. Modeling the first term $P(\mathbf{x}_{1:T}, z_{1:T}, z_c)$ elicits proper disentanglement of z_c and z_t at the latent space. Modeling the second term $P(y_{1:T}, \theta_{1:T} | z_{1:T}, z_c)$ is essentially to leverage the disentangled features to predict the labels. In what follows, we will detail how to model the two terms for better EDG.

To model $P(\mathbf{x}_{1:T}, z_{1:T}, z_c)$ for disentanglement, we can further decompose the first term by the chain rule of probability, as the following:

$$\begin{aligned} P(\mathbf{x}_{1:T}, z_{1:T}, z_c) &= P(z) P(\mathbf{x}_{1:T} | z) \\ &= \left[P(z_c) \prod_{t=1}^{T} P(z_t | z_{<t}) \right] \cdot \prod_{t=1}^{T} P(\mathbf{x}_t | z_t, z_c). \end{aligned} \tag{4.45}$$

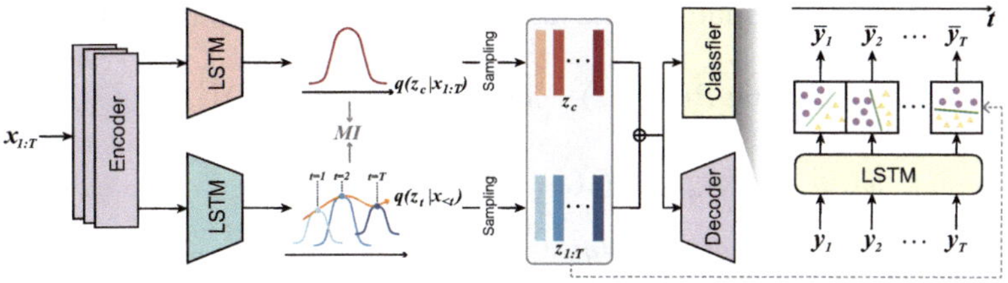

Fig. 4.14 The MISTS framework starts by encoding the input data into the latent space, where two distinct LSTMs parameterize the corresponding posteriors to obtain the invariant and dynamic latent representations, denoted as z_c and $z_{1:T}$, respectively. These representations are then passed through a decoder and an adaptive classifier to compute the reconstruction loss and classification loss, respectively. To encourage better disentanglement, mutual information (MI) terms are applied to the invariant, dynamic latent variables, and input data

Equation (4.45) shows that the generation process of domains data $\mathbf{x}_t$ at timestamp t depends on the corresponding dynamic latent representation z_t and invariant representation z_c. Moreover, the distribution of z_t is conditional on $z_{<t}$ from the history domains. For instance, z_c can represent the facial contour, while z_t denotes the visual attributes which follow fashion trends change over time. The framework is shown in Fig. 4.14.

Our objective here is to extract the latent representations given only the observed data $\mathbf{x}_{1:T}$. Therefore, we hope to learn a posterior distribution as follows:

$$Q(z_{1:T}, z_c|\mathbf{x}_{1:T}) = Q(z_c|\mathbf{x}_{1:T}) \prod_{t=1}^{T} Q(z_t|z_{<t}, \mathbf{x}_t), \tag{4.46}$$

which has a corresponding variational lower bound as follows:

$$\max_{p,q} \mathbb{E}_q \left[\log P(\mathbf{x}_{1:T}|z) - D_{\mathrm{KL}}[Q(z|\mathbf{x}_{1:T})||P(z)]\right]. \tag{4.47}$$

p and q can be parameterized by recurrent neural network, and D_{KL} denotes KL divergence. The first term denotes the reconstruction term for input data $\mathbf{x}_t$, and the second and third terms denote KL divergence which is to align the posterior distributions z_c and z_t with the corresponding prior distributions.

Although forcing the dynamic variable z_t to predict dynamic factors can guarantee that z_t contains adequate dynamic information, it fails to guarantee that z_t excludes the static information z_c. Consequently, such disentanglement results can lead to poor performance in the downstream classification tasks [171, 302]. From the information-theoretic perspective, the problem can be considered as clearly separating z_c and z_t, under the constraint that z_t and z_c should contain useful information from $\mathbf{x}_t$ in a specified domain [17, 139]. Specifically, we propose to additionally minimize the mutual information between z_t and z_c while maximizing the mutual information between the latent representation and the data observed, such that the z_t and z_c carry mutually exclusive meaningful information on input data:

$$\begin{aligned}&\max_{p,q} \mathbb{E}_q \left[\log P(\mathbf{x}_{1:T}|z) - D_{\mathrm{KL}}[Q(z|\mathbf{x}_{1:T})||P(z)]\right] \\ &\text{s.t. } p, q \in \arg\min I_q(z_t; z_c), \\ &\quad\;\; p, q \in \arg\max I_q(z_t; \mathbf{x}_t),\end{aligned} \tag{4.48}$$

where the mutual information terms are defined as

$$\begin{aligned}I_q(z_c; \mathbf{x}_t) &= \mathbb{E}_{Q(z_c, \mathbf{x}_t)} \left[\log Q(z_c|\mathbf{x}_t) - \log Q(z_c)\right], \\ I_q(z_t; \mathbf{x}_t) &= \mathbb{E}_{Q(z_t, \mathbf{x}_t)} \left[\log Q(z_t|\mathbf{x}_t) - \log Q(z_t)\right].\end{aligned} \tag{4.49}$$

$I_q(z_t; z_c)$ is defined similarly.

To address the constrained optimization problem stated in Eq. (4.48), we demonstrate that it can be transformed into the optimization of a novel valid evidence lower bound (ELBO) for the data log-likelihood. We provide a theoretical guarantee for this transformation in our paper [420], ensuring the reliability of the approach. In practice, we follow the spirit of [17, 302] to optimizating the following objective:

$$\begin{aligned}\mathcal{L}_e = \sum_{t=1}^{T} \mathbb{E}_{z_c, z_t}\big[&\log P(\mathbf{x}_t|z_c, z_t) - \alpha(D_{\mathrm{KL}}[Q(z_c|\mathbf{x}_{1:T})||P(z_c)] \\ &+ D_{\mathrm{KL}}[Q(z_t|z_{<t}, \mathbf{x}_t)||P(z_t|P(z_{<t})]) \\ &+ \beta(I_q(z_c; \mathbf{x}_x) + I_q(z_t; \mathbf{x}_t) - I_q(z_t; z_c))\big],\end{aligned} \tag{4.50}$$

To model $P(y_{1:T}, \theta_{1:T}|z_{1:T}, z_c)$ for classification, the classifier is expected to evolve accordingly, e.g., the magnitude of model parameter weights for some features will evolve gradually. We use θ_t to represent the classifier for domain t and consider θ_t as a latent variable in the category space, inspired by Qin et al. [302]. According to Eq. (4.44), the adaptive classifier θ_t can be decomposed as

$$P(y_{1:T}, \theta_{1:T}|z_{1:T}, z_c) = \prod_{t=1}^{T} P(\theta_t|\theta_{<t}) P(y_t|z_t, z_c, \theta_t),$$

where θ_t can be inferred by classifiers from previous domains and predict labels $\hat{y}_t$ given z_c and z_t. To model the dynamic variable θ_t, we parameterize $P(\theta_t|\theta_{<t})$ as a learnable categorical distribution. To optimize the prior distribution $P(\theta_t|\theta_{<t})$, the objective function can also be derived based on ELBO:

$$\begin{aligned}\mathcal{L}_c = \lambda \sum_{t=1}^{T} \mathbb{E}_{Q(\theta_t|\theta_{<t}, y_t)}\big[&\log P(y_t|z_c, z_t, \theta_t) \\ &- \alpha D_{\mathrm{KL}}[(Q(\theta_t|\theta_{<t}, y_t), P(\theta_t|\theta_{<t}))]\big],\end{aligned} \tag{4.51}$$

where $Q(w_t|w_{<t}, y_t)$ denotes the posterior distribution and $P(y_t|z_c, z_t, \theta_t)$ denotes the classification loss. Given training data D_{tr}, our proposed framework can be optimized through the final objective function $\mathcal{L}_D = \mathcal{L}_e + \mathcal{L}_c$.

Experiments To evaluate the effectiveness of MISTS, we conducted experiments on both synthetic and real-world datasets, following the setting of LSSAE [302]. Specifically, we compared our approach with invariant learning methods and the state-of-the-art EDG methods on three synthetic datasets (Circle, Sine, and Rotated MNIST) and three real-world datasets (Portraits, Caltran, and Elec). We also evaluated the results on one additional variant, Sine-C, which was created for EDG settings by Qin et al. [302].

The results of our proposed MISTS and the baseline methods are presented in Table 4.18. The experimental results reveal the significant improvement of EDG (evolving domain generalization) methods over traditional DG (domain generalization) methods. This finding aligns with our theoretical results as well as the empirical observations in the existing EDG literature. Notably, MISTS outperforms other EDG methods in terms of average accuracy across the seven datasets. These results highlight the importance of a clear separation of dynamic and invariant features and incorporating both features for successful EDG.

Beyond distribution shifts caused by evolving structures, complex data distributions can also lead to performance degradation. To address both malign and benign OOD data, the following HOOD method is proposed.

4.2.3.3 Harnessing Out-of-Distribution Examples via Augmenting Content and Style

Here, we consider OOD generalization under various key areas, including OOD detection [150, 215, 229, 246], open-set domain adaptation (DA) [241, 322], and open-set semi-supervised learning (SSL) [164, 169, 220, 323, 441]. In these problems, OOD data can generally be categorized into two types: *benign* OOD data[1] and *malign* OOD data. Benign OOD data could be potentially helpful for enhancing generalization on target distributions through DA techniques [374], but it can also be misleading if not appropriately leveraged. On the other hand, malign OOD data, from unknown classes, can harm classification results and are particularly deceptive and challenging to detect [150, 229]. When we have benign and malign OOD data at the same time, it is essential to distinguish and treat them properly. Existing methods commonly fail in two aspects: (1) aggressive exploitation of malign OOD data and (2) over-rejection of benign OOD data.

To overcome these limitations, we introduce HOOD (Fig. 4.16), which effectively handles and utilizes out-of-distribution (OOD) data across different problems. Our approach follows the structural causal model (SCM) [121, 291] to mimic the

[1] Following [30], augmented data are regarded as a type of OOD data.

Table 4.18 Experimental results (accuracy %) on synthetic and real-world datasets across different methods

Algorithm	Circle	Sine	Sine-C	Elec	RMNIST	Portraits	Caltran	Avg
ERM	49.3 ± 2.1	62.7 ± 1.1	62.3 ± 1.2	70.8 ± 0.6	41.6 ± 0.7	87.9 ± 1.4	61.2 ± 2.3	62.3
IRM	53.6 ± 2.7	62.3 ± 0.9	59.4 ± 0.7	70.5 ± 0.1	40.2 ± 0.4	87.2 ± 1.7	63.8 ± 0.8	62.4
Mixup	48.7 ± 1.5	62.2 ± 0.7	62.0 ± 1.5	70.2 ± 0.9	42.1 ± 0.4	87.8 ± 1.0	66.1 ± 1.0	62.7
LISA	48.4 ± 1.1	61.7 ± 0.5	61.3 ± 1.2	70.1 ± 1.0	41.9 ± 0.8	88.0 ± 0.9	66.3 ± 1.1	62.5
CORAL	51.2 ± 4.2	58.7 ± 1.9	60.2 ± 2.1	70.1 ± 0.7	42.3 ± 0.8	86.1 ± 1.9	65.1 ± 1.5	62.0
GroupDRO	54.3 ± 3.4	59.3 ± 0.2	59.2 ± 4.1	68.6 ± 0.6	42.8 ± 1.1	84.3 ± 1.8	63.6 ± 1.3	61.7
EWC	59.1 ± 3.9	65.4 ± 4.8	64.2 ± 4.3	69.5 ± 1.1	32.2 ± 5.7	88.7 ± 1.2	60.1 ± 3.2	62.7
SI	58.3 ± 2.7	69.7 ± 3.9	63.6 ± 4.7	69.8 ± 1.3	31.4 ± 5.3	88.7 ± 0.8	59.4 ± 3.7	63.0
DRAIN	54.5 ± 4.0	69.5 ± 2.7	**71.0 ± 1.3**	71.1 ± 0.7	44.2 ± 1.1	–	–	–
LSSAE	**63.5 ± 4.7**	68.4 ± 3.5	63.6 ± 3.7	71.0 ± 0.5	45.3 ± 1.4	88.9 ± 1.5	68.8 ± 3.4	67.0
MISTS	62.7 ± 2.7	**78.2 ± 3.3**	70.1 ± 3.2	**71.4 ± 0.8**	**47.5 ± 1.3**	**89.2 ± 1.3**	**70.2 ± 2.1**	**69.9**

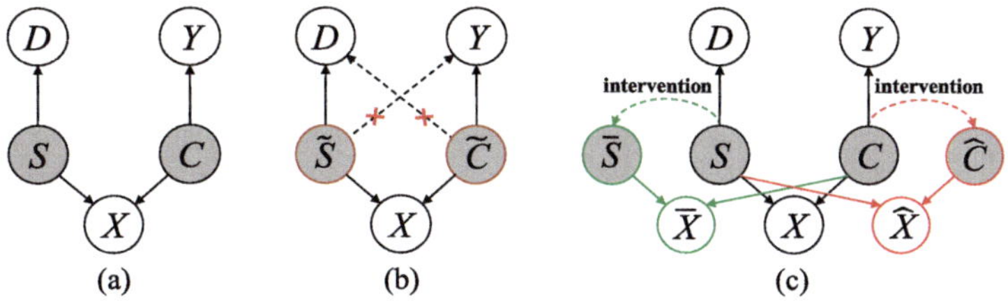

Fig. 4.15 (**a**) Illustration of the data generation. (**b**) Disentanglement of HOOD. The variables $\tilde{C}$ and $\tilde{S}$ approximate the content C and style S, respectively. The dashed lines denote the erroneous causal relations to be broken. (**c**) The augmentation approach of HOOD. The green lines and red lines are related to benign OOD data $\bar{X}$ and malign OOD data $\hat{X}$, respectively

data generation process, as shown in Fig. 4.15a. In HOOD, an image X is composed of two factors: (1) *content* C that denotes the object of interest and (2) *style* S which represents external factors like brightness and orientation. The content C relates to class label Y, while the style variable S reflects the environment D. Intuitively, OOD data with unseen content should not be used for training, as it introduces new semantic variations; meanwhile, OOD data associated with known content but different styles can still help generalization. Our HOOD achieves this by disentangling content and style features and harnessing them effectively.

To achieve this disentanglement, we optimize an approximate evidence lower bound (ELBO) [37] of the joint distribution $P(X, Y, D)$, breaking the spurious correlations [121, 155, 158, 226, 291] between content and style (Fig. 4.15b). Additionally, we perform data augmentation by independently intervening on the style and content, as illustrated in Fig. 4.15c. This helps identify style-altered data $\bar{X}$ as benign OOD data and content-altered data $\hat{X}$ as malign OOD data.

Method Our framework is shown in Fig. 4.16. To encode content information, we utilize the labels and pseudo-labels [213] for class supervision; meanwhile, domain supervision is completed by augmentation approaches that create various styles. As a result, each instance $\mathbf{x}$ is associated with a class label y and a domain label d. Two distinct encoders, g_c and g_s, parameterized by θ_c and θ_s, respectively, are employed to model the posterior distributions $q_{\theta_c}(C \mid X)$ and $q_{\theta_s}(S \mid X)$. The extracted content C and style S are then processed by two classifiers, f_c and f_s, parameterized by ϕ_c and ϕ_s, which generate predictions $q_{\phi_c}(Y \mid C)$ and $q_{\phi_s}(D \mid S)$. To further reinforce the disentanglement of C and S, we employ a decoder h parameterized by ψ.

According to the structural causal model (SCM) in Fig. 4.15a, the joint distribution $P(X, Y, D, C, S)$ can be decomposed as

$$P(X, Y, D, C, S) = P(C, S)P(Y, D \mid C, S)P(X \mid C, S). \tag{4.52}$$

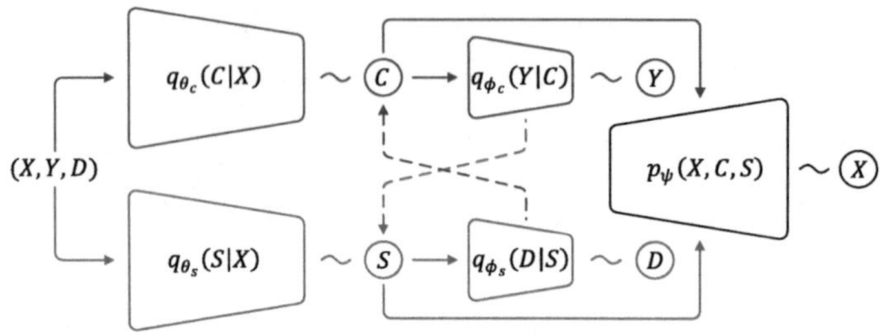

Fig. 4.16 Architecture of the HOOD. The solid lines, dashed lines, and tildes stand for inference flow, disentanglement, and latent approximation, respectively

From the SCM in Fig. 4.15a, it follows that Y and D are conditionally independent given (C, S), i.e., $Y \perp\!\!\!\perp D \mid (C, S)$, which allows us to express $P(Y, D \mid C, S)$ as

$$P(Y, D \mid C, S) = P(Y \mid C, S)P(D \mid C, S). \tag{4.53}$$

Similarly, we assume that C and S are independent, yielding $P(C, S) = P(C)P(S)$. Furthermore, since Y depends only on C and D depends only on S, we can further refine the factorization as

$$P(Y, D \mid C, S) = P(Y \mid C)P(D \mid S). \tag{4.54}$$

Spurious correlations often occur in OOD data, as highlighted by Arjovsky et al. [8]. Consequently, when variational inference relies on the factorization in Eq. (4.52), the approximated representations of content $\tilde{C}$ and style $\tilde{S}$ may both influence Y and D directly. Specifically, this results in dependencies such as $Y \leftarrow \tilde{C} \rightarrow D$ and $Y \leftarrow \tilde{S} \rightarrow D$, leading to inaccurate approximations. To mitigate these correlations, we propose leveraging the posteriors $q_{\phi_s}(D \mid C)$ and $q_{\phi_c}(Y \mid S)$ approximated by the decoders ϕ_s and ϕ_c, respectively, as denominators to $q_{\phi_c}(Y \mid C)$ and $q_{\phi_s}(D \mid S)$. By doing so, we can effectively disentangle content C from style S. Accordingly, the factorization in Eq. (4.52) can be reformulated as

$$\tilde{P}(X, Y, D, C, S) := \frac{P(C)P(S)P(Y \mid C)P(D \mid S)P(X \mid C, S)}{q_{\phi_s}(D \mid C)q_{\phi_c}(Y \mid S)}. \tag{4.55}$$

We then maximize the log-likelihood of the joint distribution $p(\mathbf{x}, y, d)$ for each data point $(\mathbf{x}, y, d)$:

$$\log p(\mathbf{x}, y, d) := \log \int_c \int_s \tilde{p}(\mathbf{x}, y, d, c, s) \mathrm{d}c\mathrm{d}s. \tag{4.56}$$

Here, lowercase letters denote specific values of the corresponding variables. Since directly integrating over the latent variables C and S is intractable, we adopt the variational inference framework [37] to derive an approximate evidence lower bound (ELBO):

$$\log p(\mathbf{x}, y, d) \geq \mathbb{E}_{(c,s)\sim q_\theta(C,S|\mathbf{x})}\left[\log \frac{\tilde{p}(\mathbf{x}, y, d, c, s)}{q_\theta(c, s \mid \mathbf{x})}\right] := \widetilde{ELBO}(\mathbf{x}, y, d). \tag{4.57}$$

Using the modified joint distribution factorization from Eq. (4.55), we obtain:

$$\begin{aligned}
\widetilde{ELBO}(\mathbf{x}, y, d) =& \mathbb{E}_{(c,s)\sim q_\theta(C,S|\mathbf{x})}\left[\log \frac{p(c)p(s)q_{\phi_c}(y \mid c)q_{\phi_s}(d \mid s)p_\psi(\mathbf{x} \mid c, s)}{q_\theta(c, s \mid \mathbf{x})q_{\phi_c}(y \mid s)q_{\phi_s}(d \mid c)}\right] \\
=& - KL(q_{\theta_c}(c \mid \mathbf{x})\|p(C)) - KL(q_{\theta_s}(s \mid \mathbf{x})\|p(S)) \\
&+ \mathbb{E}_{c\sim q_{\theta_c}(C|\mathbf{x})}\left[\log q_{\phi_c}(y \mid c) - \log q_{\phi_s}(d \mid c)\right] \\
&+ \mathbb{E}_{s\sim q_{\theta_s}(S|\mathbf{x})}\left[\log q_{\phi_s}(d \mid s) - \log q_{\phi_c}(y \mid s)\right] \\
&+ \mathbb{E}_{(c,s)\sim q_\theta(C,S|\mathbf{x})}\left[\log p_\psi(\mathbf{x} \mid c, s)\right] \qquad (4.58a) \\
=& ELBO(\mathbf{x}, y, d) - \mathbb{E}_{c\sim q_{\theta_c}(C|\mathbf{x})}\left[\log q_{\phi_s}(d \mid c)\right] \\
&- \mathbb{E}_{s\sim q_{\theta_s}(S|\mathbf{x})}\left[\log q_{\phi_c}(y \mid s)\right]. \qquad (4.58b)
\end{aligned}$$

In Eq. (4.58a), the first two terms stand for Kullback-Leibler (KL) divergence between C and S and prior distributions, respectively. Practically, the priors $p(C)$ and $p(S)$ are normally assumed to follow standard Gaussian distributions. The third and fourth terms approximate the log-likelihoods, further helping the separation of C and S. The final term denotes the reconstruction likelihood of $\mathbf{x}$.

In Eq. (4.58b), our estimated $\widetilde{ELBO}$ consists of two components: the original ELBO from Eq. (4.52) and two extra regularizations. These regularization terms explicitly maximize the log-likelihoods $\log q_{\phi_s}(d \mid c)$ and $\log q_{\phi_c}(y \mid s)$ to encourage content-style disentanglement, as depicted by the dashed lines in Fig. 4.16. By optimizing $\widetilde{ELBO}$, we can ensure that the predictor remains robust and invariant to style changes.

After disentangling content and style, we aim to leverage OOD examples through two contrasting augmentation techniques: *positive data augmentation* and *negative data augmentation*. These methods are designed to generate benign OOD data $\bar{\mathbf{x}}$ and malign OOD data $\hat{\mathbf{x}}$, respectively. To accomplish this, we adopt adversarial data augmentation [123, 271, 379]. For each data point $(\mathbf{x}, y, d)$, the perturbation e is computed by minimizing the intervention objective $\mathcal{L}(\cdot)$:

$$e = \underset{e;\|e\|_p<\epsilon}{\arg\min}\, \mathcal{L}(\mathbf{x} + e, y, d; \theta_c, \phi_c, \theta_s, \phi_s), \tag{4.59}$$

where ϵ denotes the magnitude of the perturbation e with $\mathcal{L}_p$-norm.

Table 4.19 Comparison with OOD detection baselines

OOD dataset	LSUN	DTD	CUB	Flowers	Caltech	Dogs
ID dataset	SVHN					
Likelihood	52.25±0.3	50.33±0.7	48.76±0.6	47.33±0.2	51.54±0.4	54.34±0.4
ODIN	55.72±0.2	53.32±0.5	52.70±0.4	50.47±0.7	56.41±0.4	61.16±0.3
Likelihood ratio	79.34±0.5	78.42±0.3	75.90±0.7	74.53±0.4	76.25±0.3	83.55±0.4
OpenGAN	83.77±0.4	80.36±0.5	77.49±0.8	79.26±0.5	**86.66±0.5**	86.84±0.5
HOOD	**84.10±0.6**	**80.68±0.6**	**79.24±0.5**	**80.93±0.7**	85.34±0.7	**87.58±0.8**
ID dataset	CIFAR-10					
Likelihood	54.32±0.5	52.16±0.4	50.67±0.4	49.26±0.3	53.86±0.4	56.92±0.2
ODIN	58.60±0.3	55.59±0.6	58.48±0.7	51.44±0.9	59.36±0.4	64.82±0.5
Likelihood ratio	81.41±0.6	79.77±0.5	79.35±0.8	77.17±0.7	80.67±0.5	86.76±0.3
OpenGAN	84.03±0.4	81.29±0.8	82.84±1.0	**82.32±0.4**	86.78±0.3	90.14±0.5
HOOD	**86.12±0.6**	**83.64±0.5**	**83.53±0.6**	81.56±0.8	**87.24±0.8**	**90.86±0.6**

For positive data augmentation, we have the following intervention objective:

$$\mathcal{L}_{pos} = \mathcal{L}_d(g_c(\mathbf{x}; \theta_c), g_c(\mathbf{x} + e; \theta_c)) - \mathcal{L}_{ce}(f_s(g_s(\mathbf{x} + e; \theta_s); \phi_s), d). \tag{4.60}$$

For negative data augmentation, the intervention objective is designed as

$$\mathcal{L}_{neg} = \mathcal{L}_d(g_s(\mathbf{x}; \theta_s), g_s(\mathbf{x} + e; \theta_s)) - \mathcal{L}_{ce}(f_c(g_c(\mathbf{x} + e; \theta_c); \phi_c), y). \tag{4.61}$$

To realize the adversarial augmentation, multistep PGD [258] is leveraged. Formally, the optimal $\bar{\mathbf{x}}$ and $\hat{\mathbf{x}}$ can be iteratively found through

$$\bar{\mathbf{x}}^{t+1} = \bar{\mathbf{x}}^t + \underset{e^t; \|e^t\|_p < \epsilon}{\arg\min}\ \mathcal{L}_{pos}(\bar{\mathbf{x}}^t + e^t), \hat{\mathbf{x}}^{t+1} = \hat{\mathbf{x}}^t + \underset{e^t; \|e^t\|_p < \epsilon}{\arg\min}\ \mathcal{L}_{neg}(\hat{\mathbf{x}}^t + e^t). \tag{4.62}$$

where the final iteration t is set to 15 in practice.

Experiments The experimental results for OOD detection, open-set SSL, and open-set DA are presented in Tables 4.19, 4.20, and 4.21, respectively. Compared to baseline methods, the proposed HOOD framework benefits significantly from the incorporation of both benign and malign OOD data. By adapting to benign OOD data and simultaneously detecting malign OOD data, our HOOD improves generalization performance and robustness at the same time. Hence, it outperforms many well-established baselines across most scenarios.

To further validate the disentanglement between content and style, we conduct a cross-prediction experiment using latents from different contents and styles. Specifically, we evaluate four types of input-prediction pairs: content-to-class,

Table 4.20 Comparison with open-set SSL baselines

Training dataset		CIFAR-10			CIFAR-100		
No. of labeled data		50	100	400	50	100	400
Clean Acc.	UASD	72.82±0.9	75.53±1.8	76.74±1.7	58.87±0.6	61.68±1.2	65.97±2.4
	DS3L	74.44±1.3	76.89±1.5	78.80±0.6	60.40±0.5	64.35±1.5	67.65±1.3
	MTCF	79.88±1.3	81.41±1.0	83.92±0.8	62.78±0.5	65.84±2.1	69.46±0.6
	OpenMatch	**84.10±1.1**	**85.30±0.4**	**87.92±1.0**	65.76±0.9	**68.46±0.5**	72.87±1.4
	T2T	82.74±1.2	83.56±1.4	85.97±0.8	65.16±1.2	67.58±0.9	71.96±1.1
	HOOD	83.55±1.2	84.16±1.5	86.22±2.7	**66.39±1.7**	68.03±2.6	**73.32±0.6**
Corrupted Acc.	UASD	39.36±1.2	41.38±0.7	42.66±1.8	31.55±2.0	33.39±1.7	35.20±0.8
	DS3L	39.97±0.8	42.58±0.8	44.39±0.6	33.72±0.8	34.67±0.8	36.64±0.6
	MTCF	40.16±1.2	40.58±1.1	43.33±0.7	32.72±0.8	34.33±2.3	35.53±0.6
	OpenMatch	41.38±0.7	42.90±0.6	45.79±0.8	35.98±1.3	36.47±0.7	38.56±0.6
	T2T	41.39±1.6	45.56±1.6	49.88±1.5	**41.03±1.7**	39.64±0.7	41.38±1.6
	HOOD	**44.42±1.7**	**48.38±0.9**	**50.74±0.6**	40.82±1.5	**41.65±0.9**	**43.72±2.2**

Table 4.21 Comparison with Open-set DA baselines

Dataset	Office						VisDA
Domain	A →W	A →D	D →W	W →D	D →A	W →A	Synthetic →Real
OSBP	86.5±2.0	88.6±1.4	97.0±1.0	97.9±0.9	88.9±2.5	85.8±2.5	62.9±1.3
UAN	87.7±1.2	87.0±0.8	93.5±1.3	97.2±1.6	88.4±0.7	87.8±1.6	63.8±2.4
STA	89.5±0.6	93.7±1.5	97.5±0.2	**99.5±0.2**	89.1±0.5	87.9±0.9	66.4±1.3
HOOD	**90.1±1.5**	**94.2±1.4**	**99.6±0.6**	98.3±0.9	**89.8±0.8**	**91.3±1.8**	**72.4±1.6**

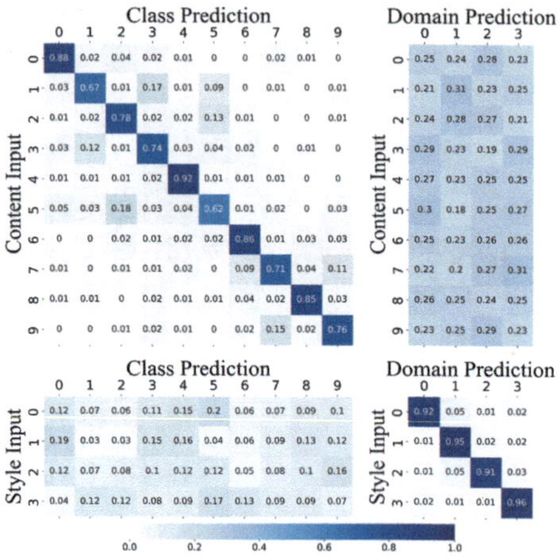

Fig. 4.17 Justification of disentanglement

content-to-domain, style-to-class, and style-to-domain. In Fig. 4.17, we can see that only content features contribute to accurate class predictions, while style features are solely relevant to domain predictions. Furthermore, the class predictor cannot identify style, and the domain predictor cannot identify content, which confirms that HOOD can effectively disentangle content and style.

Chapter 5
Advanced Topics in Trustworthy Machine Learning

As machine learning systems gain significance in practical applications, guaranteeing their reliability across various deployment contexts has become a significant problem. This section examines three essential areas in trustworthy machine learning: federated learning, graph learning, and foundation models (FMs), each offering distinct problems and potential for developing trustworthy machine learning systems.

Federated learning offers a framework for training models on decentralized devices while preserving data privacy. Nonetheless, this distributed nature presents novel vulnerabilities and challenges in maintaining model trustworthiness, including hostile assaults on adversarial attacks on local updates to heterogeneous data distributions across participants. The trustworthiness of federated learning systems requires attention to privacy protection, resilience against malicious participants, and efficacy under data heterogeneity.

Graph learning functions on interconnected data structures and has unique issues preserving trustworthiness. The relational characteristics of graph data create extra attack vectors and vulnerabilities that may undermine model dependability. Moreover, the intricate interdependencies among nodes and edges necessitate specific methodologies to guarantee resilient learning and inference, especially in adversarial perturbations or noisy graph configurations.

FMs recognized as potent general-purpose AI systems present significant hurdles in maintaining trustworthiness at scale. Their enormous capabilities and wide-ranging deployment scenarios require thorough consideration of safety, dependability, and resilience. This encompasses tackling issues associated with model alignment, generalization outside the training distribution, and alleviating inherent biases acquired from extensive training datasets.

This chapter will analyze the core issues and methodological advancements in each domain, thoroughly reviewing contemporary strategies for developing reliable AI systems. Our lecture will include theoretical theories and actual implementations,

B. Han, T. Liu, *Trustworthy Machine Learning under Imperfect Data*,
https://doi.org/10.1007/978-981-96-9396-2_5

providing insights into constructing dependable and resilient machine learning systems across these many paradigms.

5.1 Trustworthy Federated Learning

5.1.1 Introduction and Notation

Federated learning (FL) is a distributed machine learning approach that allows multiple clients to collaboratively train a model without sharing their local data. Instead of centralizing data, FL uses the computational power of edge devices (like smartphones and IoT devices) for local computations, sharing only model updates, which are then aggregated by a central server to improve the global model.

This method enhances data privacy and security since raw data remains on local devices, minimizing the risk of breaches and ensuring compliance with data protection regulations. FL is especially beneficial for sensitive data contexts, such as healthcare and finance, or when data is widely dispersed, making centralized training impractical.

Problem Formulation Formally, the global optimization problem of FL can be formulated as [260] in FL, a set of clients $\mathcal{M} = \{1, 2, \cdots, M\}$ with data distribution $\mathcal{D}_m$ with M being the total number of clients. The goal of FL is to collaboratively learn a machine learning model parameterized by $\theta \in \mathbb{R}^{|\theta|}$ and minimizing the following objective function $\mathcal{L}$:

$$\min_{\theta \in \mathbb{R}^{|\theta|}} \mathcal{L}(\theta) := \sum_{m=1}^{M} \frac{n_m}{N} \mathcal{L}_m(\theta) = \sum_{m=1}^{M} \frac{n_m}{N} \mathbb{E}_{(\mathbf{x},y)\sim\mathcal{D}_m}[\ell(f_\theta(\mathbf{x}), y)], \tag{5.1}$$

where $(\mathbf{x}, y) \in \mathcal{X} \times \mathcal{Y}$, where $\mathbf{x}$ is input in space $\mathcal{X}$ and y is its label. The model is denoted by $f_\theta(\mathbf{x}) : \mathcal{X} \to \mathbb{R}^{\mathcal{Y}}$. $\mathcal{L}_m(\theta) = \mathbb{E}_{(\mathbf{x},y)\sim\mathcal{D}_m}[\ell(f_\theta(\mathbf{x}), y)]$ is the local objective function of client m with $\ell(f_\theta(\mathbf{x}), y) = CE(f_\theta(\mathbf{x}), y)$, CE denotes the cross-entropy loss, $\frac{n_m}{N} > 0$ and $\sum_{m=1}^{M} \frac{n_m}{N} = 1$, where n_m denotes the number of samples on client m and $N = \sum_{m=1}^{M} n_m$. The foundational FL algorithm, FedAvg [260], was designed to reduce the communication frequency and promote more local computations to mitigate some of these challenges. Also, FedAvg communicates the global model to all clients, which avoids directly sharing private data. However, FL encounters two major heterogeneity challenges:

Data Heterogeneity Data heterogeneity refers to the non-Independent and independent distribution (non-IID) of data across different clients. This heterogeneity arises because each client collects data in unique contexts and environments, leading to significant variations in data characteristics such as feature distributions, label distributions, and data quality.

- Feature distribution skew (covariate shift): Each data distribution on clients $P_i(X)$ varies across clients. However, the $P_i(Y|X) = \mathcal{D}(Y|X)$ for all clients. Imagine that different users will write the same words with different styles, like stroke width and slant.
- Label distribution skew (prior probability shift): The $P_i(Y)$ varies across clients. However, the $P_i(X|Y) = \mathcal{D}(X|Y)$ for all clients. For example, different countries have their particular animal species.
- Same label, different features (concept shift): The $P_i(X|Y)$ may be different across clients, even if $P_i(Y)$ is the same.
- Same features, different label (concept shift): The $P_i(Y|X)$ may be different across clients, even if $P_i(X)$ is the same.
- Quantity skew: This is similar to imbalance learning, referring to different clients having different amounts of data.

The client drift induced by non-IID data distributions significantly compromises FedAvg's effectiveness [189, 222, 223, 256, 385]. While various strategies have been proposed to alleviate client drift—ranging from adjustments on both client and server sides—a benchmark study revealed that FedAvg often outperforms its variants, underscoring the complexity of addressing client drift across diverse scenarios [224].

System Heterogeneity System heterogeneity encompasses the variations in computational resources, communication capabilities, and data distributions across participating clients. The main factors contributing to system heterogeneity include:

- Computational Heterogeneity: Varying processing power, memory, and hardware resources among clients.
- Communication Heterogeneity: Different network conditions leading to varying communication delays.
- Data Amount Heterogeneity: Varying amounts of data across clients, resulting in different local training completion times.

To address these challenges, particularly the communication bandwidth limitations, the classical FL algorithm FedAvg proposes utilizing local updates. In each round r, the server:

- Sends the global model θ^{r-1} to a randomly selected subset of clients $\mathcal{S}^r \subseteq \mathcal{M}$
- Receives updated client models $\{\theta_m^r\}$ after local computations
- Aggregates local models based on client dataset sizes to obtain a new global model θ^r

The following sections will introduce the advanced techniques of current FL that address data and system heterogeneity.

5.1.2 Reducing Gradient Dissimilarity

Recent studies highlight a connection between client drift and the variability of gradients among clients [310, 402]. Gradient correction methods [189, 385] only marginally bridge the performance gap between federated and centralized training.

5.1.2.1 Improving FL Client Update by Feature Estimation

This study [362] presents a novel approach to refining gradient and parameter updates in FL by conceptualizing client-based local training as a problem of enhancing the "generalization impact" of the global model. The proposed framework assesses how training on individual clients can improve the server model's ability to generalize across heterogeneous client data distributions. Our analysis reveals that aligning conditional distributions among clients maximizes the efficacy of FL training.

In light of privacy constraints that restrict centralized aggregation of client data, the authors propose a decoupled architecture separating deep neural networks into distinct feature extracter and classifier. This strategy promotes alignment in the feature representation space while preserving data confidentiality. By approximating and exchanging feature distribution information across clients, the method seeks to amplify the generalization benefits of local training and mitigate gradient divergence, thereby advancing the robustness of federated learning systems.

We follow in the notations and definitions in Sect. 5.1.1, the global optimization problem of FL is formulated as [260]

$$\min_{\theta \in \mathbb{R}^{|\theta|}} \mathcal{L}(\theta) := \sum_{m=1}^{M} \frac{n_m}{N} \mathcal{L}_m(\theta) = \sum_{m=1}^{M} \frac{n_m}{N} \mathbb{E}_{(\mathbf{x},y)\sim\mathcal{D}_m}[\ell(f_\theta(\mathbf{x}), y)]. \tag{5.2}$$

To defend against limited communication bandwidth in distributed learning systems, the conventional federated learning framework FedAvg [260] employs client-side model refinement. During each communication round r, the central server distributes the current global model θ^{r-1} to a randomly selected group of participants $\mathcal{S}^r \subseteq \mathcal{M}$. These clients iteratively optimize their local models, generating updated parameters $\{\theta_m^r\}$, which are subsequently transmitted back to the server. The global model θ^r is then reconstructed through a weighted aggregation of client parameters, where weighting coefficients correspond to relative sizes of local datasets.

We propose evaluation criteria for model generalization during inference. Drawing on margin-based analysis [100, 199], we adopt a worst-case margin metric[1] to quantify model robustness across data distributions:

[1] This concept aligns with prior margin definitions [112].

Definition 5.1 (Worst-Case Margin) For a given data distribution $\mathcal{D}$, the worst-case margin for model $f_\theta(\cdot)$ is defined as $W_d(f_\theta, \mathcal{D}) = \mathbb{E}_{(\mathbf{x},y)\sim\mathcal{D}}[\inf\{d(\mathbf{x}', \mathbf{x}) : \arg\max_i f_\theta(\mathbf{x}')_i \neq y\}]$, where d denotes a predefined distance metric. The $\text{argmax}_i f_\theta(\mathbf{x}')_i \neq y$ means the $f_\theta(\mathbf{x}')$ misclassifies the $\mathbf{x}'$.

This metric evaluates the average minimum perturbation required to induce errors across the distribution. Reduced margin values correlate with higher vulnerability to misclassification, enabling quantitative assessment of model generalizability. A zero margin implies near-universal failure on the target distribution.

FL inherently prioritizes generalization across heterogeneous clients. Under the premise that broader client participation enhances model quality, we formalize a client's generalization contribution as

$$\mathbb{E}_{\Delta:\mathbf{L}(\mathcal{D}_m)}[W_d(f_{\theta+\Delta}, \mathcal{D}\backslash\mathcal{D}_m)], \tag{5.3}$$

where Δ represents simulated parameter updates[2] from applying learning algorithm $\mathbf{L}(\cdot)$ to client data $\mathcal{D}_m$. The term $\mathcal{D}\backslash\mathcal{D}_m$ denotes the aggregated data distribution excluding client m. This formulation quantifies how local training on $\mathcal{D}_m$ influences global performance on unseen data, prioritizing clients whose contributions maintain provable generalization bounds.

Definition 5.2 The conditional Wasserstein distance $C_d(\mathcal{D}, \mathcal{D}')$ between the distribution $\mathcal{D}$ and $\mathcal{D}'$:

$$\begin{aligned} C_d(\mathcal{D}, \mathcal{D}') =& \frac{1}{2}\mathbb{E}_{(\cdot,Y)\sim\mathcal{D}}[\inf_{P'\in P(\mathcal{D},\mathcal{D}'|Y)} \mathbb{E}_{(\mathbf{x},\mathbf{x}')\sim P'}[d(\mathbf{x}, \mathbf{x}')]] \\ &+ \frac{1}{2}\mathbb{E}_{(\cdot,Y)\sim\mathcal{D}'}[\inf_{P'\in P(\mathcal{D},\mathcal{D}'|Y)} \mathbb{E}_{(\mathbf{x},\mathbf{x}')\sim P'}[d(\mathbf{x}, \mathbf{x}')]]. \end{aligned}$$

Theorem 5.1 *With Δ obtained by $\mathbf{L}(\mathcal{D}_m)$, we have bounded generalization contribution as*

$$\begin{aligned} &\mathbb{E}_{\Delta:\mathbf{L}(\mathcal{D}_m)}[W_d(f_{\theta+\Delta}, \mathcal{D}\backslash\mathcal{D}_m)] \\ \geq &\mathbb{E}_{\Delta:\mathbf{L}(\mathcal{D}_m)}[W_d(f_{\theta+\Delta}, \tilde{D}_m)] - \big|\mathbb{E}_{\Delta:\mathbf{L}(\mathcal{D}_m)}[W_d(f_{\theta+\Delta}, \mathcal{D}_m) \\ &- W_d(f_{\theta+\Delta}, \tilde{D}_m)]\big| - 2C_d(\mathcal{D}_m, \mathcal{D}\backslash\mathcal{D}_m), \end{aligned}$$

where $\tilde{D}_m$ represents the dataset sampled from $\mathcal{D}_m$.

Gradient dissimilarity FL emerges from the heterogeneous nature of data, where the data distribution $\mathcal{D}_m$ for client m differs from $\mathcal{D}_k$ of client k [189].

[2] Computed per round r as $\Delta^r = \theta_T^{r-1} - \theta_0^{r-1}$, with T indicating total local training steps.

Definition 5.3 Inter-client gradient variance (CGV): [187, 189, 402]

$$\mathrm{CGV}(\mathcal{L}, \theta) = \mathbb{E}_{(\mathbf{x},y)\sim\mathcal{D}_m} ||\nabla \ell(f_\theta(\mathbf{x}), y) - \nabla \mathcal{L}(\theta)||^2.$$

It is typically assumed that CGV has an upper limit [187, 402], expressed as $\mathrm{CGV}(\mathcal{L}, \theta) \leq \sigma^2$ for a constant σ.

With representing $\nabla \ell(f_\theta(\mathbf{x}), y)$ as $\left\{\nabla_{\theta_{\text{low}}} \ell(f_\theta(\mathbf{x}), y), \nabla_{\theta_{\text{high}}} \ell(f_\theta(\mathbf{x}), y)\right\}$, the CGV can be splitted as

$$\begin{aligned}\mathrm{CGV}(\mathcal{L}, \theta) =& \mathbb{E}_{(\mathbf{x},y)\sim\mathcal{D}_m} [||\nabla \ell(f_\theta(\mathbf{x}), y) - \nabla \mathcal{L}(\theta)||^2] \\ =& \mathbb{E}_{(\mathbf{x},y)\sim\mathcal{D}_m} [||\nabla_{\theta_{\text{low}}} \ell(f_\theta(\mathbf{x}), y) - \nabla_{\theta_{\text{low}}} \mathcal{L}(\theta)||^2 \\ &+ ||\nabla_{\theta_{\text{high}}} \ell(f_\theta(\mathbf{x}), y) - \nabla_{\theta_{\text{high}}} \mathcal{L}(\theta)||^2].\end{aligned}$$

Method The generalization performance of federated models can be improved by sharing and estimating feature distributions. Thus, clients can train their local models on analogous distributions while maintaining privacy. A key challenge is accurately estimating these distributions while keeping communication and computation costs in check.

To tackle these challenges, we propose a decoupling strategy that delineates the processes of feature extraction and classification in deep neural networks. Figure 5.1 illustrates this framework, where the feature extractor network, defined by $\varphi_{\theta_{\text{low}}}(\cdot)$ and parameterized by θ_{low}, produces hidden features. We write the feature distribution $\mathcal{H}|y$ as $\mathcal{H}$ for brevity in rest of the paper. Subsequently, the classifier network, parameterized by θ_{high}, is trained using these features along with shared feature distributions. This separation mitigates gradient dissimilarity, which arises from the varied data distributions among clients.

Specifically, client m estimates its own hidden feature distribution as $\mathcal{H}_m$ using the local hidden features $h = \varphi_{\theta_{\text{low}}}(\mathbf{x})|_{(\mathbf{x},y)\sim\mathcal{D}_m}$ and sends $\mathcal{H}_m$ to the server for the global distribution approximation. Then, the server aggregates the received distributions to obtain the global feature distribution $\mathcal{H}$ and broadcasts it, being

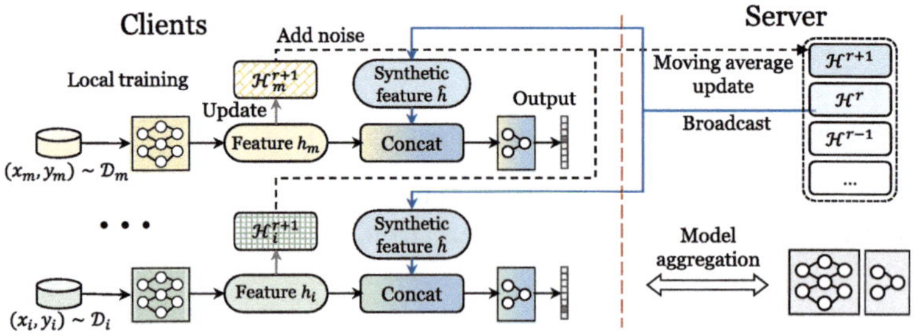

Fig. 5.1 Training process of our framework. h_m represents the hidden features of client m and $\mathcal{H}_m^r$ the distribution of h_m at the round r

similar to the model average in the FedAvg. Finally, classifier networks of all clients thus perform local training on both the local hidden features $h_{(\mathbf{x},y)\sim\mathcal{D}_m}$ and the shared $\mathcal{H}$ during the local training.

With shared $\mathcal{H}$, client m will sample $\hat{h} \sim \mathcal{H}$ and $h_m = \varphi_{\theta_{\text{low}}}(\mathbf{x})|_{(\mathbf{x},y)\sim\mathcal{D}_m}$ to train their classifier network, and then, the objective function becomes as [3]

$$\min_{\theta\in\mathbb{R}^d} \hat{\mathcal{L}}(\theta) \triangleq \sum_{m=1}^{M} \frac{n_m + \hat{n}_m}{n_m + \hat{n}_m} \mathbb{E}_{\substack{(\mathbf{x},y)\sim\mathcal{D}_m \\ \hat{h}\sim\mathcal{H}}} \left[\ell(f_\theta(\mathbf{x}), y) + \ell(f_\theta(\hat{h}), y) \right], \tag{5.4}$$

where $\hat{n}_m$ represents the size of sampling $\hat{h}$ of client m.

Theorem 5.2 *With Definition 5.3 and $\hat{n}_m$ satisfying $\frac{\hat{n}_m}{n_m+\hat{n}_m} = \frac{\hat{N}}{N+\hat{N}}$, the objective function $\hat{\mathcal{L}}(\theta)$ leads to a tighter bounded gradient dissimilarity, i.e.,*

$$\begin{aligned} CGV(\hat{\mathcal{L}}, \theta) =& \mathbb{E}_{(\mathbf{x},y)\sim\mathcal{D}_m} [||\nabla_{\theta_{\text{low}}} \ell(f_\theta(\mathbf{x}), y) - \nabla_{\theta_{\text{low}}} \mathcal{L}(\theta)||^2 \\ &+ \frac{N^2}{(N+\hat{N})^2} ||\nabla_{\theta_{\text{high}}} \ell(f_\theta(\mathbf{x}), y) - \nabla_{\theta_{\text{high}}} \mathcal{L}(\theta)||^2] \le CGV(\mathcal{L}, \theta). \end{aligned}$$

The training procedure in FedImpro is straightforward and can be integrated into existing FL frameworks with minimal adjustments. By sampling hidden features from the shared global distribution, clients can reduce the dissimilarity in gradients, leading to improved convergence rates and overall model performance.

Experiments We conducted experiments using four widely adopted datasets in the FL community: CIFAR-10, FMNIST, SVHN, and CIFAR-100. We trained ResNet-18 on CIFAR-10, FMNIST, and SVHN, and ResNet-50 on CIFAR-100, with different configurations of local epochs and client numbers.

Table 5.1 presents the best test accuracy achieved by various FL algorithms, including our proposed `FedImpro`. Notably, `FedImpro` consistently outperformed other methods across different datasets and non-IID settings. For instance, on CIFAR-10 with a non-IID degree of 0.1, `FedImpro` achieved an accuracy of 88.45%, significantly higher than FedAvg's 83.65%.

Different from feature perspective, FedGAMMA proposed to improve FL through the sharpness aware minimization. FedGAMMA seeks a global flat landscape with high performance.

5.1.2.2 FL with Global Sharpness Aware Minimization

Federated learning (FL) allows decentralized clients to train models jointly while maintaining data privacy. However, a key challenge in this framework is client

[3] We reuse f here for brevity; the input of f can be the input $\mathbf{x}$ or the hidden feature $h = \varphi_{\theta_{\text{low}}}(\mathbf{x})$.

Table 5.1 Best test accuracy (%) of all experimental results. "Cent." means centralized training; "Acc." test accuracy

Dataset	Cent. Acc.	FL setting α	E	M	FL test accuracy FedAvg	FedProx	SCAFFOLD	FedNova	FedDyn	FedImpro
CIFAR-10	92.53	0.1	1	10	83.65±2.03	83.22±1.52	82.33±1.13	84.97±0.87	84.73±0.81	**88.45±0.43**
		0.05	1	10	75.36±1.92	77.49±1.24	33.6±3.87	73.49±1.42	77.21±1.25	**81.75±1.03**
		0.1	5	10	85.69±0.57	85.33±0.39	84.4±0.41	86.92±0.28	86.59±0.39	**88.10±0.20**
		0.1	1	100	73.42±1.19	68.59±1.03	59.22±3.11	74.94±0.98	75.29±0.95	**77.56±1.02**
		Average			79.53	78.66	64.89	80.08	80.96	**83.97**
FMNIST	93.7	0.1	1	10	88.67±0.34	88.92±0.25	87.81±0.36	87.97±0.41	89.01±0.25	**90.83±0.19**
		0.05	1	10	82.73±0.98	83.66±0.82	76.16±1.29	81.89±0.91	83.20±1.19	**86.42±0.71**
		0.1	5	10	87.6±0.52	88.41±0.38	88.44±0.29	87.66±0.62	88.50±0.52	**89.87±0.21**
		0.1	1	100	90.12±0.19	90.39±0.12	88.24±0.31	90.40±0.18	90.57±0.21	**90.98±0.15**
		Average			87.28	87.85	85.16	86.98	87.82	**89.53**
SVHN	95.27	0.1	1	10	88.20±1.21	87.04±0.89	83.87±2.15	88.48±1.31	90.82±1.09	**92.37±0.82**
		0.05	1	10	80.67±1.92	82.39±1.35	82.29±1.81	84.01±1.48	84.12±1.28	**90.25±0.69**
		0.1	5	10	86.32±1.19	86.05±0.72	83.14±1.51	88.10±0.91	89.92±0.50	**91.58±0.72**
		0.1	1	100	92.42±0.21	92.29±0.18	92.06±0.20	92.44±0.31	92.82±0.13	**93.42±0.15**
		Average			86.90	86.94	85.34	88.26	89.42	**91.91**
CIFAR-100	74.25	0.1	1	10	69.38±1.02	69.78±0.91	65.74±1.52	69.52±0.75	69.59±0.52	**70.28±0.33**
		0.05	1	10	63.80±1.29	64.75±1.25	61.49±2.16	64.57±1.27	64.90±0.69	**66.60±0.91**
		0.1	5	10	68.39±1.02	68.71±0.88	68.67±1.20	67.99±1.04	68.52±0.41	**68.79±0.52**
		0.1	1	100	53.22±1.20	54.10±1.32	23.77±3.21	55.40±0.81	54.82±0.81	**56.07±0.75**
		Average			63.70	64.34	54.92	64.37	64.46	**65.44**

drift, caused by non-IID data distributions, iterative local updates, and partial client participation. These factors contribute to inconsistencies between local and global model updates. Many existing methods predominantly leverage empirical risk minimization (ERM) but fail to consider how the global loss landscape influences generalization. To mitigate this issue, we introduce FedGAMMA, an innovative FL algorithm that leverages global sharpness aware minimization to encourage a smoother loss landscape, thereby enhancing model effectiveness.

Problem Definition Utilizing the relation between the loss landscape of global model and its generalization ability, SAM [111] is introduced in FL [47, 305] which results in a novel (SAM-FL) formulation:

$$\begin{aligned} &\min_{\theta \in \mathbb{R}^d} L_{\text{SAM}}(\theta), \\ &\text{where } L_{\text{SAM}}(\theta) \triangleq \max_{\|\epsilon\|_p \leq \gamma} F(\theta + \epsilon), \\ &\text{and } F(\theta + \epsilon) \triangleq \sum_{m=1}^{M} \frac{n_m}{N} \mathcal{L}_i(\theta + \epsilon), \end{aligned} \tag{5.5}$$

where $\gamma \geq 0$ is a predefined constant controlling the radius of the perturbation, $||\cdot||_p$ is p-norm, which is often a l_2-norm, and $\mathcal{L}_i(\cdot)$ is the empirical risk associated with finite local training samples for client i. In order to minimize $L_{\text{SAM}}(\theta)$ in SAM-FL, the inner maximization should be approximated efficaciously, which in turn enables applying SGD directly to the outside SAM-FL objective. Following the same step of centralized SAM optimizer [111], the general inner maximization term can be estimated with $\epsilon^*(\theta)$ as follows:

$$\epsilon^*(\theta) \approx \gamma \text{sign}\left(\frac{1}{M}\sum_{i=1}^{M} \nabla \ell_i(\theta)\right) \frac{\frac{1}{M}\sum_{i=1}^{M} \nabla \ell_i(\theta)}{||\frac{1}{M}\sum_{i=1}^{M} \nabla \ell_i(\theta)||}, \tag{5.6}$$

where sign$(\cdot)$ denotes element-wise signum function. After getting the approximation of the inner maximization, $F(\theta + \epsilon^*(\theta))$ can be solved via efficient SGD-type methods.

Method To more accurately estimate $\epsilon^*(\theta)$ for the global model at each optimization step and mitigate the client-drift issue arising from non-IID data, multiple local updates, and partial participation, we draw inspiration from variance reduction techniques [189] to approximate the true global gradient while enhancing global flatness with reduced bias. Our approach provides a more precise solution to (5.6) when the model is trained effectively.

In each communication round, we maintain a stateful control variance vector c at the server and an individual counterpart c_i for each client. To closely emulate sharpness aware minimization on the global model, a key step is accurately estimating $\frac{1}{M}\sum_{i=1}^{M} \nabla \ell_i(\theta+\epsilon)$. To achieve this, we introduce a periodic signposting

Table 5.2 Train and test accuracies on CIFAR-10 (1000 communication rounds). *Italics* and **bold** denote the best in train or test

	ResNet-18 (BN based)						ResNet-18 (GN based)					
	I.I.D		Dir(0.6)		Dir(0.1)		I.I.D		Dir(0.6)		Dir(0.1)	
Algorithm	Train	Test	Train	Test	Train	Test	Train	Test	Train	Test	Train	Test
FedAvg	99.90	91.11	98.95	90.47	92.21	84.27	99.48	89.60	96.82	87.41	84.39	77.00
SCAFFOLD	99.99	94.23	99.99	93.86	97.71	90.71	99.99	91.99	99.98	91.05	91.44	83.12
FedCM	98.83	92.31	94.67	89.86	93.41	87.33	97.85	91.45	97.31	91.36	93.95	82.86
FedSAM	99.97	94.17	99.72	93.56	95.25	89.27	99.75	91.19	98.02	89.07	85.21	78.24
FedGAMMA	99.94	**95.56**	99.83	**95.06**	97.28	**92.04**	99.98	**93.42**	99.86	**91.93**	93.54	**85.91**

Table 5.3 Train and test accuracies on CIFAR-100 (1000 communication rounds). *Italics* and **bold** denote the best in train or test

	ResNet-18 (BN based)						ResNet-18 (GN based)					
	I.I.D		Dir(0.6)		Dir(0.1)		I.I.D		Dir(0.6)		Dir(0.1)	
Algorithm	Train	Test	Train	Test	Train	Test	Train	Test	Train	Test	Train	Test
FedAvg	89.79	64.76	89.26	64.63	81.80	57.18	78.53	58.24	77.61	57.47	66.24	48.03
SCAFFOLD	*99.97*	76.36	*99.97*	75.43	*99.69*	73.89	*99.97*	67.06	*99.96*	66.45	*98.27*	59.06
FedCM	95.93	73.69	93.71	70.45	83.34	66.73	88.46	66.79	94.57	65.86	69.33	54.15
FedSAM	99.93	75.63	99.69	75.53	97.71	71.11	99.64	60.98	98.32	59.34	88.41	55.06
FedGAMMA	99.94	**79.57**	99.90	**79.33**	98.62	**75.72**	99.96	**70.54**	98.22	**70.49**	86.49	**62.20**

mechanism, where c serves as a global directional estimate from the previous round. This allows local updates to be adjusted based on the periodic signposting and each client's recorded last update c_i in subsequent rounds. By integrating global information into each local training process, our method ensures a stronger global influence. This adjustment enables each client's local optimizer to efficiently explore flatter regions in the global model, naturally aligning local ϵ with global perturbations ϵ. Consequently, the discrepancy between local updates and the actual global update in each communication round is reduced, strengthening the robustness of the global sharpness aware minimization process.

Experiments In Tables 5.2 and 5.3, we present a comparative analysis of training and testing accuracy across multiple baselines (FedAvg, SCAFFOLD, FedCM, FedSAM) and our proposed FedGAMMA algorithm. Evaluations are conducted on two datasets (CIFAR-10, CIFAR-100), three partition strategies (I.I.D, Dir(0.6), Dir(0.1)), and two network architectures (BN-based ResNet-18, GN-based ResNet-18). As expected, FedAvg yields the lowest performance due to its naive averaging of selected client updates. SCAFFOLD, incorporating variance reduction techniques, demonstrates some effectiveness in addressing non-IID challenges. FedCM, leveraging a client-momentum approach, exhibits varying performance across different settings, as it requires careful tuning of hyperparameters such as the global learning rate and alpha. Although SCAFFOLD and FedCM show improvements over FedAvg by mitigating client drift, their primary focus remains on ERM-FL.

On the other hand, FedSAM shares conceptual similarities with our approach by considering the relationship between the loss landscape and generalization ability. However, it applies sharpness aware minimization independently at the client level, leading to biased estimates of ϵ in the SAM-FL formulation (5.5) due to client drift. In contrast, our proposed FedGAMMA achieves superior performance by enabling each client to simulate global flatness optimization directly, effectively addressing the client-drift problem and solving SAM-FL in a more principled manner.

5.1.3 Exploiting Virtual Data

Different from mitigating client drift and the variability of gradients among clients, another perspective to alleviate the influences of data heterogeneity is to consider modifying training datasets. As a promising direction to directly tackle the root cause of client drift—data heterogeneity is the incorporation of external or virtual data.

5.1.3.1 Virtual Homogeneity Learning with Virtual Data

Maintaining different types of data in collaborative learning through sharing user-defined data is considered an effective strategy [256, 479]. However, recent studies have shown that there is a trade-off between protecting privacy and improving performance [358]. Most traditional methods rely on detecting outliers to account for distribution differences or using a process called distillation to refine the sample.

A recent research framework addresses these differences by using synthetic datasets containing common and unrelated elements instead of traditional datasets. This technology helps avoid privacy risks and streamline workflows.

The new method is based on phase transition theory and allows to align object representations between virtual objects and virtual objects. Such optimization not only improves the accuracy of the model but also increases the training speed. Experimental results show that our framework improves the model performance in different learning scenarios. This allows large problems to be solved in a distributed manner. This study explored the following key questions. Can research institutions use shared datasets to reduce data variation and preserve privacy?

To solve this problem, virtual homogeneity theory (VHL) [358] provides a unique set of parameters. This asset pool is separate from client assets, reducing changes in asset allocation. VHL organizes object space by connecting virtual and real objects using planning techniques. This makes the model more robust and efficient. Experiments at different levels show that VHL can improve learning speed and processing power in different ways.

Building on the notations in Sect. 5.1.1, VHL operates at both data and feature levels; traditional solutions distribute consistent datasets to clients, assuming that the shared data reflects the real-world distribution. However, locality is often

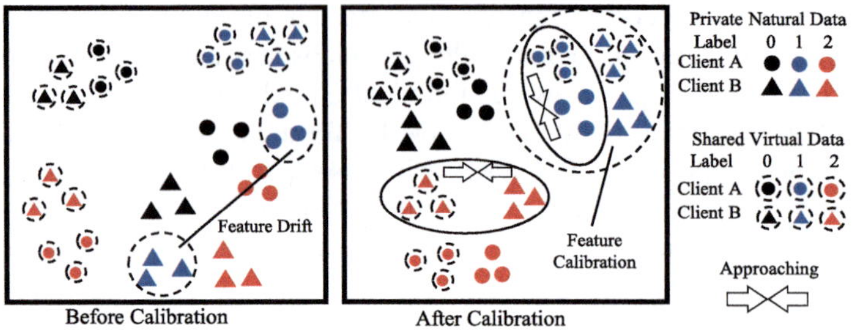

Fig. 5.2 Feature distribution of different clients

compromised by privacy constraints. VHL solves this problem by coordinating feature distributions through a calibration mechanism (see Fig. 5.2). Synthetic datasets no longer rely on natural data but allow clients to fine-tune local models without exposing sensitive information.

VHL offers a set of candidate regularizers with different, ineffective, or modified GAN structures than other candidates for [190].

By explaining the differences, VHL demonstrates the differences in the distribution methods to optimize for new users. Each client's objective function minimizes

$$\begin{aligned}&\mathbb{E}_{(\mathbf{x},y)\sim P_k}[\ell\left(f\circ\phi\left(\mathbf{x}\right),y\right)]+\mathbb{E}_{(\mathbf{x},y)\sim P_v}[\ell\left(f\circ\phi\left(\mathbf{x}\right),y\right)]\\&\quad+\lambda\mathbb{E}_y[d\left(P_k\left(\phi\left(\mathbf{x}\right)|y\right),P_v\left(\phi\left(\mathbf{x}\right)|y\right)\right)],\end{aligned}\tag{5.7}$$

where P_v represents a virtual distribution (e.g., Gaussian), ϕ denotes the feature extractor, and $P_k(\phi(\mathbf{x})|y)$ and $P_v(\phi(\mathbf{x})|y)$ correspond to label-conditioned feature distributions of natural and synthetic data, respectively. The divergence metric $d(\cdot)$ ensures ϕ produces domain-invariant features by penalizing mismatches between P_k and P_v.

Theoretical analysis links VHL's penalty to a bounded statistical margin, guaranteeing robust generalization. A derived theorem establishes that minimizing this penalty constrains the margin, thereby reducing overfitting risks.

CIFAR-10 under specified non-IID rules using latent Dirichlet distribution (LDA), FMNIST, SVHN and CIFAR-100 analysis were performed on the ResNet model. FedAvg, FedProx, STEIER, and FedNova show the superiority of VHL in terms of compilation speed and accuracy (Tables 5.4, 5.5). The results show improved performance, especially in the effectiveness of communication.

Furthermore, the distribution of real data improves the robustness of FL by maintaining a realistic model in contrast to the limitations of the aggregated data. The methods of analyzing data to resolve mismatches are described in the following sections.

Table 5.4 Experiment results with/without VHL. The left is the results of CIFAR-10, while the right is the results of FMNIST. "ROUND" means the communication rounds that need to attain the accuracy

	ACC ↑	ROUND ↓	Speedup ↑
	w/ (w/o) VHL		
Centralized training ACC = **92.88%** (92.53%)			
$\alpha = 0.1$, $E = 1$, $M = 10$ (Target ACC = 79%)			
FedAvg	**87.82** (79.98)	**128** (287)	×**2.2** (× 1.0)
FedProx	**87.30** (83.56)	**128** (188)	×**2.2** (× 1.5)
SCAFFOLD	**84.87** (83.58)	**90** (291)	×**3.2** (× 1.0)
FedNova	**87.56** (81.35)	**128** (351)	×**2.2** (× 0.8)
$\alpha = 0.05$, $E = 1$, $M = 10$ (Target ACC = 69%)			
FedAvg	**79.23** (69.02)	**112** (411)	×**3.7** (× 1.0)
FedProx	**80.84** (78.66)	**151** (201)	×**2.7** (× 2.0)
SCAFFOLD	**55.73** (38.55)	Nan (Nan)	**Nan** (Nan)
FedNova	**80.59** (64.78)	**247** (Nan)	×**1.66** (Nan)
$\alpha = 0.1$, $E = 5$, $M = 10$ (Target ACC = 84%)			
FedAvg	**89.93** (84.79)	**91** (255)	×**2.8** (× 1.0)
FedProx	**86.41** (82,18)	**255** (Nan)	×**1.0** (Nan)
SCAFFOLD	**87.27** (86.20)	**45** (66)	×**5.7** (× 2.0)
FedNova	**90.24** (86.09)	**67** (127)	×**3.8** (× 1.0)
$\alpha = 0.1$, $E = 1$, $M = 100$ (Target ACC = 49%)			
FedAvg	**70.20** (49.61)	**385** (957)	×**2.5** (× 1.0)
FedProx	**73.90** (49.97)	**325** (842)	×**2.9** (× 1.1)
SCAFFOLD	**59.66** (52.24)	**479** (664)	×**2.0** (× 1.4)
FedNova	**61.59** (46.53)	**554** (Nan)	×**1.7** (Nan)

	ACC ↑	ROUND ↓	Speedup ↑
	w/ (w/o) VHL		
Centralized training ACC = **93.80%** (93.70%)			
$\alpha = 0.1$, $E = 1$, $M = 10$ (Target ACC = 86%)			
FedAvg	**92.05** (86.81)	**52** (119)	×**2.3** (× 1.0)
FedProx	**90.68** (87.12)	**31** (135)	×**3.8** (× 0.9)
SCAFFOLD	**90.27** (86.21)	**14** (143)	×**8.5** (× 0.8)
FedNova	**91.88** (86.99)	**52** (83)	×**2.3** (× 1.4)
$\alpha = 0.05$, $E = 1$, $M = 10$ (Target ACC = 78%)			
FedAvg	**89.06** (78.57)	**53** (425)	×**8.0** (× 1.0)
FedProx	**87.76** (81.96)	**30** (41)	×**14.2** (× 10.4)
SCAFFOLD	**80.68** (76.08)	**58** (Nan)	×**7.3** (Nan)
FedNova	**87.25** (79.06)	**30** (538)	×**14.2** (× 0.8)
$\alpha = 0.1$, $E = 5$, $M = 10$ (Target ACC = 87%)			
FedAvg	**91.52** (87.45)	**51** (278)	×**5.5** (× 1.0)
FedProx	**88.27** (86.07)	**74** (Nan)	×**3.8** (Nan)
SCAFFOLD	**91.82** (87.10)	**20** (105)	×**13.9** (× 2.7)
FedNova	**91.86** (87.53)	**51** (193)	×**5.5** (× 1.4)
$\alpha = 0.1$, $E = 1$, $M = 100$ (Target ACC = 90%)			
FedAvg	**91.14** (90.11)	**436** (658)	×**1.5** (× 1.0)
FedProx	**91.37** (90.71)	**283** (491)	×**2.3** (× 1.3)
SCAFFOLD	**87.91** (85.99)	Nan (Nan)	Nan (Nan)
FedNova	**88.34** (87.09)	Nan (Nan)	Nan (Nan)

Table 5.5 Experiment results with/without VHL. The left is the results of SVHN, while the right is the results of CIFAR-100

	ACC ↑	ROUND ↓	Speedup ↑
	w/ (w/o) VHL		
Centralized training ACC = 95.01% (**95.27%**)			
$\alpha = 0.1$, $E = 1$, $M = 10$ (Target ACC = 88%)			
FedAvg	**93.49** (88.56)	**75** (251)	×**3.3** (× 1.0)
FedProx	**91.70** (86.51)	**271** (Nan)	×**0.9** (Nan)
SCAFFOLD	**87.54** (80.61)	Nan (Nan)	Nan (Nan)
FedNova	**93.35** (89.12)	**75** (251)	×**3.3** (× 1.0)
$\alpha = 0.05$, $E = 1$, $M = 10$ (Target ACC = 82%)			
FedAvg	**92.26** (82.67)	**94** (357)	×**3.8** (× 1.0)
FedProx	**89.30** (78.57)	**320** (Nan)	×**1.1** (Nan)
SCAFFOLD	**83.89** (74.23)	**147** (Nan)	×**2.4** (Nan)
FedNova	**91.82** (82.22)	**128** (741)	×**2.8** (× 0.5)
$\alpha = 0.1$, $E = 5$, $M = 10$ (Target ACC = 87%)			
FedAvg	**90.52** (87.92)	**145** (131)	×**0.9** (× 1.0)
FedProx	**87.20** (78.43)	**351** (Nan)	×**0.4** (Nan)
SCAFFOLD	**88.04** (81.07)	**210** (Nan)	×**0.6** (Nan)
FedNova	**90.99** (88.17)	**75** (162)	×**1.7** (× 0.8)
$\alpha = 0.1$, $E = 1$, $M = 100$ (Target ACC = 89%)			
FedAvg	**92.05** (89.44)	**362** (618)	×**1.7** (× 1.0)
FedProx	**92.08** (89.51)	**356** (618)	×**1.7** (× 1.0)
SCAFFOLD	89.21 (**89.55**)	968 (**643**)	× 0.6 (×**1.0**)
FedNova	**92.01** (82.08)	**676** (Nan)	×**0.9** (Nan)

	ACC ↑	ROUND ↓	Speedup ↑
	w/ (w/o) VHL		
Centralized training ACC = 71.90 % (**74.25 %**)			
$\alpha = 0.1$, $E = 1$, $M = 10$ (Target ACC = 67%)			
FedAvg	**70.04** (67.95)	**384** (497)	×**1.3** (× 1.0)
FedProx	**68.29** (65.29)	**617** (Nan)	×**0.8** (Nan)
SCAFFOLD	**67.88** (67.14)	**294** (766)	×**1.7** (× 0.6)
FedNova	**69.58** (68.26)	**384** (472)	×**1.3** (× 1.1)
$\alpha = 0.05$, $E = 1$, $M = 10$ (Target ACC = 62%)			
FedAvg	**65.61** (62.07)	**354** (514)	×**1.5** (× 1.0)
FedProx	**64.39** (61.52)	**482** (Nan)	×**1.1** (Nan)
SCAFFOLD	**60.67** (59.04)	Nan (Nan)	Nan (Nan)
FedNova	**66.45** (60.35)	**320** (Nan)	×**1.6** (Nan)
$\alpha = 0.1$, $E = 5$, $M = 10$ (Target ACC = 69%)			
FedAvg	**69.85** (69.81)	**327** (283)	×**0.9** (× 1.0)
FedProx	**63.83** (62.62)	Nan (Nan)	Nan (Nan)
SCAFFOLD	69.43 (**70.68**)	291 (**171**)	× 1.0 (×**1.7**)
FedNova	68.86 (**70.05**)	Nan (**292**)	Nan (×**1.0**)
$\alpha = 0.1$, $E = 1$, $M = 100$ (Target ACC = 48%)			
FedAvg	**53.45** (48.33)	**717** (967)	×**1.3** (× 1.0)
FedProx	**52.68** (48.14)	**717** (955)	×**1.3** (× 1.0)
SCAFFOLD	**54.93** (51.63)	**656** (827)	×**1.5** (× 1.2)
FedNova	**53.50** (48.12)	**797** (967)	×**1.2** (× 1.0)

5.1.3.2 Federated Learning with Feature Distillation

One simple and promising solution to data heterogeneity has been viewed as sharing client data [256, 479]. Nonetheless, other research draw attention to the difficulty of striking a balance between performance enhancement and strict privacy protection. In this FL context, The FedFed framework was introduced based on three questions that became increasingly specific over time:

- **Q1:** *Can sharing partial features help address the challenges associated with heterogeneity in federated learning?*
- **Q2:** *How can data be partitioned into performance-robust and performance-sensitive features?*
- **Q3:** *What are the potential risks if performance-sensitive features inadvertently reveal private information?*

By dividing data features into performance-robust and performance-sensitive categories, with no overlap, the privacy-performance trade-off for question 1 can be addressed. The features that are robust in performance, which hold the majority of the data's information, are kept confidential. On the other hand, clients share the performance-sensitive that are necessary to improve model generalization. Second, ideal features should only preserve the bare minimum of information needed for generalization, according to the information bottleneck principle [369]. Therefore, the performance-sensitive features, which embody this crucial information, should be exchanged among clients to address data heterogeneity. However, sharing these performance-sensitive features inevitably poses privacy risks. To mitigate this, we can apply conventional privacy protection techniques, such as adding random noise to these features.

Building upon the insights gained from addressing the three key questions, the method FedFed (**Fed**erated **fe**ature **d**istillation) was proposed [429]; the features are divided into performance-sensitive ($\mathbf{x}_s$) and performance-robust ($\mathbf{x}_r$) categories. To protect $\mathbf{x}_s$, the method shares the protected features $\mathbf{x}_p$ by applying random noise $\mathbf{n}$ to $\mathbf{x}_s$, i.e., $\mathbf{x}_p = \mathbf{x}_s + \mathbf{n}$, where $\mathbf{n}$ is drawn from the distribution $\mathcal{N}(\mathbf{0}, \sigma_s^2\mathbf{I})$ and σ_s denotes the intensity of Gaussian noise. Extensive experiments validate the effectiveness of FedFed in enhancing model performance.

Method There are two primary ways to describe the distinctions between IB and FedFed: FedFed's goal is to keep the discarded characteristics as similar to the original features as possible, which aids in protecting the majority of private data. However, IB prioritizes the preservation of only the most important information and aims to minimize the difference between the original features and the ones that are kept. FedFed, on the other hand, eliminates information directly in the data space, while IB does it in the representation space. More specifically, the goal of feature distillation can be formulated as follows:

$$\min_{Z} I(X; Y|Z), \ s.t. \ I(X; X - Z|Z) \geq I_{\mathsf{FF}}, \tag{5.8}$$

where X and Y are the input and label, respectively. Z is the desired representation extracted from input X. I_{FF} is a constant to balance the information between representation Z and input X.

An explicit competition mechanism is introduced in the data space to make feature distillation more tractable and ensure that the performance-robust features closely resemble the original data. Specifically, the preserved features, denoted as $\mathbf{x}_r$, are modeled explicitly by $q_\theta(\mathbf{x})$, while the discarded features $\mathbf{x}_r$ are modeled implicitly as $z_\theta(\mathbf{x}) \triangleq \mathbf{x} - q_\theta(\mathbf{x})$. The objective is then formulated as:

$$\min_{\theta,\mathbf{w}_k} -\mathbb{E}_{(\mathbf{x},\mathbf{y})\sim P(\mathbf{X}_k,Y_k)}\left[\ell_{\mathrm{CE}}(f(z_\theta(\mathbf{x});\mathbf{w}_k),y)\right], \; s.t.||z_\theta(\mathbf{x})||_2^2 \le \rho, \tag{5.9}$$

where $q_\theta(\mathbf{x})$ stands for a generative model, $\mathbf{w}_k$ is a classifier to ensure the rest information is useful, while ρ represents a tunable hyperparameter. $\mathbf{x} - q_\theta(\mathbf{x})$ is shared to server for data heterogeneity mitigating. All clients share the protected $\mathbf{x}_p$ by DP, enabling a global dataset constructed on the server side. Using this shared dataset, clients can train their own classifiers $f_{\phi_k}(\cdot)$, parameterized by ϕ_k, with both private and shared data:

$$\min_{\phi_k} \mathbb{E}_{(\mathbf{x},y)\sim P(\mathbf{X}_k,Y_k)}\left[\ell_{CE}(f_{\phi_k}(\mathbf{x}),y) + \mathbb{E}_{(\mathbf{x}_p,y)\sim P(\mathbf{h}(\mathbf{X}_k),Y_k)}\ell_{\mathrm{CE}}(f_{\phi_k}(\mathbf{x}_p),y)\right], \tag{5.10}$$

where $f_{\phi_k}(\cdot)$ is trained for aggregating the global model, $(\mathbf{x}_p, y)$ are the protected $\mathbf{x}_e$ received from all clients, and $\mathbf{h}(\mathbf{x}_k) := \mathbf{z}(\mathbf{x}_k) + \mathbf{n}$, with noise n.

Experiments Table 5.6 displays the CIFAR-10 experimental findings. Model accuracy is consistently improved by the FedFed framework in a variety of setups. Moreover, it improves the rate at which various algorithms converge. Table 5.7 demonstrates that, in comparison to LDA with $\alpha = 0.1$, two particular forms of heterogeneous partitions cause a more noticeable performance reduction. Significant progress is shown by FedFed, proving its resistance to non-IID partitions. In order to evaluate FedFed's resilience to noise selection, we examined the effects of Laplacian noise when using differential privacy (DP) to improve privacy protection. Table 5.8 presents the findings for Laplacian noise, which emphasizes FedFed's resilience.

5.1.4 *Enhancing FL Robustness*

Except for considering the origin data heterogeneity in federated learning, the unique distributed learning paradigm also aggregates the potential adversarial risks from different client or participants, which is a major concern for some safety-critical applications in real-world scenarios. Thus, how to construct a robust model under the federated learning framework is an important research question.

Table 5.6 Top-1 accuracy results with and without FedFed under varying levels of heterogeneity, local epochs, and client numbers on CIFAR-10

	Centralized training ACC = 95.48% w/(w/o) FedFed							
	ACC ↑	Gain ↑	Round ↓	Speedup ↑	ACC ↑	Gain ↑	Round ↓	Speedup ↑
	$\alpha = 0.1$, $E = 1$, $K = 10$ (Target ACC =79%)				$\alpha = 0.05$, $E = 1$, $K = 10$ (Target ACC =69%)			
FedAvg	**92.34**(79.35)	12.99 ↑	**39**(284)	×**7.3**(×1.0)	**90.02**(69.36)	20.66 ↑	**44**(405)	×**9.2**(×1.0)
FedProx	**92.12**(83.06)	9.06 ↑	**62**(192)	×**4.6**(×1.5)	**90.73**(78.98)	11.75 ↑	**48**(203)	×**8.4**(×2.0)
SCAFFOLD	**89.66**(83.67)	5.99 ↑	**34**(288)	×**8.4**(×1.0)	**81.04**(37.87)	43.17 ↑	**37**(None)	×**10.9**(None)
FedNova	**92.23**(80.95)	11.28 ↑	**33**(349)	×**8.6**(×0.8)	**91.21**(65.08)	26.13 ↑	**32**(None)	×**12.7**(None)
	$\alpha = 0.1$, $E = 5$, $K = 10$ (Target ACC =85%)				$\alpha = 0.1$, $E = 1$, $K = 100$ (Target ACC =49%)			
FedAvg	**93.24**(83.79)	9.45 ↑	**17**(261)	×**15.4**(×1.0)	**84.06**(49.72)	34.34 ↑	**163**(967)	×**5.9**(×1.0)
FedProx	**91.39**(82.32)	8.97 ↑	**76**(None)	×**3.4**(None)	**87.01**(50.01)	37.00 ↑	**127**(831)	×**7.6**(×1.2)
SCAFFOLD	**92.34**(85.31)	7.03 ↑	**15**(66)	×**17.0**(×4.0)	**79.60**(52.76)	26.84 ↑	**171**(627)	×**5.7**(×1.5)
FedNova	**92.85**(86.21)	6.64 ↑	**31**(120)	×**8.4**(×2.2)	**86.64**(45.97)	40.67 ↑	**199**(None)	×**4.9**(None)

"Round": communication rounds arriving at the target accuracy. "None": not attaining the target accuracy. Results in "(·)" indicate training without FedFed

Table 5.7 Experimental results of various non-IID data partitioning approaches

	Test accuracy w/(w/o) FedFed			
Partition method	FedAvg	FedProx	SCAFFOLD	FedNova
$\alpha = 0.1$	**92.34**(79.35)	**92.12**(83.06)	**89.66**(83.67)	**92.23**(80.95)
$\#C = 2$	**89.23**/42.54	**88.17**/58.45	**84.43**/46.82	**89.54**/45.42
Subset	**90.29**/39.53	**89.11**/32.87	**89.92**/35.26	**90.00**/38.52

Table 5.8 Experiment results with different noise adding on CIFAR-10

	Test accuracy on different noise with FedFed			
Noise type	FedAvg	FedProx	SCAFFOLD	FedNova
Gaussian noise	92.34	92.12	89.66	92.23
Laplacian noise	92.30	91.36	91.24	91.73

5.1.4.1 Combating Exacerbated Heterogeneity for Robust Models in Federated Learning

Federated learning [187, 260] has garnered heightened interest owing to apprehensions over data privacy and governance challenges. The investigation of adversarial resilience in federated systems was prompted by the fact that training on local clients makes them more vulnerable to adversarial attacks [124]. Adversarial training has been the focus of recent research regarding its integration into federated learning. Progress in adversarially resilient federated systems is impeded by the unanticipated decrease in robustness that results from the merging of two paradigms, a significant issue in the algorithmic aspect. The later stages of training in federated adversarial training (FAT), which employs adversarial training within the framework of federated learning based on FedAvg, demonstrate a substantial decrease in robust accuracy in contrast to centralized scenarios (Fig. 5.3).

In [493], the deterioration of robustness is examined, and it is demonstrated that it may be the result of *augmented heterogeneity* generated by adversarial training with local clients. FAT allocates training data to each client, resulting in independent adversarial training for each client that is unaffected by the data of other clients, in contrast to centralized adversarial training. Due to this, the adversarial examples generated by the inner maximization of adversarial training [258] often lean toward each local distribution. A previous study discovered that local training in federated learning has an optimization bias as a result of the heterogeneity of data across clients. The heterogeneity in federated optimization can be exacerbated by the adversarial data produced by biased local models, which complicates the convergence to a robust optimum.

Method A novel learning framework, termed *Slack Federated Adversarial Training* (SFAT) [493], is suggested to tackle the challenge of integrating adversarial training with federated learning using a α-slack mechanism. SFAT allocates client-

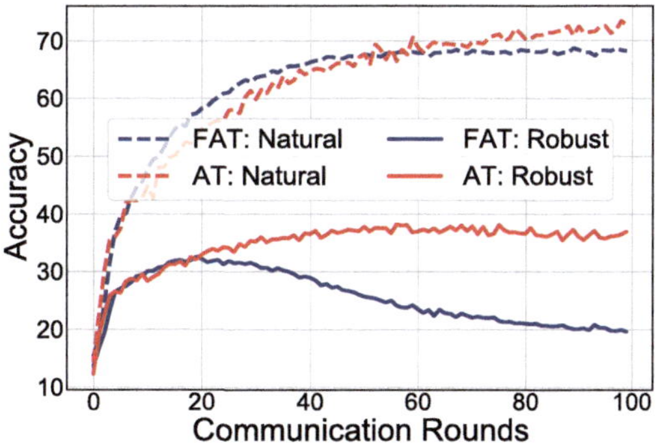

Fig. 5.3 Comparison between centralized AT and FAT in terms of the robust accuracy and the natural accuracy

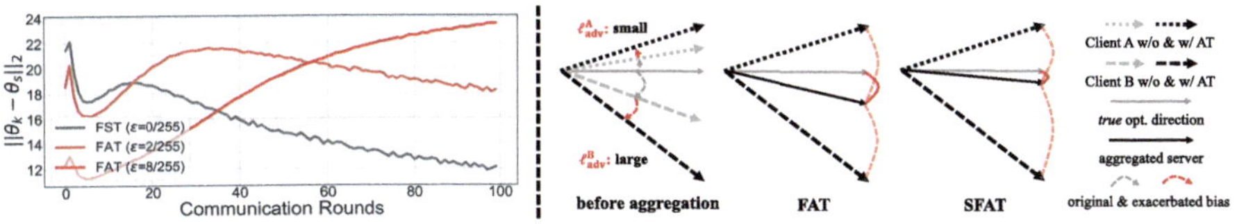

Fig. 5.4 Left panel: heterogeneity metric employing client drift with various adversarial generation using CIFAR-10. Right panel: a comparison of FAT with our SFAT regarding the aggregation of local clients. The local antagonistic generation would amplify the loss and are noted to intensify the heterogeneity. In contrast to FAT, SFAT selectively adjusts the weights of clients with modest or substantial adversarial training losses to mitigate the issue during aggregation, adhering to our α-slack technique that transforms the original aim into a lower bound

specific slack during aggregation to increase the influence of clients with minimal adversarial training loss and simultaneously reduce the impact of clients with substantial losses. This approach mitigates increased heterogeneity and alleviates the decline in robustness. The right panel of Fig. 5.4 illustrates the implementation of a laxity in the inner maximization to reduce the exacerbation of optimization bias. The inner maximization of adversarial training is formally decomposed into $M = |\mathcal{M}|$ independent populations that correspond to the M clients in the framework of federated learning. This decomposition is then relaxed to a lower bound denoted by α as follows:

$$\begin{aligned}
\mathcal{L}_{\mathrm{AT}} &= \frac{1}{N}\sum_{n=1}^{N}\max_{\tilde{\mathbf{x}}_n\in\mathcal{B}_\epsilon[\mathbf{x}_n]}\ell(f_\theta(\tilde{\mathbf{x}}_n), y_n) = \sum_{m=1}^{M}\frac{N_m}{N}\underbrace{\left(\frac{1}{N_m}\sum_{i=1}^{N_m}\max_{\tilde{\mathbf{x}}_i\in\mathcal{B}_\epsilon[\mathbf{x}_i]}\ell(f_\theta(\tilde{\mathbf{x}}_i^m), y_i^m)\right)}_{\mathcal{L}_m} \\
&\geq (1+\alpha)\sum_{m=1}^{\widehat{M}}\frac{N_{\phi(m)}}{N}\mathcal{L}_{\phi(m)} + (1-\alpha)\sum_{m=\widehat{m}+1}^{M}\frac{N_{\phi(m)}}{N}\mathcal{L}_{\phi(m)} \\
&\doteq \mathcal{L}^{\alpha}(\widehat{M}), \qquad \text{s.t. } \alpha\in[0,\ 1),\ \widehat{M}\leq\frac{m}{2},
\end{aligned} \tag{5.11}$$

where $\mathcal{B}_\epsilon[\mathbf{x}_n]$ indicates the perturbation ball of data sample $\mathbf{x}$ constrained by l_p-norm not larger than ϵ, $\tilde{\mathbf{x}}$ indicates the adversarial data, and $\phi(\cdot)$ is a function that associates the index with the original population arranged in ascending order according to $\{\frac{N_m}{N}\mathcal{L}_m\}$. It is important to note that $\phi(\cdot)$ is a nonparametric auxiliary operation that does not influence gradient back-propagation. The concept is to mitigate the original objective to reduce the intensification of heterogeneity, akin to standard continuous relaxation methods (e.g., Gumbel-SoftMax) in discrete optimization. The inverse function of α-slack is also examined in the experiments, which seeks an upper limit but further intensifies the heterogeneity. The α-slack mechanism is implemented as follows:

$$\min \mathcal{L}_{\text{SFAT}} = \min_{f_\theta \in \mathcal{H}} \frac{1}{\sum_m^M N_m} \sum_{m=1}^{M} W_m N_m \cdot \underbrace{\left(\frac{1}{N_m} \sum_{n=1}^{N_m} \max_{\tilde{\mathbf{x}}_n^m \in \mathcal{B}_\epsilon[\mathbf{x}_n^m]} \ell(f_\theta(\tilde{\mathbf{x}}_n^m), y_n^m) \right)}_{\mathcal{L}_m}, \tag{5.12}$$

where W_m indicates the weight assigned to the m-th client according to an ascending sort of weighted client losses compared with the $\widehat{M}$-th one, which can be $(1+\alpha) \cdot \mathbf{1}(\frac{M}{N}\mathcal{L}_m \leq \mathcal{L}_{\text{sorted}}[\widehat{M}]) + (1-\alpha) \cdot \mathbf{1}(\frac{M}{N}\mathcal{L}_m > \mathcal{L}_{\text{sorted}}[\widehat{M}])$ that corresponds to the relaxed version of loss in Eq. (5.11).

For simplicity, without loss of generality, the slack mechanism can be transformed as the weight adjustment of those selected clients with smaller adversarial training losses and adopt the following to guarantee the derivation of the lower limit:

$$\begin{aligned} W_m = ((1+\alpha)/(1-\alpha) \cdot \mathbf{1}(\frac{N_m}{N}\mathcal{L}_m \leq \mathcal{L}_{\text{sorted}}[\widehat{M}]) + \\ 1 \cdot \mathbf{1}(\frac{N_m}{N}\mathcal{L}_m > \mathcal{L}_{\text{sorted}}[\widehat{M}]))/((\sum_{m=1}^{M} W_m) + 2\alpha/(1-\alpha)). \end{aligned} \tag{5.13}$$

After receiving the global model parameters from the server, each client will conduct adversarial training on its local dataset. Simultaneously, the client loss associated with hostile situations is documented, which suggests that the modest local bias resulting from the excessive adversarial construction is not significant. The client transmits its model parameters and loss to the server when the training steps satisfy the necessary criteria. On the server side, after collecting the model parameters $\{\theta_m\}_{m=1}^M$ and the losses $\{\mathcal{L}_m\}_{m=1}^M$ from all clients, it will first arrange $\{\frac{N_m}{N}\mathcal{L}_m\}_{m=1}^M$ in ascending order to identify the top-$\widehat{M}$ clients. The α-slack approach will be used to aggregate global model parameters. The top-$\widehat{M}$ clients' parameters will be upweighted by a ratio of $(1+\alpha)/(1-\alpha)$, while the parameters of the remaining clients will be downweighted.

Experiments Table 5.9 on *CIFAR-10* demonstrates that SFAT outperforms FAT on non-IID data in terms of both natural and robust accuracy. FedProx and Scaffold, which are specifically engineered to address the numerous challenges

Table 5.9 Test accuracy on non-IID settings with different federated methods (Mean±Std)

Setting		Non-IID				
CIFAR-10		Natural	FGSM	PGD-20	CW_∞	AA
FedAvg	FAT	58.13±0.68%	40.06±0.62%	32.56±0.01%	30.88±0.37%	29.17±0.03%
	SFAT	**63.36±0.07%**	**44.82±0.32%**	**37.14±0.03%**	**33.39±0.61%**	**31.66±0.70%**
FedProx	FAT	59.95±0.45%	41.44±0.15%	33.83±0.01%	31.65±0.36%	30.11±0.09%
	SFAT	**62.04±0.47%**	**44.21±0.08%**	**36.64±0.11%**	**32.62±0.20%**	**31.83±0.15%**
Scaffold	FAT	61.44±1.37%	42.85±0.76%	34.08±0.05%	32.56±0.02%	31.03±0.08%
	SFAT	**63.16±0.96%**	**45.55±0.50%**	**37.33±0.02%**	**34.82±0.04%**	**33.32±0.01%**
CIFAR-100		Natural	FGSM	PGD-20	CW_∞	AA
FedAvg	FAT	34.63±0.56%	19.92±0.28%	15.40±0.20%	13.23±0.03%	12.23±0.01%
	SFAT	**35.65±0.54%**	**20.23±0.44%**	**16.24±0.16%**	**13.53±0.02%**	**12.45±0.03%**
FedProx	FAT	31.93±0.43%	19.06±0.17%	15.30±0.08%	12.93±0.02%	12.01±0.04%
	SFAT	**34.87±0.24%**	**20.54±0.08%**	**16.09±0.10%**	**13.35±0.12%**	**12.44±0.20%**
Scaffold	FAT	39.98±0.02%	24.30±0.04%	19.34±0.07%	16.49±0.12%	15.29±0.08%
	SFAT	**44.13±0.05%**	**25.32±0.94%**	**20.22±0.07%**	**16.96±0.17%**	**15.80±0.10%**
SVHN		Natural	FGSM	PGD-20	CW_∞	AA
FedAvg	FAT	**91.52±0.28%**	88.13±0.18%	68.98±0.11%	68.04±0.15%	66.59±0.04%
	SFAT	91.26±0.01%	**88.27±0.02%**	**72.04±0.32%**	**69.96±0.16%**	**68.89±0.27%**
FedProx	FAT	91.00±0.08%	87.65±0.15%	68.48±0.04%	67.16±0.02%	65.76±0.18%
	SFAT	**91.19±0.06%**	**88.15±0.01%**	**71.84±0.30%**	**69.88±0.35%**	**68.84±0.37%**
Scaffold	FAT	90.82±0.87%	87.89±0.66%	69.51±0.84%	68.12±0.88%	67.19±0.54%
	SFAT	**90.93±0.76%**	**88.27±0.45%**	**71.77±0.38%**	**69.49±0.67%**	**68.37±0.48%**

associated with federated learning, may enhance the efficacy of the model in FAT in comparison to FedAvg. SFAT enhances performance by mitigating the additional variability that adversarial training introduces. The results of *CIFAR-10* across three federated optimization approaches are consistent with the similar improvements observed in *CIFAR-100* and *SVHN*, as illustrated in Table 5.9. Many comparisons have shown that SFAT is consistently effective.

Apart from adversarial data samples reducing the model's robustness, label deficiency further degrades its performance. To tackle these challenges and enhance robustness in label-deficient scenarios, federated semi-supervised learning (FSSL) has been proposed.

5.1.4.2 Federated Models Training Under Extremely Label Deficiency

Label scarcity is a problem in decentralized contexts with restricted resources as well [184, 248]. A lot of current approaches combine supervised and unsupervised learning techniques. These methods frequently use pseudo-labeling techniques or consistency when applied to one single global model. However, a problem known as gradient conflict is frequently encountered by models trained with multiple learning objectives [442]. Gradient conflict occurs when the model's two-part gradient's cosine similarity to the objective function is negative during optimization.

In a federated semi-supervised learning (FSSL) setting, this can both happen in optimization objectives and data samples. This arises when labeled and unlabeled data aims diverge and only certain samples are annotated. When one model optimizes two objectives in FSSL, gradient conflicts inevitably occur. The problem of label scarcity has been tackled with a dual-model architecture called **Twin-sight** [499]. By utilizing complementary insights from both labeled and unlabeled data, this system improves mutual guiding. The approach adapts traditional semi-supervised learning techniques to the FL setting.

Problem Definition In this section, we focus on a traditional situation in which T clients have fully labeled data, but the rest of B clients only have data without label information. The set of $\mathcal{M}^L = \{m_t\}_{t=1}^{T}$ and $\mathcal{M}^U = \{m_b\}_{b=1}^{B}$ corresponds to the fully labeled clients and fully unlabeled, respectively. The full set of clients is represented by $\mathcal{M} = \{m_k\}_{k=1}^{K} = \mathcal{M}^L \cup \mathcal{M}^U$. The dataset of the t-th client is denoted as $\mathcal{D}_t^L = \{(\mathbf{x}_t^i, y_t^i)\}_{i=1}^{N_t} \sim P(X_t, Y_t)$, while the dataset for the n-th client in $\mathcal{M}^U$ is $\mathcal{D}_b^U = \{\mathbf{x}_b^i\}_{b=1}^{N_b} \sim P(X_b)$, representing the data without annotations for b_s.

Method The twin-model paradigm consists of two separate models, one supervised model with parameter θ_s is trained by conventional cross-entropy loss, and another unsupervised model is parameterized by θ_u. The unsupervised model is trained using a fine-grained downstream classification task, specifically instance discrimination. This objective is consistent across all clients:

$$\ell^u(\theta_u) = -\log \frac{\exp\left(\text{sim}\left(f_{\theta_u}(;\mathbf{x}_i), f_{\theta_u}(\mathbf{x}_{\hat{i}})\right)/\tau\right)}{\sum_{z=1}^{2N} \mathbb{I}_{[z\neq i]} \exp\left(\text{sim}\left(f_{\theta_u}(;\mathbf{x}_i), f_{\theta_u}(\mathbf{x}_z)\right)/\tau\right)}, \tag{5.14}$$

where $\text{sim}(\cdot,\cdot)$ gets the similarity between two samples, $\mathbf{x}_{\hat{i}}$ and $\mathbf{x}_i$ are the positive sample pairs, $f_{\theta_u}(\cdot)$ represents the unsupervised model, and τ denotes the temperature hyperparameter. It is also important to note that the label information is not valid for clients sampled from the unlabeled subset $\mathcal{M}^U = \{m_b\}_{b=1}^{B}$. Therefore, a surrogate loss $\ell_t^s(\cdot)$ is introduced to train the supervised model on unlabeled data for client m_t. This surrogate loss can be expressed as:

$$\ell_t^s(\theta_s) := -\mathbb{I}\left[\sigma(\tilde{y}) > t_u\right]\sigma(\tilde{y}) \log f_{\theta_s}(\mathbf{x}_i), \tag{5.15}$$

where $\mathbb{I}(\cdot)$ is an indicator function, $\sigma(\cdot)$ selects the maximum value from a given vector, and t_u is a threshold hyperparameter used to select predictions with high confidence. How to make two models gain from each other is the next challenge. To solve this problem, the Twin-sight architecture is completed by introducing a Twin-sight loss. To retain neighborhood relationships between data features from supervised and unsupervised models, the Twin-sight loss function $\ell_a(\cdot)$ is used. Loss is defined as

$$\min_{\theta_s,\theta_u} \ell_a(\theta_s, \theta_u) := d(\mathbf{N}(f_{\theta_s}(\mathbf{x})), \mathbf{N}(f_{\theta_u}(\mathbf{x}))), \tag{5.16}$$

where the difference between two matrices is measured by d, which can be the Frobenius norm ℓ_F, and $\mathbf{N}(\cdot)$ is the neighborhood relation function.

As a result, the objective function for labeled data, $\ell^l(\cdot)$, is formulated as:

$$\ell^l(\theta_s, \theta_u) = \underbrace{\ell_m(\theta_s)}_{\substack{Supervised\ model \\ on\ \text{fully labeled}\ clients}} + \underbrace{\lambda_u \ell^u(\theta_u)}_{\substack{Unsupervised\ model \\ on\ \text{fully labeled}\ clients}} + \underbrace{\lambda_d \ell_a(\theta_s, \theta_u)}_{Twin-sight\ loss}. \tag{5.17}$$

Similarly, unlabeled data can be incorporated by using the loss function $\ell^u(\cdot)$:

$$\ell^u(\theta_s, \theta_u) = \underbrace{\ell_t^s(\theta_s)}_{\substack{Supervised\ model \\ on\ \text{fully unlabeled}\ clients}} + \underbrace{\lambda_u \ell^u(\theta_u)}_{\substack{Unsupervised\ model \\ on\ \text{fully unlabeled}\ clients}} + \underbrace{\lambda_d \ell_a(\theta_s, \theta_u)}_{Twin-sight\ loss}, \tag{5.18}$$

where λ_u and λ_d are hyperparameters used to adjust the relative weight of the different loss components.

Experiments Table 5.10 shows the Twin-sight experimental findings on SVHN and FMNIST. A ratio of $\gamma_u = 60\%$ was obtained by randomly selecting six of the ten clients to be fully unlabeled and the remaining four to be fully labeled. Throughout the testing procedure, a constant random seed was used in the tests. For every FL system communication round, five clients were chosen ($S = 50\%$). Both Twin-sight and baseline techniques fail to meet FedAvg's upper bound. Nonetheless, our new method, Twin-sight, outperforms all baseline alternatives, indicating its efficacy in resolving gradient conflict.

To assess the robustness of our method, we performed experiments with varying ratios of unlabeled clients, specifically from $\gamma_u = 90\%$ down to $\gamma_u = 40\%$. The results, presented in Fig. 5.5, indicate that Twin-sight's performance is influenced by the proportion of unlabeled clients. Nonetheless, in every situation, the strategy consistently performs better than the baseline alternatives. It became shown that general performance drops when the percentage of unlabeled customers increases. Twin-sight outperforms other methods even when there is a higher percentage of unlabeled consumers. Furthermore, Twin-sight consistently outperforms other methods across various ratios of labeled to fully unlabeled clients, highlighting the importance of labeled data in facilitating Twin-sight interactions and boosting overall performance.

Table 5.10 The performance of Twin-sight with state-of-the-art (SOTA) methods on the SVHN and FMNIST datasets, using parameters of $\alpha = 0.1$ and $K = 10$

Method	No. Fully labeled clients/fully unlabeled clients		SVHN		FMNIST	
	Labeled clients (T)	Unlabeled clients (B)	Acc ↑	Round ↓	Acc ↑	Round ↓
FedAvg-upper bound	10	0	87.34		88.98	
FedAvg-lower bound	4	0	51.10	70	72.46	172
FedProx-lower bound	4	0	49.22	None	70.71	None
FedAvg+FixMatch	4	6	58.68	**35**	67.52	None
FedProx+FixMatch	4	6	45.58	None	63.20	None
FedAvg+Freematch	4	6	<u>59.74</u>	<u>45</u>	63.10	None
FedProx+Freematch	4	6	50.91	None	69.62	None
Fed-Consist	4	6	56.87	103	68.51	None
RSCFed	4	6	54.50	69	<u>76.58</u>	**88**
Twin-sight (ours)	4	6	**62.94**	125	**79.95**	<u>140</u>

The performance of FedAvg-lower bound is the target accuracy. "Round" refers to the communication round required to reach the target accuracy. "None" indicates that this method did not attain the target accuracy throughout the entire training period. The bold indicates the best result, while the underlined represents the runner-up

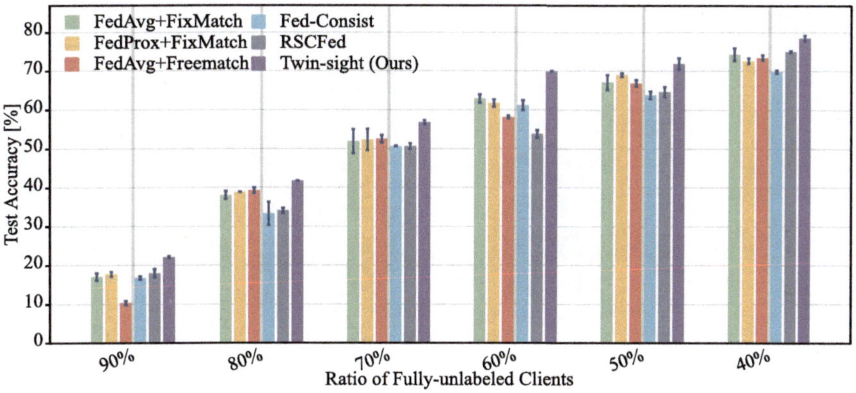

Fig. 5.5 The effects of different ratios of unlabeled clients (α), comparing five baseline methods with Twin-sight

5.1.5 Communication Heterogeneity

Because of geographic dispersion and diverse participanting devices, FL systems also face challenges from client resource disparities and communication inefficiencies [260].

Decentralized FL (DFL) frameworks often struggle to optimize varying bandwidth capacities and adapt to dynamic network topologies, leading to suboptimal performance and unstable convergence [140, 148, 187, 298]. Additionally, traditional centralized FL architectures create communication bottlenecks and vulnerabilities [260], while decentralized approaches must overcome limitations in bandwidth utilization and topology unpredictability to ensure reliable and efficient model training [140, 255].

5.1.5.1 Decentralized FL with Sparsified and Adaptive Communication

GossipFL [357] is a framework for building decentralized federated languages (DFL) using bandwidth efficient approaches, combining model compression and adaptive language selection. However, the frameworks consider dynamic, bandwidth-sensitive communication topologies and gossip matrices to maintain training efficiency without sacrificing convergence. This chapter discusses the design, algorithms, and empirical results of GossipFL, which provides performance improvements over conventional federated algorithms.

The study investigates (1) methods for bandwidth-constrained DFL training while preserving convergence speed and (2) strategies to optimize client bandwidth allocation for enhanced communication efficiency.

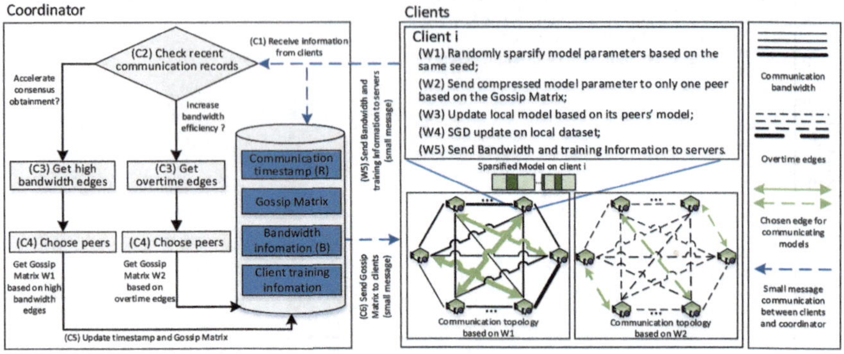

Fig. 5.6 GossipFL's system architecture features two roles: coordinator and client [357]

Method GossipFL [357] reduces communication costs by enabling clients to share compressed models with one peer per iteration. A dynamic peer selection algorithm prioritizes clients' real-time bandwidth availability, optimizing network resource allocation. GossipFL [357] replaces centralized servers with a decentralized, gossip-driven protocol. A coordinator oversees training coordination without directly aggregating model parameters, minimizing overhead and improving privacy. Key innovations include communication traffic reduction and intelligent peer selection for bandwidth optimization. The architecture is illustrated in Fig. 5.6.

GossipFL models customer interactions as a dynamic graph where all vertices correspond to participants and edges define active communication channels. The topology is adaptively evolved based on the gossip matrices generated by the coordinators governing connection patterns. During data transmission, clients exchange data with others. A bandwidth-sensitive selection mechanism optimizes bandwidth utilization by matching customers and partners to increase throughput. As shown in Table 5.11, GossipFL saves 180.20 MB of data transfering. This efficiency persisted in non-IID scenarios, maintaining accuracy while using substantially less bandwidth than competitors. Participants employ dimension reduction through selective parameter nullification before peer exchanges, discarding substantial portions of model elements. Coordinated randomization ensures mask consistency across nodes through seed synchronization. This framework implements a performance-optimized routing matrix that dynamically prioritizes high-speed connections. The adaptive structures are continuously updated in real time with bandwidth measurements and topological changes. Overall, GossipFL proposes parameter sparsification, intelligent partner selection, and adaptive network optimization. Proven convergence features and practical validation make it a robust solution for decentralized kitchens.

Experiments The evaluations include three network scales (14, 32, and 100 nodes) and realistic bandwidth conditions (0.2–330 Mbits/s) using the MNIST and CIFAR10 datasets. Further evaluation is conducted on FedAvg, S-FedAvg,

Table 5.11 Theoretical communication traffic (MB) and time (second) to reach the target accuracy on IID data on 14 clients

Algorithm	MNIST-CNN (98%)		CIFAR10-CNN (60%)		ResNet-20 (80%)	
	Traffic (MB)	Time (s)	Traffic (MB)	Time (s)	Traffic (MB)	Time (s)
FedAvg	2994.95	1497.77	–	–	222.24	111.14
S-FedAvg	840.74	420.45	–	–	172.34	86.18
D-PSGD	39645.14	33037.61	48777.69	40648.08	16446.06	13705.05
DCD-PSGD	34238.98	28532.48	93334.25	77778.54	15075.55	12562.96
CHOCO-SGD	288.32	240.27	581.58	484.65	228.41	190.34
GossipFL (ours)	**180.20**	**120.46**	**272.02**	**181.85**	**114.20**	**76.34**

Table 5.12 Top-1 validation accuracy on IID data on 14 clients

Algorithm	MNIST-CNN	CIFAR10-CNN	ResNet-20
FedAvg	98.89%	56.89%	87.24%
S-FedAvg	98.88%	53.85%	86.15%
D-PSGD	99.28%	**82.754%**	**90.42%**
DCD-PSGD	99.13%	66.70%	88.01%
CHOCO-SGD	**99.30%**	82.31%	90.03%
GossipFL (ours)	99.23%	82.46%	90.3%

Table 5.13 Top-1 validation accuracy on non-IID data on 14 clients

Algorithm	MNIST-CNN	CIFAR10-CNN
FedAvg	96.79%	46.08%
FedNova	96.68%	37.67%
FedProx	95.83%	37.54%
S-FedAvg	96.56%	41.06%
D-PSGD	98.22%	10.94%
DCD-PSGD	97.69%	29.04%
CHOCO-SGD	94.72%	10.77%
GossipFL (ours)	**99.26%**	**61.10%**

D-PSGD, DCD-PSGD, and CHOCO-SGD under both shared and separate data sharing.

The main advantages of GossipFL are maintaining competitive accuracy as shown in Tables 5.12 and 5.13, the main advantages of GossipFL are maintaining competitive accuracy while significantly reducing latency (98.5% bandwidth reduction compared to D-PSGD). The resilience to node failures and the efficient utilization of nodes demonstrate the practical feasibility of federated systems among multiple nodes.

5.1.5.2 Communication-Efficient Personalized FL via Decentralized Sparse Training

To tackle personal federated learning issues, we introduce Dis-PFL, a decentralized framework that utilizes a peer-to-peer communication protocol. This approach incorporates personalized sparse masks to create tailored, sparse local models directly on client devices.

To tailor models for every user, we include individualized masks into the PFL problem, in contrast to the general FL specified in Sect. 5.1.1. The following is the target formulation:

$$\min_{\theta\in\mathbb{R}^d; p_1,\cdots,p_M} \mathcal{L}(\theta, p_m) := \sum_{m=1}^{M} \frac{n_m}{N} \mathbb{E}_{(\mathbf{x},y)\sim D_m} \ell(\theta \odot p_m; \mathbf{x}, y), \tag{5.19}$$

The customized sparse binary mask for the m-th client is represented by $p_m \in \{0, 1\}^d$, and the Hadamard product between two vectors is indicated by $\odot$. The matching weight in the global model is still active in the m-th personalized model if its value in p_m is 1, and it is inactive if its value is 0.

Based on each client's computing capacity, memory limitations, and communication budget (expressed as c_m, distinct sparsity masks are generated). According to the Erdos-Renyi kernel (ERK) scheme [103], the mask initialization assigns greater sparsity to layers with more parameters and retains more weights for layers with less parameters. Since each client only uses specific parts of the same dense global model, the local mask p_k is multiplied element-wise when a weighted average of the intersection of the active parameters from the received models is obtained. We use local SGD to further improve communication efficiency, which allows for several local training steps before obtaining the updated sparse model.

Two crucial elements are present during local training: (**i**) since each client only maintains a sparse model, certain weight coordinates are set to zero before the forward pass in order to reduce the computational cost at this step. (**ii**) In the backward phase, p_m masks the stochastic gradient so that gradients corresponding to trimmed parameters are not calculated, further lowering the computational cost.

We update the sparsity mask using a similar method as [103]. The pruning rate α_t is dynamically altered using cosine annealing. Each client used gradient information to restore weights with the greatest relevance while pruning α_t-fraction of the lowest-magnitude weights per layer in order to better adjust to their individual features.

Experiments In our experiments across three datasets and two widely used non-IID partitioning strategies, Dis-PFL consistently outperforms both centralized and decentralized baseline models. We observe that some methods perform poorly with Dirichlet-based partitions but do better in more extreme non-IID settings. We attribute this discrepancy to the limited local data available for each class, restricting the generalization ability of these methods. In contrast, Dis-PFL demonstrates superior performance across various partitioning schemes, proving its robustness. Furthermore, as illustrated in Table 5.14, while both Dis-PFL and SubFedAvg use client-specific masks, our method achieves a significant performance edge.

Table 5.14 Table illustrating performance of different methods

Task	Methods	Dir Part Acc	Path Part Acc	Comm (MB)	FLOP (1e12)
CIFAR-10	Local	61.55±0.2	86.48±0.2	–	8.3
	FedAvg	78.07±0.5	54.53±0.6	446.9	8.3
	FedAvg-FT	81.20±0.5	84.96±0.2	446.9	8.3
	D-PSGD	79.02±0.4	58.07±0.5	446.9	8.3
	D-PSGD-FT	83.90±0.2	90.87±0.2	446.9	8.3
	Ditto	74.68±0.2	87.73±0.1	446.9	8.3
	FOMO	64.68±0.2	88.24±0.1	446.9	8.3
	SubFedAvg	76.70±0.2	88.30±0.2	278.8	**4.7**
	Dis-PFL	**85.70±0.2**	**91.05±0.2**	**223.4**	7.0
CIFAR-100	Local	29.23±0.2	52.46±0.2	–	8.3
	FedAvg	41.72±0.5	33.24±0.6	448.7	8.3
	FedAvg-FT	49.19±0.5	63.53±0.7	448.7	8.3
	D-PSGD	41.87±0.4	35.42±0.2	448.7	8.3
	D-PSGD-FT	51.42±0.4	67.24±0.1	448.7	8.3
	Ditto	38.26±0.2	54.02±0.3	448.7	8.3
	FOMO	28.39±0.1	52.74±0.1	448.7	8.3
	SubFedAvg	43.91±0.2	60.67±0.1	346.6	**5.7**
	Dis-PFL	**53.48±0.3**	**68.64±0.4**	**224.3**	7.0
Tiny-Imagenet	Local	6.76±0.2	17.68±0.3	–	66.6
	FedAvg	12.30±0.3	10.40±0.3	450.7	66.6
	FedAvg-FT	14.80±0.2	28.30±0.2	450.7	66.6
	D-PSGD	12.13±0.5	16.50±0.4	450.7	66.6
	D-PSGD-FT	15.50±0.3	28.60±0.3	450.7	66.6
	Ditto	15.69±0.2	24.55±0.3	450.7	66.6
	FOMO	5.20±0.4	9.39±0.3	450.7	66.6
	SubFedAvg	12.18±0.4	19.73±0.5	290.9	**40.2**
	Dis-PFL	**16.95±0.4**	**31.71±0.4**	**225.3**	54.5

5.1.6 One-Shot Federated Learning

The limited bandwidth in typical Internet environments, often ranging from 1 to 10 MB/s, results in significant communication delays, especially when dealing with large neural networks [187, 356, 357, 359, 360, 360]. For instance, transmitting a ResNet-50 model with 25.56 million parameters can take over 100 seconds, while a model like GPT-3, with 175 billion parameters, could require nearly 200 hours for a single communication round. Current FL techniques attempt to mitigate these communication costs by avoiding the gradient synchronization typical of traditional distributed training, yet the necessity for hundreds or thousands of communication rounds still renders the overall time unacceptable [187, 260].

One-shot federated learning (OFL) has become a promising learning paradigm, enabling the training of a global server model via a single communication round. In

OFL, the server model is aggregated by distilling knowledge from all client models (the ensemble), which are also responsible for synthesizing samples for distillation. In this regard, advanced works show that the performance of the server model is intrinsically related to the quality of the synthesized data and the ensemble

5.1.6.1 One-Shot FL Through Data and Ensemble Co-Boosting

In this chapter, we introduce Co-Boosting, an innovative framework designed to iteratively refine both the data and the model. At the heart of Co-Boosting is the interaction between synthesized data and an ensemble model, where each continually enhances the other. This cyclical process allows for continuous improvement, driving both the quality of the data and the model's performance.

What sets Co-Boosting apart from other methods is its ability to deliver substantial improvements without altering the local training processes or requiring additional data or model transfers. It also supports heterogeneous client architectures, making it an adaptable solution that fits various environments. By focusing on this interplay between data generation and model optimization, Co-Boosting offers a robust and efficient approach for enhancing machine learning systems.

Method Specifically, we explore how hard sample generation can enhance data quality while keeping the ensemble fixed and, conversely, how improving the ensemble can boost the quality of the synthetic dataset. These two processes are naturally intertwined and work together to improve each other.

To optimize both the ensemble $A_{\mathbf{w}}$ and the synthetic dataset D_{syn}, we address the following optimization problem:

$$\min_{\mathbf{w}} \frac{1}{|D_{\text{syn}}|} \sum_{\{\mathbf{x}_s, y_s\} \in D_{\text{syn}}} \max_{\delta \in S_{\text{per}}} \ell_{\text{CE}}\Big(\sum_{k=1}^{n} w_k f_{\theta_k}(\mathbf{x}_s + \delta), y_s\Big), \tag{5.20}$$

where $\mathbf{w}$ is the assigned weight vector for each client, y_s is the label of the sample $\mathbf{x}_s$, and δ represents the perturbation within the set S_{per}. This optimization can be tackled in an adversarial manner, allowing data and ensemble improvements to occur simultaneously. This mutual enhancement ultimately leads to a higher-performing global model, achieved through the distillation process described in Eq. (5.20). Each time one element is refined, the other benefits, resulting in a continuous loop of improvement. This synergistic approach in Co-Boosting leads to progressive enhancements in both data quality and model performance. Ultimately, the global server model derived from this process surpasses competing methods by consistently leveraging these intertwined improvements.

Experiments To evaluate the performance of our method, we conduct a series of experiments across various non-IID settings. As shown in Table 5.15, unlike other methods, including FedAvg, which rely on simple logit averaging and aim to distill knowledge in a static manner, Co-Boosting introduces a dynamic mutual

Table 5.15 Test accuracy of the server model

Method	α	FedAvg	FedDF	F-ADI	F-DAFL	DENSE	Co-Boosting
MNIST	0.05	46.35±0.98	80.73±1.08	80.12±1.76	78.49±1.36	81.06±1.12	**93.93±0.69**
	0.1	75.68±0.82	87.91±0.92	85.92±0.82	87.44±0.61	87.83±1.38	**94.44±1.02**
	0.3	78.97±0.82	**97.66±0.10**	96.34±0.83	96.36±0.98	96.96±0.53	97.25±0.44
FMNIST	0.05	20.07±1.98	44.73±0.40	42.25±2.01	41.66±0.83	44.77±1.87	**50.62±1.13**
	0.1	46.61±1.94	68.40±1.80	63.19±1.99	67.81±0.93	69.43±1.94	**74.86±1.70**
	0.3	60.13±2.62	**83.14±0.47**	74.80±1.72	78.68±0.49	81.31±0.83	83.11±1.28
SVHN	0.05	39.41±2.19	60.79±0.52	56.58±1.05	59.38±1.19	60.24±1.31	**65.40±0.86**
	0.1	46.22±1.92	68.98±0.63	66.33±1.69	67.77±0.34	68.30±1.01	**72.88±1.19**
	0.3	72.61±2.06	79.78±0.55	76.75±1.47	78.01±1.02	78.73±0.84	**81.31±1.09**
CIFAR-10	0.05	17.49±2.51	37.53±0.67	36.94±1.70	37.82±1.30	38.37±1.08	**47.20±0.81**
	0.1	27.54±1.80	49.63±0.80	47.19±0.97	46.32±0.97	47.80±1.21	**57.09±0.94**
	0.3	46.39±2.37	67.18±0.60	60.60±1.32	65.89±1.69	66.77±1.55	**70.24±1.56**
CIFAR-100	0.05	6.45±0.92	16.07±0.54	13.75±1.01	15.79±0.21	16.17±1.33	**19.24±1.42**
	0.1	10.28±1.70	22.07±0.43	19.44±1.66	20.99±1.17	22.21±1.41	**23.59±1.27**
	0.3	15.22±2.08	30.71±0.53	26.14±1.37	28.79±1.25	30.33±1.24	**31.30±1.30**

enhancement mechanism. This process iteratively improves both the synthesized data and the ensemble, resulting in a much more powerful ensemble model compared to FedENS. In conclusion, the strength of Co-Boosting lies in its ability to continuously generate high-quality data and refine ensemble performance, leading to a global server model that consistently outperforms others.

5.1.6.2 One-Shot Federated Learning with Progressive Model Fusion

This work [361] explores the one-shot federated learning (OFL) framework, which reduces communication overhead by transmitting local models only once. While efficient, OFL risks convergence instability due to limited model aggregation. Key limitations include restricted feature extraction from isolated local training and excessive server storage demands for multiple client models.

Existing FL methods reduce gradient synchronization but still require hundreds of communication rounds. This motivates the core research question: how can FL achieve high performance under *minimal communication budgets* without added computational or storage burdens?

Method To address OFL's limitations, [361] proposes *FuseFL*, which tackles the *isolation problem* (spurious correlations in non-IID data) via cross-client intermediate feature augmentation. Neural networks are modeled as

$$F = \Lambda \circ H^K \circ H^{K-1} \circ \cdots \circ H^k \circ \cdots \circ H^1, \quad \text{for } 1 \le k \le K, \tag{5.21}$$

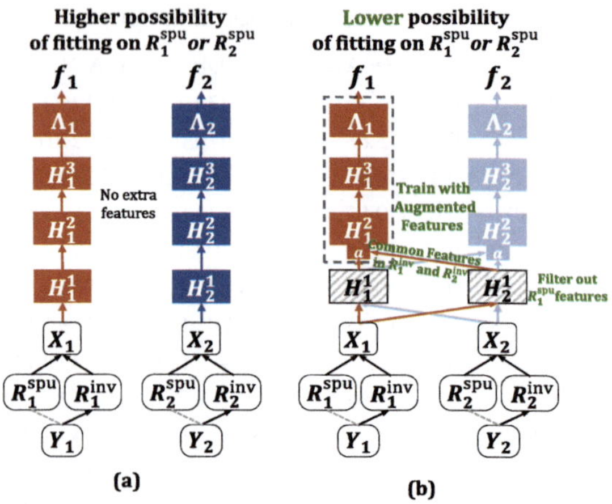

Fig. 5.7 Structure equation model [291] of FL. R_i^{spu} represents the spurious features for i-th class while R_i^{inv} invariant features for i-th class. (**a**) Isolated training and ensemble. (**b**) Federated fusion

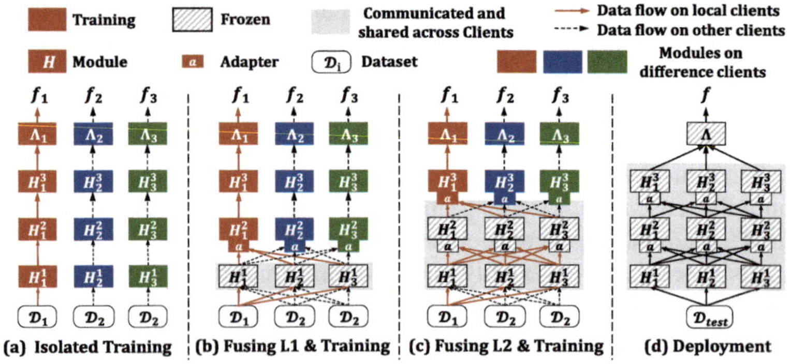

Fig. 5.8 Λ represents the classifier and f the neural network. And H_m^k denotes the modular layer on m-th client and the k-th layer. (**a**) During the first phase, each neural network block is trained independently across clients. (**b**) Subsequently, the initial blocks (L1) are exchanged between clients and synchronized, after which their parameters remain fixed. (**c**) An analogous procedure is applied to the third blocks (L3), mirroring the collaborative training and fusion steps outlined for L1. (**d**) In FuseFL's inference stage, original client-specific blocks (depicted as larger units) process inputs, while compact adaptors (smaller units) merge features from preceding modules

where Λ is the classifier and H_k denotes modular layers and $H^i \circ H^j(\cdot)$ means $H^i(H^j(\cdot))$. Using structural equation modeling (SEM), FuseFL identifies invariant features while mitigating spurious ones, as illustrated in Fig. 5.7.

FuseFL proposes a multistage training network (Fig. 5.8) as follows:

- The local client model is trained independently.
- Intermediate buildings are matched to clients with the help of the clients themselves (e.g., at the center of convolutional layers).
- The top layer is frozen for immutability protection.
- Couplers were selected to increase the robustness of the model. This approach relies on shared semantic knowledge to reduce noise caused by error recovery.

Experiments The MNIST [211], CIFAR-10 [204], and SVHN [278] estimators train FuseFL on non-IID segmentation based on latent Dirichlet allocation. CIFAR-10 test accuracy ($\alpha = 0.5$) outperforms FedAvg and FedDF (Table 5.16). Random and shoddy FuseFL designs now make FL laugh while holding their own.

We also check how well FuseFL works when there are many clients. You can see the results in Table 5.17. The findings show that FuseFL does a great job and beats older methods like FedAvg and FedDF as we add more clients.

To sum up, our experiments show that FuseFL can solve the data imbalance issue in federated learning. It performs better than traditional communication methods and does this at a low computational cost.

5.2 Trustworthy Graph Learning

5.2.1 Introduction and Notation

Graph neural networks (GNNs) [120, 193, 455] demonstrate efficacy across diverse applications, including social networks, knowledge inferencing, and drug discovery. Unlike traditional neural networks that operate on regular structured data such as images or sequences, GNNs can effectively learn from and model complex relationships between entities represented as nodes and edges in a graph structure. These networks address the fundamental challenge of applying deep learning to irregular graph-structured data by implementing the *message-passing* framework. Each node iteratively aggregates information from the neighborhood to refine its representation. This aggregation mechanism allows GNNs to acquire local and global knowledge.

By incorporating structural and semantic information, KGs (knowledge graphs) organize and represent real-world information in a graphical style. KGs organize information into entities (nodes) linked by semantic relationships (edges), forming a comprehensive, interconnected network of facts and concepts. KGs represent relational data, enabling link prediction to infer new facts [27, 61, 180]. Existing methods fall into two categories: semantic models, which learn low-dimensional embeddings to capture local evidence [39, 94, 383, 463], and structural models, which explore KG structures using recurrent neural networks or GNNs [84, 304, 318, 324, 365, 376, 457, 490].

Notation Let $\mathcal{V} = \{v_i\}_{i=1}^{N}$ denote the collection of nodes and $\mathcal{E} = \{e_{ij}\}_{ij=1}^{M}$ indicate the collection of edges. An undirected graph is defined as $\mathcal{G} = (A, \mathbf{X})$, where $A_{ij} = 1$ indicates an edge between v_i and v_j. Each node's D-dimension node feature is denoted as $\mathbf{X}_{[i,:]} \in \mathbb{R}^D$, and its label $y_i \in \mathcal{Y}$ indicates the node category. The symbol $I(\cdot;\cdot)$ represents the mutual information between two variables.

A knowledge graph is defined as $\mathcal{G} = (\mathcal{V}, \mathcal{R}, \mathcal{E})$, where $\mathcal{V}$ is the set of entities, $\mathcal{R}$ is the set of relations, and $\mathcal{E}$ is the set of edges. A query of a knowledge graph is defined as $(e_q, r_q, \cdot)$, where e_q and r_q represent the query

Table 5.16 Accuracy of different methods across $\alpha = \{0.1, 0.3, 0.5\}$ on different datasets. Ensemble means ensemble learning with local trained models, which is an upper bound of all previous methods but impractical in FL due to the large memory costs and the weak scalability of clients. Thus, we highlight the best results in **bold font** except ensemble

Dataset	MNIST			FMNIST			CIFAR-10			SVHN			CIFAR-100			Tiny-Imagenet		
Method	α=0.1	α=0.3	α=0.5	α=0.1	α=0.3	α=0.5	α=0.1	α=0.3	α=0.5	α=0.1	α=0.3	α=0.5	α=0.1	α=0.3	α=0.5	α=0.1	α=0.3	α=0.5
FedAvg	48.24	72.94	90.55	41.69	82.96	83.72	23.93	27.72	43.67	31.65	61.51	56.09	4.58	11.61	12.11	3.12	10.46	11.89
FedDF	60.15	74.01	92.18	43.58	80.67	84.67	40.58	46.78	53.56	49.13	73.34	73.98	28.17	30.28	36.35	15.34	18.22	27.43
Fed-DAFL	64.38	74.18	93.01	47.14	80.59	84.02	47.34	53.89	58.59	53.23	76.56	78.03	28.89	34.89	38.19	18.38	22.18	28.22
DENSE	66.61	76.48	95.82	50.29	83.96	85.94	50.26	59.76	62.19	55.34	79.59	80.03	32.03	37.32	42.07	22.44	28.14	32.34
Ensemble	86.81	96.76	97.22	67.71	87.25	89.42	57.5	77.35	79.91	65.29	88.31	85.7	35.69	49.41	53.39	30.85	39.43	45.8
FuseFL $K=2$	97.02	**98.43**	**98.54**	83.15	**89.94**	89.47	70.85	81.41	**84.34**	76.88	**91.07**	**90.87**	34.07	**45.12**	46.12	**29.28**	31.11	**34.34**
FuseFL $K=4$	**97.19**	98.34	98.29	83.05	84.58	**90.50**	**73.79**	**84.58**	81.15	78.08	89.63	89.34	**36.86**	42.79	**49.30**	27.63	**33.04**	34.28
FuseFL $K=8$	96.66	98.35	98.16	**83.2**	88.57	88.24	70.46	80.70	74.99	**80.31**	88.88	89.94	34.97	39.08	40.73	25.21	32.59	33.82

Table 5.17 Accuracy of different methods across $M = \{5, 10, 20, 50\}$ on CIFAR-10

Dataset	CIFAR-10			
Method	$M = 5$	$M = 10$	$M = 20$	$M = 50$
FedAvg	43.67	38.29	36.03	23.01
FedDF	53.56	54.44	43.15	29.52
Fed-DAFL	55.46	56.34	45.98	29.41
Fed-ADI	58.59	57.13	46.45	27.45
DENSE	62.19	61.42	52.71	28.51
Ensemble	79.91	77.25	59.69	55.63
FuseFL $K = 2$	**84.34**	73.71	62.85	**42.18**
FuseFL $K = 4$	81.15	**78.28**	62.57	37.08
FuseFL $K = 8$	74.99	67.35	**63.19**	28.28

entity and relation, respectively, and $\cdot$ represents the target entity. A GNN for KG reasoning is defined as $f_\theta(\hat{\mathcal{G}}^L)$, where θ represents the model parameters and $\hat{\mathcal{G}}^L$ denotes the *propagation path*, consisting of entity sets $\{\mathcal{V}^0, \mathcal{V}^1, \ldots, \mathcal{V}^L\}$ at each propagation step. The objective is to design adaptive propagation paths $\hat{\mathcal{G}}^L$ that allow the model to accurately predict the target entity e_a for queries $(e_q, r_q, \cdot)$. A *query-dependent propagation path* for a query $(e_q, r_q, \cdot)$ is defined as $\hat{\mathcal{G}}^L_{e_q,r_q} = \{\mathcal{V}^0_{e_q,r_q}, \mathcal{V}^1_{e_q,r_q}, \ldots, \mathcal{V}^L_{e_q,r_q}\}$.

Downstream Tasks The node classification task is to predict the label Y of each node via a parameterized model $f_\theta(\cdot)$. Similarly, the graph classification task aims to predict the label of the graph based on the overall node representation. On the other hand, the link prediction problem entails detecting the existence of query edges that are not present in A. The binary labels of the query edges are denoted as Y, where 0 signifies negative edges and 1 signifies positive edges.

5.2.2 Robust Graph Learning

5.2.2.1 Strengthening and Defending Graph Reconstruction Attack

GNNs [120, 193, 455] are effective in various applications like social networks and drug discovery. However, data privacy remains a challenge. The model inversion attack (MIA) [113, 345, 464] exploits trained models and publicly available data to extract sensitive information. Graph reconstruction attack (GRA) [147, 470] is a specialized MIA targeting GNNs, exposing sensitive relational data.

While the research of MIA is comprehensive in the image domain, understanding and addressing GRA is still in its early stages, leaving GNN-based systems vulnerable. To address this, Zhou et al. [485] systematically analyze GRA by modeling the encoding and decoding processes of GNNs as a Markov chain, treating attacks as chain approximations [485], as shown in Fig. 5.9.

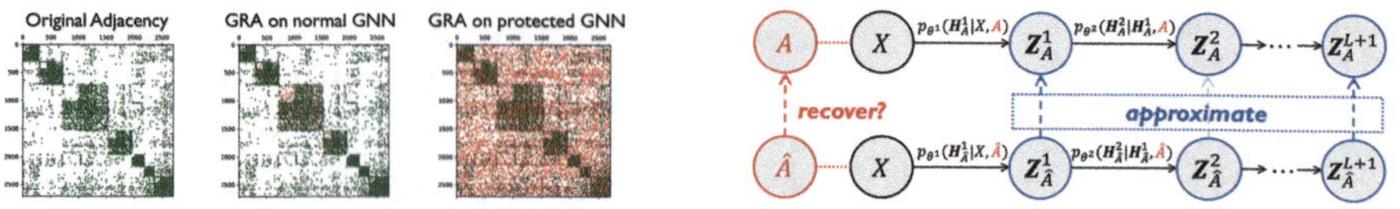

(a) Recovered adjacency on Cora dataset. (b) Modeling the GRA problem Markov chain

Fig. 5.9 Adjacency recovery in GRA attack (**a**) and Markov chain modeling in GRA problem (**b**)

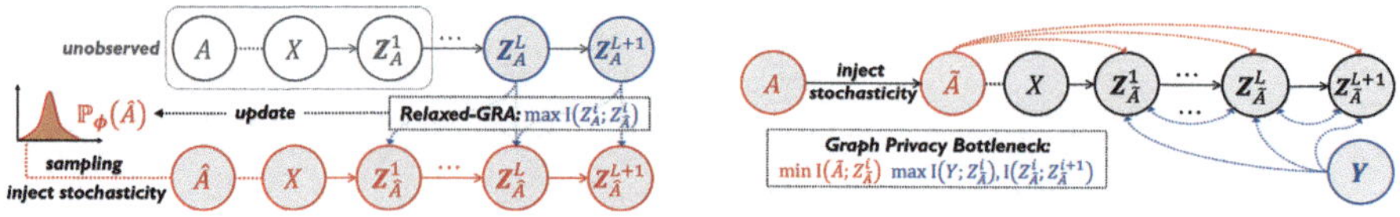

(a) The MC-GRA attacking framework. (b) The MC-GPB defending framework.

Fig. 5.10 Depictions of the two suggested approaches for boosting (**a**) and defending (**b**) the GRA

Based on this formulation, this work proposes two methods: (1) the Markov chain-based graph reconstruction attack (MC-GRA) to estimate the transformation process in attacks and (2) the Markov chain-based graph privacy bottleneck (MC-GPB) to limit mutual dependencies and reduce privacy leakage, as shown in Fig. 5.10. Empirically, both of the proposed methods outperform the baselines across six datasets.

Problem Formulation This study focuses on the privacy risk of GNNs. GRA aims to recover the original linking relations of the training graph. GRA is conducted post hoc, leveraging the target model and any prior knowledge $\mathcal{K} = \{\mathbf{X}, Y, \mathbf{H}_\mathbf{A}, \hat{Y}\}$, where $\mathbf{H}_\mathbf{A}$ denotes the representation of a node and $\hat{Y}$ indicates the classification outputs.

Method The GRA problem is defined as the approximation of the original Markov chain, known as the ORI-chain, by the attack chain, designated as the GRA-chain. This work develops attack and defense strategies, MC-GRA and MC-GPB, by drawing on the information bottleneck principle [336, 370, 405].

$$\begin{aligned} &\text{ORI-chain:} Z^0 \xrightarrow[\theta^1]{\mathbf{A}} Z^1_\mathbf{A} \xrightarrow[\theta^2]{\mathbf{A}} Z^2_\mathbf{A} \rightarrow \cdots \xrightarrow[\theta^{L+1}]{\mathbf{A}} Z^{L+1}_\mathbf{A}, \\ &\text{GRA-chain:} Z^0 \xrightarrow[\theta^1]{\hat{\mathbf{A}}} Z^1_{\hat{\mathbf{A}}} \xrightarrow[\theta^2]{\hat{\mathbf{A}}} Z^2_{\hat{\mathbf{A}}} \rightarrow \cdots \xrightarrow[\theta^{L+1}]{\hat{\mathbf{A}}} Z^{L+1}_{\hat{\mathbf{A}}}, \end{aligned} \tag{5.22}$$

where $Z^0 = \mathbf{X}$, $Z^i_\mathbf{A} = \mathbf{H}^i_\mathbf{A}$ for $i = 1, \cdots, L$ and $Z^{L+1}_\mathbf{A} = \hat{Y}_\mathbf{A}$. The MC-GRA is a maximin game that seeks to maximize the approximation of the forward processes of the two Markov chains while reducing the complexity of each transition with $\hat{\mathbf{A}}$ to prevent trivial solutions by constraining the density. The adaptive capability of MC-GRA derives from its utilization of any existing knowledge collection, enabling its application across all combinations of prior knowledge.

$$\text{MC-GRA: } \hat{\mathbf{A}}^* = \arg\max_{\hat{\mathbf{A}}} \underbrace{\alpha_p I(\mathbf{H}_{\mathbf{A}}; \mathbf{H}^i_{\hat{\mathbf{A}}})}_{\text{propagation approximation}} + \underbrace{\alpha_o I(Y_{\mathbf{A}}; Y_{\hat{\mathbf{A}}}) + \alpha_s I(Y; Y_{\hat{\mathbf{A}}})}_{\text{outputs approximation}} - \underbrace{\alpha_c \mathbf{H}(\hat{\mathbf{A}})}_{\text{complexity}} . \tag{5.23}$$

The MC-GPB architecture emphasizes the reduction of adjacency information in learned representations, which may jeopardize accuracy. The principal objective is to minimize conditional mutual information, achieving an equilibrium between accuracy and privacy. This approach improves privacy through randomization, allowing GNNs to train securely and efficiently. The defense objectives are

$$\text{MC-GPB:} \theta^* = \arg\min_{\theta} \sum_{i=1}^{L} \underbrace{-I(Y; \mathbf{H}^i_{\mathbf{A}})}_{\text{accuracy}} + \underbrace{\beta^i_p I(\mathbf{A}; \mathbf{H}^i_{\mathbf{A}})}_{\text{privacy}} + \sum_{i=1}^{L-1} \underbrace{\beta^i_c I(\mathbf{H}^i_{\mathbf{A}}; \mathbf{H}^{i+1}_{\mathbf{A}})}_{\text{complexity}} . \tag{5.24}$$

Experiments Compared with the baseline information recovery as shown in Tables 5.18 and 5.19, the empirical study evaluates the performance of MC-GRA and MC-GPB using the AUC metric to assess the mutual information between the ground truth adjacency and the approximated adjacency, as shown in Tables 5.20 and 5.21. A higher AUC indicates greater information recovery during the attack, while a lower AUC reflects stronger privacy protection by GNNs.

MC-GRA, as indicated by the experimental results, achieves the best results in all six datasets with various settings of prior knowledge sets, and more prior knowledge generally brings a higher attack AUC. In addition, the learnable MC-GRA brings significantly and consistently better results than the previous non-learnable methods, particularly with the more difficult datasets, namely, Brazil and AIDS, where a maximum promotion of 22.8% and 15.5% can be attained, respectively.

MC-GPB is able to defend all the GRA attack methods, namely, it can counter all attack strategies employed by GRA. In the Polblogs dataset, MC-GPB attains a 13.4% average decrease in privacy leakage, incurring a 1.1% fall in accuracy. Besides, as the results are shown in Table 5.22, MC-GPB also effectively defends MC-GRA, indicating consistent and significant reductions in attack AUC. The above results verify the effectiveness of MC-GPB in defending all kinds of GRA methods, showing the potential to employ MC-GPB in practical applications, such as recommendation systems and social networks.

Furthermore, this empirical study performs MC-GRA and MC-GPB using diverse GNN architectures and assess their generalizability. The results are pre-

Table 5.18 Quantitative analysis of $I(\mathbf{A}; Z)$

Dataset	$I(\mathbf{A}; \mathbf{X})$	$I(\mathbf{A}; \mathbf{H_A})$	$I(\mathbf{A}; \hat{Y}_\mathbf{A})$	$I(\mathbf{A}; Y)$
Cora	<u>0.781</u>	0.766	0.712	**0.815**
Citeseer	**0.881**	0.760	0.743	<u>0.779</u>
Polblogs	–	**0.763**	<u>0.772</u>	0.705
USA	–	**0.850**	<u>0.826</u>	0.728
Brazil	–	**0.758**	<u>0.732</u>	0.613
AIDS	0.521	**0.584**	<u>0.561</u>	0.536

A higher AUC denotes more privacy information leakages

Table 5.19 An ensemble study on the extent of prior knowledge incorporated through different combinations of graph features

X	$\mathbf{H_A}$	$\hat{Y}_\mathbf{A}$	Y	Cora	Citeseer	Polblogs	USA	Brazil	AIDS
✓	✓			0.781	0.881	0.763	0.850	0.758	0.521
✓		✓		0.781	0.881	0.772	0.826	0.732	0.521
✓			✓	0.849	0.907	0.705	0.728	0.613	0.522
✓	✓	✓		0.781	0.881	0.763	0.848	0.756	0.521
✓	✓		✓	0.849	0.907	0.779	0.850	0.743	0.522
✓		✓	✓	0.842	0.907	0.785	0.842	0.730	0.522
✓	✓	✓	✓	0.849	0.907	0.781	0.852	0.717	0.522

sented as Tables 5.23 and 5.24. The empirical findings indicate that both methods are model agnostic, allowing for generalization across various types of GNNs with differing layers. A deeper model (more layers L) typically enhances privacy protection (reducing $I(\mathbf{A}; \mathbf{H}_\mathbf{A}^L)$). Nonetheless, this may result in significant accuracy deterioration due to the recognized over-smoothing phenomenon of GNNs during message propagation. Moreover, it has been observed that a more powerful model is more susceptible to GRA, hence posing a greater danger of privacy breaches in practice.

5.2.2.2 Harmonious Graph Adversarial Learning

In addition to the privacy vulnerability of GNNs, Chen et al. [65] extend the discussion to adversarial vulnerabilities. It has been widely demonstrated the vulnerabilities of GNNs to adversarial attacks, i.e., small, deliberate perturbations made to the input data [183, 350]. Research indicates that even moderate alterations to a graph's existing topology or node features—known as graph modification attacks (GMA)—can substantially deteriorate the performance of GNNs [55, 80, 427, 450, 497]. However, because altering the original graph is often prohibitively costly in real-world applications, recent studies have increasingly focused on graph injection attacks (GIA), where an adversary simply introduces a few malicious nodes to compromise the network [351, 384, 386, 495].

Although empirical findings on GIA have been promising, the reasons behind their rapid growth and potential drawbacks remain unclear. To address this uncer-

Table 5.20 Outcomes of MC-GRA with standard GNNs. The improvements are calculated with respect to the results in Table 5.19

X	$\mathbf{H_A}$	$\hat{Y}_\mathbf{A}$	Y	Cora	Citeseer	Polblogs	USA	Brazil	AIDS
✓	✓			0.864 (10.6% ↑)	0.912 (3.5% ↑)	0.831 (8.9% ↑)	0.883 (3.8% ↑)	0.771 (1.7% ↑)	0.574 (10.1% ↑)
✓		✓		0.839 (7.4% ↑)	0.902 (2.3% ↑)	0.836 (8.2% ↑)	0.913 (10.5% ↑)	0.800 (9.2% ↑)	0.567 (8.8% ↑)
✓			✓	0.896 (5.5% ↑)	0.918 (1.2% ↑)	0.837 (18.7% ↑)	0.825 (13.3% ↑)	0.753 (22.8% ↑)	0.574 (9.9% ↑)
✓	✓	✓		0.866 (10.8% ↑)	0.921 (4.5% ↑)	0.839 (9.9% ↑)	0.878 (3.5% ↑)	0.776 (2.6% ↑)	0.572 (9.7% ↑)
✓	✓		✓	0.905 (6.5% ↑)	0.930 (2.5% ↑)	0.832 (6.8% ↑)	0.878 (3.5% ↑)	0.758 (2.0% ↑)	0.603 (15.5% ↑)
✓		✓	✓	0.897 (5.6% ↑)	0.928 (2.3% ↑)	0.839 (6.8% ↑)	0.870 (3.3% ↑)	0.758 (3.7% ↑)	0.567 (8.6% ↑)
✓	✓	✓	✓	0.904 (6.4% ↑)	0.931 (2.6% ↑)	0.853 (9.2% ↑)	0.870 (1.9% ↑)	0.760 (5.9% ↑)	0.588 (12.6% ↑)

Table 5.21 Results of GRA with MC-GPB protected GNNs. Relative improvements and reductions are calculated with respect to the results in Table 5.18

MI	Cora	Citeseer	Polblogs	USA	Brazil	AIDS
$I(\mathbf{A}; \mathbf{H_A})$	0.706 (7.8% ↓)	0.750 (1.3% ↓)	0.724 (5.1% ↓)	0.716 (15.8% ↓)	0.745 (1.7% ↓)	0.564 (3.4% ↓)
$I(\mathbf{A}; \hat{Y}_\mathbf{A})$	0.704 (0.1% ↓)	0.730 (1.7% ↓)	0.705 (8.7% ↓)	0.587 (28.9% ↓)	0.692 (5.5% ↓)	0.559 (0.4% ↓)
$I(\mathbf{A}; \mathbf{H}^1_{\hat{\mathbf{A}}})$	0.625 (9.9% ↓)	0.691 (9.8% ↓)	0.506 (26.3% ↓)	0.300 (64.5% ↓)	0.609 (25.1% ↓)	0.514 (10.6% ↓)
Acc.	0.734 (3.0% ↓)	0.602 (4.4% ↓)	0.830 (1.1% ↓)	0.391 (16.8% ↓)	0.808 (5.1% ↑)	0.668 (0.0% ↑)

Table 5.22 Outcomes of MC-GRA utilizing MC-GPB safeguarded GNNs. Relative reductions are calculated with respect to the results in Table 5.20

X	$\mathbf{H_A}$	$\hat{Y}_\mathbf{A}$	Y	Cora	Citeseer	Polblogs	USA	Brazil	AIDS
✓	✓			0.816 (5.5% ↓)	0.871 (4.4% ↓)	0.748 (9.9% ↓)	0.841 (4.7% ↓)	0.752 (2.4% ↓)	0.503 (12.3% ↓)
✓		✓		0.817 (9.7% ↓)	0.843 (6.5% ↓)	0.707 (15.4% ↓)	0.844 (7.5% ↓)	0.747 (6.6% ↓)	0.458 (19.2% ↓)
✓			✓	0.892 (0.4% ↓)	0.888 (3.2% ↓)	0.699 (16.4% ↓)	0.738 (10.5% ↓)	0.700 (7.0% ↓)	0.490 (14.6% ↓)
✓	✓	✓		0.804 (7.1% ↓)	0.894 (2.9% ↓)	0.706 (15.8% ↓)	0.754 (14.1% ↓)	0.636 (16.7% ↓)	0.546 (3.7% ↓)
✓	✓		✓	0.890 (1.6% ↓)	0.881 (5.2% ↓)	0.731 (12.1% ↓)	0.808 (5.6% ↓)	0.705 (6.9% ↓)	0.507 (15.9% ↓)
✓		✓	✓	0.858 (4.3% ↓)	0.903 (2.6% ↓)	0.791 (5.7% ↓)	0.768 (11.7% ↓)	0.656 (13.4% ↓)	0.511 (9.8% ↓)
✓	✓	✓	✓	0.864 (4.4% ↓)	0.891 (4.2% ↓)	0.757 (11.2% ↓)	0.853 (1.9% ↓)	0.637 (16.1% ↓)	0.547 (6.9% ↓)

Table 5.23 MC-GRA with different architectures on Cora

	GCN			GAT			GraphSAGE		
$\mathcal{K}$	$L=2$	$L=4$	$L=6$	$L=2$	$L=4$	$L=6$	$L=2$	$L=4$	$L=6$
$\{\mathbf{X}, Y\}$	0.895	0.892	0.878	0.883	0.878	0.876	0.889	0.872	0.840
$\{\mathbf{X}, Y, \mathbf{H_A}\}$	0.904	0.900	0.884	0.897	0.885	0.874	0.892	0.8881	0.873
$\{\mathbf{X}, Y, \mathbf{H_A}, \hat{Y}\}$	0.905	0.895	0.892	0.913	0.887	0.879	0.909	0.893	0.865
Acc.	0.792	0.661	0.248	0.637	0.651	0.630	0.614	0.443	0.145

Table 5.24 MC-GPB with different architectures on Polblogs

	GCN			GAT			GraphSAGE		
MI	$L=2$	$L=4$	$L=6$	$L=2$	$L=4$	$L=6$	$L=2$	$L=4$	$L=6$
$I(\mathbf{A}; \mathbf{H_A})$	0.724	0.790	0.810	0.901	0.808	0.854	0.805	0.808	0.813
$I(\mathbf{A}; \hat{Y}_\mathbf{A})$	0.705	0.650	0.650	0.654	0.623	0.673	0.803	0.668	0.652
$I(\mathbf{A}; \mathbf{H}_{\hat{\mathbf{A}}})$	0.506	0.577	0.532	0.542	0.656	0.536	0.599	0.769	0.468
Acc.	0.830	0.822	0.512	0.855	0.880	0.869	0.830	0.869	0.801

tainty, we analyze both the benefits and drawbacks of GIA by directly comparing it with GMA under a unified framework. Our theoretical analysis demonstrates that, in the absence of any defense mechanisms, GIA can be demonstrably more damaging than GMA because of its greater flexibility. This flexibility allows GIA to effectively translate GMA-type perturbations into tailored GIA perturbations, which can then be further optimized to increase their adversarial impact (Fig. 5.11a). However, invoking the no free lunch principle, we also discover that the efficacy of GIA largely depends on its capacity to severely disrupt the homophily inherent in the original graph. Homophily—the tendency for nodes with similar features or labels to connect—is a critical factor underpinning the performance of most current GNN models [27, 159, 195, 261, 430]. Consequently, extensive degradation of homophily compromises the utility of GIA in assessing robustness, as non-robust models can readily counteract or even nullify the effects of GIA by leveraging this very degradation.

In particular, given that GIA attacks disrupt homophily, it is natural to develop a defensive strategy aimed at restoring this property, which we refer to as *homophily defenders*. We theoretically show that the homophily defenders are robust against GIA attacks. Empirically, the performance of GIA can be even lower than GMA when against homphily defenders that are simply implemented via edge pruning techniques [459] (Fig. 5.11b). Consequently, neglecting the impact on homophily not only renders GIA ineffective but also restricts the utility of GIA in evaluating the robustness of GNNs.

Method To ensure that GIA remains effective for evaluating robust GNNs, it is essential to maintain the homophily of the original graph during the attack. To this end, we introduce a new constraint to GIA—homophily unnoticeability—which compels GIA to preserve the inherent homophily. This constraint acts as an additional safeguard alongside the conventional unnoticeability constraints in graph

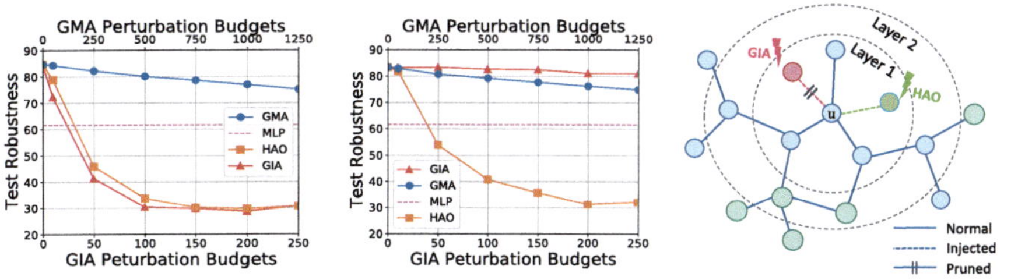

(a) Attack without defense (b) Attack with defense (c) GIA at node u

Fig. 5.11 We conduct a unified evaluation of GIA and GMA, where a lower test robustness score signifies a more effective attack. In the scenario without any defenses (panel **a**), GIA consistently outperforms GMA. However, when defenses are introduced (panel **b**), GIA without homophily-aware optimization (HAO) falls short compared to GMA, whereas GIA with HAO yields the best performance. Panel (**c**) illustrates that GIA can damage the homophily of the original data. In cases where GIA is executed without HAO, the malicious nodes (red) and edges disrupt homophily significantly, making them easy targets for homophily defenders to detect and prune. On the other hand, GIA with HAO is designed to preserve homophily by injecting adversarial nodes and edges that closely resemble legitimate ones (dark green), thereby evading detection and effectively inflicting damage

adversarial learning. Building on this idea, we propose the Harmonious Adversarial Objective (HAO) for GIA (Fig. 5.11c). HAO offers a novel, differentiable method to enforce the homophily constraint by regulating the shift in homophily distribution during the attack. As a result, adversarial modifications become less detectable by homophily defenders while still effectively compromising the model (Fig. 5.11b).

Experiments Tables 5.25 and 5.26 present the performance of various GIA methods for nontargeted and targeted attacks, respectively. In each defense category, the top-performing attack is highlighted in bold, while the runner-up is underlined.

Performance of Nontargeted Attacks Table 5.25 shows that incorporating HAO boosts the performance of every attack across all datasets by as much as 30%, underscoring its effectiveness and versatility. Notably, even when combined with a random injection strategy like PGD, HAO can degrade robust models to a level comparable with—or even worse than—a simple MLP that ignores relational information. Moreover, adaptive injection strategies generally outperform the earlier PGD and TDGIA methods by a significant margin in most scenarios, further highlighting their suitability when paired with HAO.

Performance of Targeted Attack on Large-Scale Graphs The results are given in Table 5.26. We can find that HAO improves the targeted GIA attack performance of every attack method across all datasets by as much as 15%, demonstrating the universal benefits of HAO. Additionally, when with HAO, adaptive injection strategies further elevate attack performance.

Table 5.25 Performance of nontargeted attacks against different models

		Cora (↓)			Citeseer (↓)			Computers (↓)			Arxiv (↓)		
	HAO	Homo	Robust	Combo	Homo	Robust	Combo	Homo	Robust	Combo	Homo	Robust	Combo
Clean		85.74	86.00	87.29	74.85	75.46	75.87	93.17	93.17	93.32	70.77	71.27	71.40
PGD		83.08	83.08	85.74	74.70	74.70	75.19	84.91	84.91	91.41	68.18	68.18	71.11
PGD	✓	52.60	62.60	77.99	69.05	69.05	73.04	79.33	79.33	87.83	55.38	62.89	68.68
MetaGIA†		83.61	83.61	85.86	74.70	74.70	75.15	84.91	84.91	91.41	68.47	68.47	71.09
MetaGIA†	✓	49.25	69.83	76.80	68.04	68.04	71.25	78.96	78.96	90.25	57.05	63.30	69.97
AGIA†		83.44	83.44	85.78	74.72	74.72	75.29	85.21	85.21	91.40	68.07	68.07	71.01
AGIA†	✓	47.24	**61.59**	**75.25**	70.24	70.24	71.80	75.14	75.14	86.02	59.32	65.62	69.92
TDGIA		83.44	83.44	85.72	74.76	74.76	75.26	88.32	88.32	91.40	64.49	64.49	70.97
TDGIA	✓	56.95	73.38	79.45	**60.91**	**60.91**	72.51	**74.77**	**74.77**	90.42	49.36	**60.72**	**63.57**
ATDGIA		83.07	83.07	85.39	74.72	74.72	75.12	86.03	86.03	91.41	66.95	66.95	71.02
ATDGIA	✓	**42.18**	70.30	76.87	61.08	61.08	**71.22**	80.86	80.86	**84.60**	**45.59**	63.30	64.31
MLP		61.75			65.55			84.14			52.49		

↓The lower number indicates better attack performance. †Runs with SeqGIA framework on Computers and Arxiv

Table 5.26 Performance of targeted attacks against different models

	HAO	Computers (↓)			Arxiv (↓)			Aminer (↓)			Reddit (↓)		
		Homo	Robust	Combo	Homo	Robust	Combo	Homo	Robust	Combo	Homo	Robust	Combo
Clean		92.68	92.68	92.83	69.41	71.59	72.09	62.78	66.71	66.97	94.05	97.15	97.13
PGD		88.13	88.13	91.56	69.19	69.19	71.31	53.16	53.16	56.31	92.44	92.44	93.03
PGD	✓	71.78	71.78	85.81	**36.06**	**37.22**	69.38	34.62	34.62	39.47	56.44	86.12	**84.94**
MetaGIA†		87.67	87.67	91.56	69.28	69.28	71.22	48.97	48.97	52.35	92.40	92.40	93.97
MetaGIA†	✓	70.21	71.61	85.83	38.44	38.44	48.06	41.12	41.12	45.16	**46.75**	90.06	90.78
AGIA†		87.57	87.57	91.58	66.19	66.19	70.06	50.50	50.50	53.69	91.62	91.62	93.66
AGIA†	✓	**69.96**	**71.58**	85.72	38.84	38.84	68.97	35.94	35.94	42.66	80.69	88.84	90.44
TDGIA		87.21	87.21	91.56	63.66	63.66	71.06	51.34	51.34	54.82	92.19	92.19	93.62
TDGIA	✓	71.39	71.62	**77.15**	42.56	42.56	42.53	25.78	25.78	29.94	78.16	**85.06**	88.66
ATDGIA		87.85	87.85	91.56	66.12	66.12	71.16	50.87	50.87	53.68	91.25	91.25	93.03
ATDGIA	✓	72.00	72.53	78.35	38.28	40.81	**39.47**	**22.50**	**22.50**	**28.91**	64.09	89.06	88.91
MLP		84.11			52.49			32.80			70.69		

↓The lower number indicates better attack performance. †Runs with SeqGIA framework

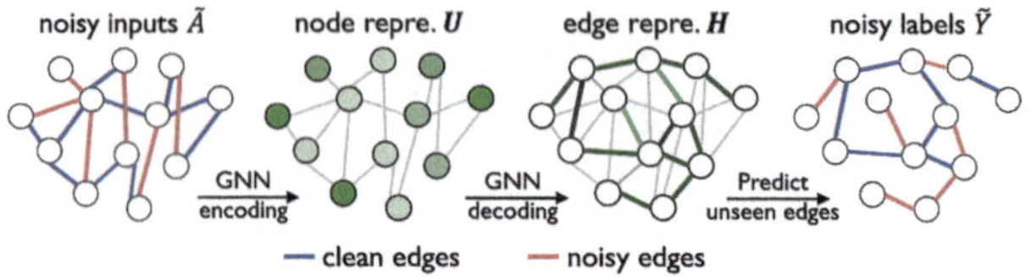

Fig. 5.12 Bilateral edge noise in link prediction. The GNN receives the graph $\mathcal{G} = (\tilde{\mathbf{A}}, \mathbf{X})$ as input, produces the edge representation **H**, and predicts the existing of unobserved edges with labels $\tilde{Y}$

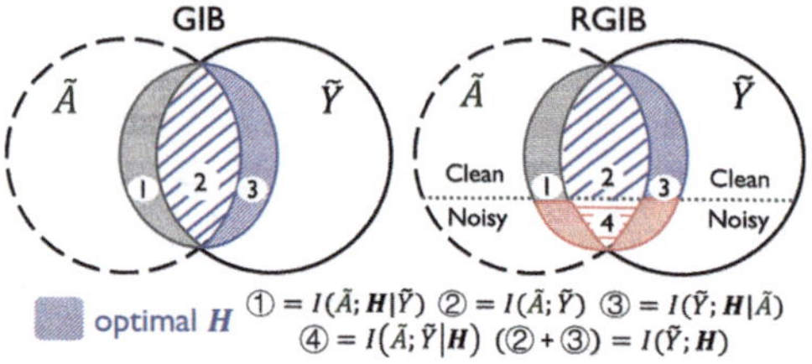

Fig. 5.13 The fundamental GIB and the suggested RGIB. Observation RGIB balances signals ①, ③, ④ to identify the best **H** and eliminate the noisy signals highlighted in red

5.2.2.3 Robust Representation Against the Bilateral Noise

While HAO enhances our understanding of GNN architectures and their functionalities, it does not completely resolve the issues associated with graph noise. As illustrated in Fig. 5.12, noise in links of nodes results in negative impacts in downstream tasks, especially link prediction. An information-theoretic framework designed to mitigate noise in graph connections is presented by Zhou et al. [486]: the Robust Graph Information Bottleneck (RGIB). Based on the Graph Information Bottleneck (GIB), RGIB effectively identifies and handles noise components in input data and breaks down the mutual dependencies between data features, label information, and learned representations significantly enhancing model performance in complex environments. This facilitates the attainment of resilient representations that diminish the impact of bilateral edge noise, as can be seen in Fig. 5.13. To incorporate RGIB into real techniques, this work proposes two specific realizations, RGIB-SSL and RGIB-REP, as shown in Fig. 5.14. RGIB and its implementations significantly enhance the resilience of GNN models under diverse noise conditions.

Problem Formulation Link prediction is the task of finding potential connections that are likely to exist but aren't currently shown in the adjacency matrix A. Each potential connection is assigned a binary label Y, where $Y = 1$ indicates a likely existing edge and $Y = 0$ indicates a likely nonexisting edge.

Method The main objective of RGIB is to construct graph representations that can be robust to bilateral link noise. The approach is formulated to address disturbances in both the graph configuration $\tilde{\mathbf{A}}$ and the label domain $\tilde{Y}$. This method enhances model robustness by building on the foundation laid by Graph Information Bottleneck (GIB) [405] through two primary mechanisms: a) examining and

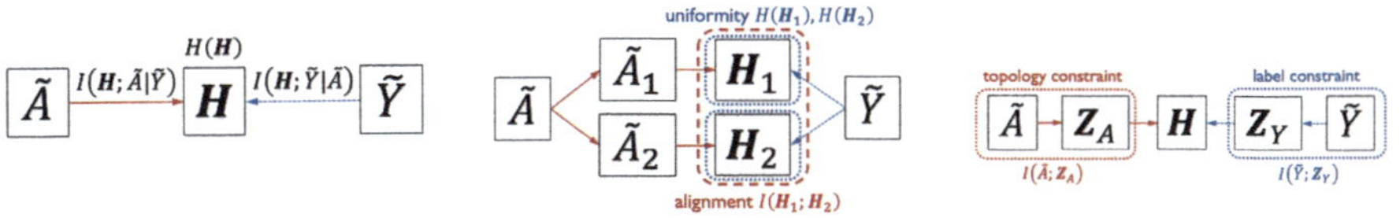

(a) RGIB (Equation (5.25)) (b) RGIB-SSL (Equation (5.26)) (c) RGIB-REP (Equation (5.27))

Fig. 5.14 Digrams illustrating the RGIB concept (**a**) and its two realizations (**b**, **c**). RGIB-SSL (**b**) employs a comparative method utilizing data perspectives after augmentation, allowing the model to sustain consistent performance despite the presence of noisy input signals. RGIB-REP (**c**) employs dual reparameterization variables $\mathbf{Z}_A$ and $\mathbf{Z}_Y$ to simultaneously enhance the reliability of both network connections and label assignments, resulting in more resilient graph representations

optimizing the interconnections among input features, label assignments, and learned embeddings and b) creating noise-resistant information metrics that mitigate the effects of data corruption on the learning process:

$$\begin{aligned} \min \text{RGIB} &\triangleq -I(\mathbf{H}; \tilde{Y}), \quad \text{s.t. } \gamma_{\mathbf{H}}^{-} < \mathbf{h}(\mathbf{H}) < \gamma_{\mathbf{H}}^{+}, \\ & I(\mathbf{H}; \tilde{Y}|\tilde{\mathbf{A}}) < \gamma_Y, \; I(\mathbf{H}; \tilde{\mathbf{A}}|\tilde{Y}) < \gamma_{\mathbf{A}}. \end{aligned} \tag{5.25}$$

Here, $\mathbf{h}(\mathbf{H})$ denotes the entropy of $\mathbf{H}$. $\gamma_{\mathbf{H}}^{+}$ and $\gamma_{\mathbf{H}}^{-}$ are hyperparameters that can avoid over-fitting in high information mutation and representation collapse in low information mutation. γ_Y and $\gamma_{\mathbf{A}}$ are upper bound of mutual information between edges and labels and between edges and adjacency noise, respectively. This work proposes two practical realizations of RGIB, RGIB-SSL, and RGIB-REP, which enhance this framework by mitigating the effects of noisy inputs and labels. Both of them show benefits in addressing bilateral edge noise and alleviating overfitting from irrelevant or noisy data.

RGIB-SSL employs self-supervised regularization to preserve the integrity of graph representations, avert collapse, and accurately capture links among graph edges. This is accomplished through the utilization of a loss function that incorporates supervised mutual information, uniformity, and alignment components, with hyperparameters serving as modulators. Through the utilization of contrastive learning, hybrid graph augmentation, self-adversarial alignment, and uniformity losses, RGIB-SSL can enhance the robustness of representations. Adaptive upgrades in graph-based activities are made easier with the help of this technology, which offers a feasible approximation of mutual information aims. The optimization objective formulation of RGIB-SSL is shown as follows:

$$\begin{aligned} \min \text{RGIB-SSL} &\triangleq -\underbrace{\lambda_s(I(\mathbf{H}_1; \tilde{Y}) + I(\mathbf{H}_2; \tilde{Y}))}_{\text{supervision}} \\ & -\underbrace{\lambda_u(\mathbf{h}(\mathbf{H}_1) + \mathbf{h}(\mathbf{H}_2))}_{\text{uniformity}} - \underbrace{\lambda_a I(\mathbf{H}_1; \mathbf{H}_2)}_{\text{alignment}}, \end{aligned} \tag{5.26}$$

where $\mathbf{H}_1, \mathbf{H}_2$ denotes aligned representations of two different adjacency noises and $\lambda_s, \lambda_u, \lambda_a$ indicates the balanced weights. Every aspect in the formulation fulfills a specific function: the supervision term amplifies the informational quality of representations, the uniformity term inhibits representation degradation, and the alignment term guarantees coherence among various augmented perspectives of the graph.

RGIB-REP demonstrates explicitly reparameterization on both topology and label spaces. It introduces latent variables $\mathbf{Z_A}$ and $\mathbf{Z}_Y$ to extract task-relevant information from the noisy adjacency $\tilde{\mathbf{A}}$ and labels $\tilde{Y}$ while discarding noise. The objective of RGIB-REP objective consists of three terms: 1) supervision, $I(\mathbf{H}; \mathbf{Z}_Y)$; 2) topology constraint, $I(\mathbf{Z_A}; \tilde{\mathbf{A}})$; and 3) label constraint, $I(\mathbf{Z}_Y; \tilde{Y})$. $\lambda_\mathbf{A}, \lambda_Y$ are balanced weights. These components are bounded and simplified into regularization terms plus a classification loss, making the optimization process computationally feasible:

$$\min \text{RGIB-REP} \triangleq \; -\underbrace{\lambda_s I(\mathbf{H}; \mathbf{Z}_Y)}_{\text{supervision}} + \underbrace{\lambda_A I(\mathbf{Z_A}; \tilde{\mathbf{A}})}_{\text{topology constraint}} + \underbrace{\lambda_Y I(\mathbf{Z}_Y; \tilde{Y})}_{\text{label constraint}}, \tag{5.27}$$

Experiments This part performs empirical analyses on six prominent datasets and evaluates our methods against three categories of baselines. As seen in Table 5.27, the RGIB demonstrates state-of-the-art results across various bilateral noise ratios. In particular, RGIB shows significant improvement in Cora and Citeseer, achieving a 12.9% increase in AUC relative to the second-best methods.

RGIB-SSL and RGIB-REP are general to diverse graph scenarios and show strengths in different density of graphs. In sparsely connected networks like Cora and Citeseer, RGIB-SSL exhibits enhanced robustness, adeptly addressing the difficulties associated with restricted connectivity. Conversely, RGIB-REP shows efficacy in high connectivity density graphs, such as Facebook and Chameleon, where its capacity to manage numerous node connections is very advantageous.

While RGIB tackles the overarching issue of noise-resistant representation learning in graphs, a more particular and equally significant obstacle arises in situations when only positive and unlabeled samples are accessible. This prompts an examination of positive-unlabeled (PU) learning within graph settings, where the traditional premise of independent sample creation is inherently compromised by the interrelated characteristics of graph data.

5.2.2.4 Unraveling the Impact of Heterophilic Structures on Graph Positive-Unlabeled Learning

Positive-unlabeled (PU) learning [92, 93, 95], which trains a classifier using only positive and unlabeled samples, has proven highly effective for image and text applications. Recently, there has been growing interest in extending PU learning

Table 5.27 Comparison of methods with GCN based on AUC under both input and label noise. The **boldface** signifies the optimal outcomes, while the underlined denotes the suboptimal results

	Cora			Citeseer			Pubmed			Facebook			Chameleon			Squirrel		
Method	20%	40%	60%	20%	40%	60%	20%	40%	60%	20%	40%	60%	20%	40%	60%	20%	40%	60%
Standard	0.811	0.742	0.697	0.786	0.738	0.709	0.887	0.875	0.864	**0.983**	0.952	0.944	0.962	0.950	0.927	0.943	0.941	0.939
DropEdge	0.802	0.742	0.730	0.764	0.740	0.709	0.871	0.848	0.835	0.981	0.968	0.947	0.957	0.955	0.941	0.944	0.938	0.937
NeuralSparse	0.819	0.732	0.729	0.777	0.740	0.715	0.891	0.873	0.863	**0.983**	0.964	0.946	0.960	0.950	0.940	0.950	0.931	0.930
PTDNet	0.805	0.756	0.739	0.780	0.742	0.728	0.887	0.873	0.862	0.973	0.967	0.949	0.961	0.951	0.942	0.949	0.933	0.930
Co-teaching	0.820	0.748	0.703	0.753	0.724	0.713	0.894	0.876	0.864	0.982	0.953	0.948	0.960	0.952	0.948	0.946	0.935	0.937
Peer loss	0.819	0.747	0.702	0.742	0.735	0.710	0.896	0.882	0.857	0.981	0.954	0.943	0.954	0.953	0.927	0.946	0.935	0.929
Jaccard	0.814	0.750	0.702	0.747	0.732	0.711	0.887	0.880	0.851	0.979	0.958	0.943	0.950	0.954	0.934	0.944	0.933	0.924
GIB	0.820	0.749	0.715	0.751	0.739	0.712	0.890	0.873	0.854	0.977	0.961	0.942	0.955	0.956	0.932	0.947	0.933	0.930
VIB	0.821	0.781	0.722	0.770	0.812	0.719	0.893	0.883	0.850	0.970	0.964	0.950	0.953	0.956	0.949	0.943	0.940	0.929
PRI	0.798	0.733	0.698	0.757	0.745	0.702	0.890	0.880	0.849	0.960	0.962	0.951	0.951	0.950	0.949	0.938	0.941	0.930
SupCon	0.824	0.782	0.749	0.755	0.746	0.730	0.885	0.872	0.853	0.959	0.951	0.930	0.956	0.953	0.947	0.947	0.935	0.930
GRACE	0.787	0.694	0.693	0.763	0.724	0.684	0.892	0.875	0.859	0.890	0.887	0.832	0.898	0.899	0.895	0.939	0.938	0.936
RGIB-REP	0.831	0.797	0.759	0.788	0.752	0.731	0.902	0.883	0.865	**0.983**	**0.977**	0.952	**0.972**	**0.962**	**0.952**	**0.951**	**0.946**	**0.943**
RGIB-SSL	**0.893**	**0.855**	**0.834**	**0.869**	**0.843**	**0.814**	**0.923**	**0.892**	**0.870**	**0.983**	0.971	**0.964**	0.966	0.959	0.950	0.950	0.943	**0.943**

to graph data (e.g., transaction networks and pandemic prediction) [284, 412, 437], where a limited number of nodes are labeled positive and the rest remain unlabeled.

Most PU methods assume that samples are generated independently—a condition that often fails in graphs where nodes are interdependent [407, 480]. Our work shows that one major challenge in graph-based PU learning is the presence of *heterophilic structures*, where positive and negative nodes frequently connect. When only positive and unlabeled data are available, the challenge becomes even more significant [252, 297].

In these situations, heterophilic edges tend to blend features from both positive and negative nodes, disrupting the clear separation needed for an accurate class prior estimate (CPE) [409]. This often leads to an inflated estimation of the proportion of positive nodes. In addition, mixing nodes with different labels interrupts the normal flow of messages among same-class nodes, making it significantly harder to accurately uncover the hidden labels of unlabeled nodes during PU classification.

To address this challenge, we introduce graph PU Learning with label propagation loss (GPL). GPL reduces the disruptive effects of heterophily by assigning lower weights to edges connecting nodes of different classes. By focusing exclusively on observed positive nodes, our approach preserves the strong connections among similar nodes while minimizing interference from heterophilic links. This strategy leads to better class prior estimates and improves PU classification performance.

Problem Formulation We have an undirected graph $\mathcal{G} = (\mathcal{V}, \mathcal{E})$ with adjacency matrix A, node features $\mathbf{x}_{[i,:]} \in \mathbb{R}^D$, and binary labels $y_i \in \{-1, +1\}$. A portion of these nodes is known to be positive (set $\mathcal{P}$), while the rest (set $\mathcal{U}$) is unlabeled, consisting of both unobserved positives and negatives. We aim to train a binary classifier $f_\theta(\mathbf{x}, \mathcal{G}_\mathbf{x}^A)$ that approximates $P(y = +1 \mid \mathbf{x}_{[i,:]}, \mathcal{G}_\mathbf{x}^A)$ for unlabeled data $\mathcal{U}$. In contrast to standard PU methods assuming i.i.d. samples, we must consider the interdependent structure introduced by A.

Method The *graph PU learning with label propagation loss* (GPL) method has the following formulation:

$$\begin{aligned} & \min_\theta \mathcal{L}_{\text{GNN}}(\hat{\pi}_\text{p}, w, \hat{A}) \\ \text{s.t.} \quad & \hat{\pi}_\text{p} = \mathcal{E}_{\text{CPE}}(\hat{A}) \\ & \hat{A} = \arg\min_A \mathcal{L}_{\text{LPL}}(A)^{(K)}, \end{aligned} \tag{5.28}$$

the proposed *label propagation loss* (LPL), i.e., $\mathcal{L}_{\text{LPL}}$ will optimize the heterophilic structure A into $\hat{A}$. Equation 5.28 effectively adds a mechanism to dampen heterophily in standard graph PU learning, addressing the challenges we mentioned.

To further reduce these heterophilic structures, we use the classic label propagation algorithm (LPA),[4] which operates exclusively on observed positive nodes, outlined as follows.

Let $\mathbf{E}^{(0)} = [P(y=+1|\mathbf{x}_{[i,:]}, \mathcal{G}_{\mathbf{x}}^A), P(y=-1|\mathbf{x}_{[i,:]}, \mathcal{G}_{\mathbf{x}}^A)]_{i\in\mathcal{V}}$ be a vector of the nodes' initial class-posterior probabilities, and under LPA, it is updated as follows:

$$\begin{aligned}\mathbf{E}^{(k)} &= \alpha\mathbf{E}^{(k-1)} + (1-\alpha)D_{dia}^{-1}A\mathbf{E}^{(k-1)},\\ \mathbf{E}^{(k)} &= [P(y=+1|\mathbf{x}_{[i,:]}, \mathcal{G}_{\mathbf{x}}^A)^{(k)}, P(y=-1|\mathbf{x}_{[i,:]}, \mathcal{G}_{\mathbf{x}}^A)^{(k)}]_{i\in\mathcal{V}}\\ &= [\mathbf{E}_i^{(k)}]_{i\in\mathcal{V}}.\end{aligned} \tag{5.29}$$

Here, $0 < \alpha < 1$ is a chosen parameter, $k \in K$ denotes the number of label-propagation iterations, and D_{dia} is the diagonal degree matrix of A. We employ LPA as a proxy to mitigate heterophily, referring to this approach as *label propagation loss* (LPL), defined as follows:

$$\begin{aligned}\hat{A} &= \arg\min_A \mathcal{L}_{\mathrm{LPL}}(A)^{(K)}\\ &= \arg\min_A \frac{1}{n_{\mathcal{P}}} \sum_{\mathbf{x}_{[i,:]}\in\mathcal{P}} \log(P(y=-1|\mathbf{x}_{[i,:]}, \mathcal{G}_{\mathbf{x}}^A)^{(K)}).\end{aligned} \tag{5.30}$$

By incorporating LPL, we address Eq. (5.28) through a bi-level framework that integrates both *class-prior estimator* (CPE) and PU classification.

Class-Prior Estimator Using multiple layers of graph convolution, each node is mapped to a positive posterior probability $z \in [0, 1]$, expressed as $z = f_\theta(\mathbf{x}, \mathcal{G}_{\mathbf{x}}^A)$ where w are the trainable parameters of the GNN model f_θ. For a probability density function P_θ and model f_θ, we introduce the following cumulative distribution function $\mathbb{Q}_\theta^A(z) = \int_{f_\theta(\mathbf{x},\mathcal{G}_{\mathbf{x}}^A)\geq z} P_\theta(\mathbf{x}, \mathcal{G}_{\mathbf{x}}^A)\mathrm{d}x$, which reflects the probability that the classifier f_θ assigns the pair $(\mathbf{x}, \mathcal{G}_{\mathbf{x}}^A)$ a value exceeding z in the transformed representation. For each probability density distribution $P_{\mathrm{p}}(\mathbf{x}, \mathcal{G}_{\mathbf{x}}^A)$, $P_{\mathrm{n}}(\mathbf{x}, \mathcal{G}_{\mathbf{x}}^A)$, and $P_{\mathrm{u}}(\mathbf{x}, \mathcal{G}_{\mathbf{x}}^A)$, we can define $\mathbb{Q}_{\mathrm{p}}^A(z)$, $\mathbb{Q}_{\mathrm{n}}^A(z)$, and $\mathbb{Q}_{\mathrm{u}}^A(z)$, respectively. Once the graph structure is updated to $\hat{A}$ following Eq. (5.30), the class prior can be estimated as

$$\hat{\pi}_{\mathrm{p}} = \min_{c\in[0,1]} \frac{\mathbb{Q}_{\mathrm{u}}^{\hat{A}}(c)}{\mathbb{Q}_{\mathrm{p}}^{\hat{A}}(c)} = \mathcal{E}_{\mathrm{CPE}}(\hat{A}). \tag{5.31}$$

PU Classification Starting with a training set of positive nodes $\mathcal{P}$ and unlabeled nodes $\mathcal{U}$, along with an estimated class prior $\hat{\pi}_{\mathrm{p}}$, we begin by ranking the nodes in $\mathcal{U}$ according to their positive class-posterior probabilities $P(y_i = +1 \mid x_i, \mathcal{G}_{x_i}^A)$. At

[4] A well-known method that iteratively refines node labels across the edges of a graph.

each training epoch, we create a temporary subset of likely negative nodes, $\mathcal{U} \setminus \mathcal{S}^{\mathcal{G}}_{\hat{\pi}_\mathrm{p}}$, by removing $\mathcal{S}^{\mathcal{G}}_{\hat{\pi}_\mathrm{p}}$—the top $\hat{\pi}_\mathrm{p}$ fraction of $\mathcal{U}$ with the largest positive probabilities. We then update the GNN classifier by minimizing the following loss function on the optimized graph structure $\hat{A}$:

$$\mathcal{L}_{\mathrm{GNN}}(\hat{\pi}_\mathrm{p}, w, \hat{A}) = \frac{1}{|\mathcal{P} \cup \mathcal{S}^{\mathcal{G}}_{\hat{\pi}_\mathrm{p}}|} \sum_{\mathbf{x}_{[i,:]} \in \mathcal{P} \cup \mathcal{S}^{\mathcal{G}}_{\hat{\pi}_\mathrm{p}}} \mathcal{L}\Big(f_\theta\big(\mathbf{x}_{[i,:]}, \mathcal{G}^{\hat{A}}_{\mathbf{x}_{[i,:]}}\big), +1\Big)$$
$$+ \frac{1}{|\mathcal{U} \setminus \mathcal{S}^{\mathcal{G}}_{\hat{\pi}_\mathrm{p}}|} \sum_{\mathbf{x}_{[i,:]} \in \mathcal{U} \setminus \mathcal{S}^{\mathcal{G}}_{\hat{\pi}_\mathrm{p}}} \mathcal{L}\Big(f_\theta\big(\mathbf{x}_{[i,:]}, \mathcal{G}^{\hat{A}}_{\mathbf{x}_{[i,:]}}\big), -1\Big).$$

Experiments We evaluated our approach on several real-world datasets whose edge homophily ratios span from strongly homophilic to strongly heterophilic. As shown in Table 5.28, our GPL method consistently outperforms other techniques, achieving the highest classification accuracy across all tested settings.

Moreover, we evaluated its efficacy in class-prior estimation (CPE) and examined how the method differentiates between heterophilic and homophilic edges by inspecting the learned weight distribution. In the CPE task (Table 5.29), our approach shows reduced estimation errors as the heterophily ratio rises, suggesting that GPL effectively tackles heterophily. Furthermore, Fig. 5.15 illustrates that during training, the weights for heterophilic edges steadily decrease, while those for homophilic edges remain high. Overall, these observations confirm that GPL mitigates the negative effects of heterophilic structures, thereby boosting the performance of graph PU learning.

PU learning concentrates on situations involving solely positive and unlabeled input, whereas graph-based learning encounters a significant challenge: the presence of noisy labels across all categories. In contrast to PU learning's particular emphasis on handling unlabeled data, situations with noisy labels must tackle the dissemination of label inaccuracies throughout the graph structure. This disparity requires varied methodologies for effective learning, especially in addressing the interrelated characteristics of graph data.

5.2.2.5 Mitigating Label Noise on Graphs via Topological Sample Selection

Noisy labels are a frequent challenge in real-world settings [88, 269], often impeding the training speed and generalization of deep neural networks [239, 313]. This issue is particularly severe in graph data, where labeling errors can easily spread along the network's edges—contrasting with the independent and identically distributed (i.i.d.) conditions typical of image data—thus intensifying the need for robust GNN training methods [81, 225, 411].

Table 5.28 Mean F1 score ± std over different datasets. The best model is bolded. The methods marked with an asterisk (*) rely on class prior knowledge

	Cora	Pubmed	Citeseer	Wiki-CS	Cornell	Chameleon	Squirrel	Actor	Wisconsin	Texas
Hete. ratio h	**0.19**	**0.20**	**0.26**	**0.35**	**0.70**	**0.77**	**0.78**	**0.78**	**0.79**	**0.89**
#Nodes $\|\mathcal{V}\|$	2,708	19,717	3,327	11,701	183	2,277	5,201	7,600	251	183
#Edges $\|\mathcal{E}\|$	13,264	108,365	12,431	431,206	298	36,101	217,073	30,019	515	325
GCN	25.2±5.2	19.7±3.1	25.7±3.1	35.4±4.1	1.4±2.9	0.7±0.6	0.8±0.7	0.6±0.3	0.9±1.8	1.3±3.0
MLP	16.6±8.1	8.0±9.0	4.2±8.4	5.4±8.4	11.3±9.1	18.8±6.9	14.8±6.4	17.6±8.1	15.9±8.9	11.3±9.2
GCN+TED	80.1±0.8	75.4±0.4	70.0±1.4	80.6±0.5	13.9±5.5	17.5±3.5	22.8±1.3	27.6±1.0	19.6±6.1	11.9±5.0
MLP+TED	24.4±9.2	15.6±19.1	10.6±9.5	16.2±7.5	29.2±9.5	26.5±4.1	31.4±2.6	43.3±6.7	40.4±9.1	41.0±7.9
GCN+NNPU*	76.7±0.9	76.5±2.3	66.2±1.1	71.1±0.6	11.7±6.3	28.2±9.2	15.5±12.6	25.2±21.7	15.5±8.6	22.8±19.7
MLP+NNPU*	25.2±5.6	34.2±2.3	12.9±9.8	24.4±5.1	32.9±4.9	21.0±3.7	32.1±5.6	43.2±2.6	35.6±5.2	40.1±6.2
LSDAN*	63.5±4.1	69.6±0.4	47.0±19.	80.4±0.9	26.8±4.3	30.2±8.5	25.5±9.3	38.4±9.2	25.5±8.6	41.8±9.7
GRAB	80.4±0.2	71.6±0.3	69.7±0.4	79.4±1.0	30.6±9.0	15.5±2.2	29.0±8.4	25.2±0.7	39.5±9.6	40.2±8.2
PU-GNN*	79.8±0.3	73.0±0.4	69.5±0.4	80.3±1.8	28.3±3.9	31.5±6.3	29.5±4.8	32.6±8.4	27.8±8.2	42.3±9.4
GPL	**81.9**±0.5	**79.1**±0.4	**74.1**±0.9	**82.3**±0.8	**37.9**±1.3	**36.2**±1.4	**37.7**±5.8	**48.0**±4.2	**45.2**±2.7	**46.3**±3.2
	↑ 2.2%	↑ 3.4%	↑ 5.8%	↑ 2.1%	↑ 15.2%	↑ 14.5%	↑ 17.4%	↑ 10.9%	↑ 12.5%	↑ 10.2%

Table 5.29 In the first row, we present the absolute estimation error with respect to the true class prior across multiple datasets. The results, reported as the mean absolute error over five runs, highlight the best-performing model (i.e., the smallest estimation error)

	Cora	Pubmed	Citeseer	Wiki-CS	Cornell	Chameleon	Squirrel	Actor	Wisconsin	Texas
Hete. ratio h	**0.19**	**0.20**	**0.26**	**0.35**	**0.70**	**0.77**	**0.78**	**0.78**	**0.79**	**0.89**
#True π_p	0.1779	0.2496	0.1179	0.1293	0.2887	0.1294	0.1113	0.1485	0.3073	0.3835
KM	0.09±0.02	0.08±0.04	0.10±0.05	0.06±0.03	0.21±0.04	0.18±0.05	0.39±0.10	0.16±0.06	0.19±0.08	0.31±0.05
DEDPUL	0.05±0.03	0.05±0.05	0.07±0.04	0.05±0.02	0.19±0.05	0.16±0.04	0.32±0.09	0.14±0.07	0.17±0.06	0.25±0.07
MPE	0.04±0.03	0.04±0.02	0.08±0.04	0.02±0.01	0.13±0.06	0.10±0.03	0.36±0.17	0.08±0.02	0.14±0.06	0.26±0.03
ReMPE	0.03±0.03	0.04±0.01	0.06±0.03	0.02±0.02	0.14±0.07	0.12±0.05	0.34±0.12	0.11±0.04	0.12±0.05	0.25±0.03
BBE	0.03±0.01	0.05±0.01	0.04±0.02	0.03±0.01	0.27±0.01	0.12±0.01	0.10±0.08	0.11±0.01	0.23±0.05	0.27±0.02
TED	0.02±0.01	0.04±0.00	0.02±0.01	0.02±0.01	0.18±0.04	0.25±0.03	0.13±0.04	0.25±0.10	0.16±0.02	0.26±0.03
GRAB	0.07±0.13	0.08±0.09	0.06±0.08	0.09±0.12	0.35±0.26	0.18±0.07	0.36±0.19	0.30±0.07	0.37±0.18	0.33±0.20
GPL	**0.01±0.01**	**0.008±0.01**	**0.01±0.01**	**0.01±0.00**	**0.04±0.03**	**0.07 ± ±0.01**	**0.02±0.01**	**0.02±0.01**	**0.03±0.02**	**0.08±0.03**
	↓ 0.01	↓ 0.03	↓ 0.01	↓ 0.01	↓ 0.09	↓ 0.03	↓ 0.08	↓ 0.06	↓ 0.09	↓ 0.17

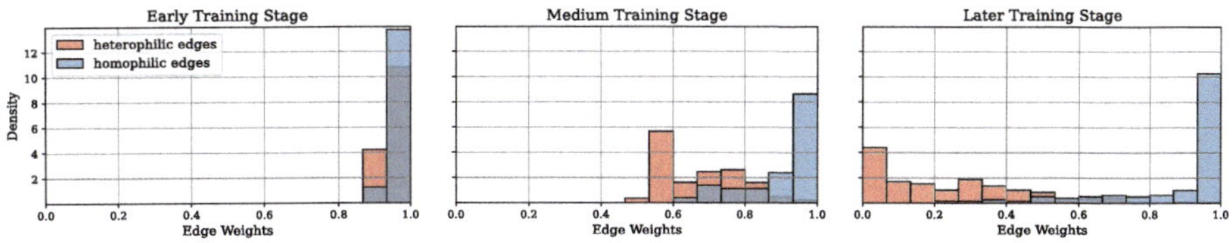

Fig. 5.15 This graph illustrates how the distributions of edge weights for both heterophilic and homophilic connections change throughout the GPL training process on the Cornell dataset

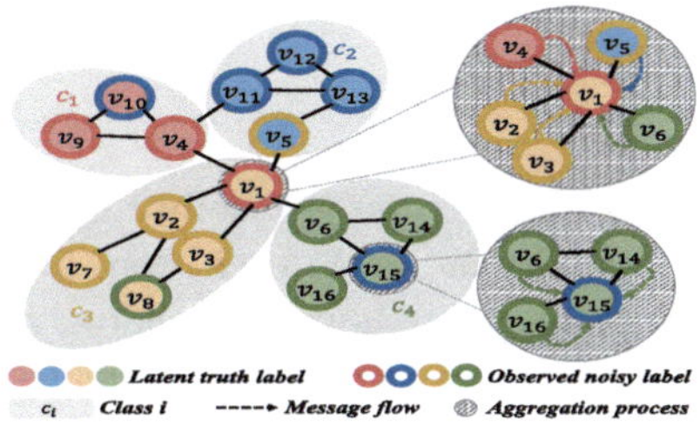

Fig. 5.16 Illustration of nodes with noisy labels in different topological contexts: v_1 is a mislabeled node situated near the class boundaries, while v_{15} is a mislabeled node located further away from those boundaries

For i.i.d. datasets, one effective strategy to counter label noise is sample selection [136, 182]. This approach capitalizes on the model's tendency to learn from correctly labeled data first—the so-called memorization effect—to sift out and retain only reliable instances (such as those exhibiting low loss or high confidence). By creating a cleaner training set, this method significantly reduces the detrimental impact of noisy labels.

Directly applying sample selection techniques to graph data with noisy labels often falls short because they fail to account for the underlying topological structure. As illustrated in Fig. 5.16, boundary-near nodes, which are highly informative as they connect diverse classes, are difficult to identify because they are rare, lose discriminative power when aggregating information from heterogeneous neighbors. Until recently, no existing methods have successfully utilized graph topology to improve sample selection.

We now propose a *topological sample selection* (TSS) approach that identifies likely clean samples in graphs from noisily labeled graph. TSS begins by prioritizing nodes situated far from class boundaries, as they tend to possess more reliable labels and support more accurate model representations. After selecting nodes far from class boundaries, TSS then focus on nodes near class boundaries. During this phase, the *class-conditional betweenness centrality* (CBC) metric is employed to capture the intricate flow of information between classes. By iteratively refining the selection process, TSS gradually guides the model toward data that is both cleaner and more informative.

Problem Formulation The dataset D is composed of two parts: a noisy training set $\tilde{D}_{\text{tr}} = \{(A, \mathbf{x}_{[i,:]}, \tilde{y}_i)\}_{i=1}^{n_{\text{tr}}}$, derived from a noisy distribution $P_{\tilde{D}}$, and a clean test set D_{te}, sampled from a clean distribution P_D. Our aim is to train a GNN $f_\theta : (A, \mathbf{x}) \rightarrow \mathbf{Y}$ using only $\tilde{D}_{\text{tr}}$.

Method We introduce a *class-conditional betweenness centrality* (CBC) measure that use the personalized PageRank matrix:

$$\boldsymbol{\pi} = \alpha\big(\mathbf{I} - (1-\alpha)\hat{A}\big)^{-1} \quad (\boldsymbol{\pi} \in \mathbb{R}^{n\times n}).$$

We compute CBC for each node v_i by measuring how often random walks—launched between nodes of different classes within $\mathcal{G}$—pass through v_i:

$$\mathbf{Cb}_i := \frac{1}{n(n-1)} \sum_{\substack{v_u \neq v_i \neq v_v, \\ \tilde{y}_u \neq \tilde{y}_v}} \frac{\boldsymbol{\pi}_{u,i}\,\boldsymbol{\pi}_{i,v}}{\boldsymbol{\pi}_{u,v}}, \tag{5.32}$$

where $\boldsymbol{\pi}_{u,i}$ is the probability that an α-discounted random walk, initiated at v_u, terminates at $\mathbf{v}_i$. At each step, the walk either remains at its current node with probability α or hops to a randomly chosen neighbor with probability $1-\alpha$.

We propose *topological sample selection* (TSS) to better identify informative samples in noisy datasets. TSS operates under an "easy-to-hard" curriculum guided by CBC and is framed as a series of training objectives $\langle \tilde{Q}_\lambda \rangle$, controlled by a pace parameter $0 \leq \lambda \leq 1$. Initially, $\tilde{Q}_\lambda$ targets nodes that are simpler to learn—generally those located far from class boundaries. As λ increases, nodes closer to the boundaries are gradually incorporated into the selection. This enables TSS to collect a confident node set $P_{\hat{D}}$, which are from the noisy distribution $P_{\tilde{D}}$. TSS is formally defined:

$$\hat{Q}_\lambda(z) \propto W_\lambda(z)\, P_{\hat{D}}(z), \tag{5.33}$$

such that

$$\int_z \hat{Q}_\lambda(z)\, dz = 1,$$

where z is the set of confident nodes under $\hat{Q}_\lambda(z)$. In definition, there are a series of training objectives $\langle \hat{Q}_\lambda \rangle$ parameterized by λ, where each $\hat{Q}_\lambda$ reweights the confident distribution $P_{\hat{D}}(z)$ with z representing a confident node. And $0 \leq W_\lambda(z) \leq 1$ be the weight for z at iteration λ.

The following properties hold:

(1) The entropy $H(\hat{Q}_\lambda)$ increases monotonically with λ.
(2) For confident nodes, the weighting function $W_\lambda(z)$ grows as λ increases.

Experiments We looked at both how robust and how effective CBC is in practice. In Fig. 5.17b, CBC remains worked even with high label noise ratio and the results that are very similar to the perfect label which showed in Fig. 5.17a. Conversely, Fig. 5.18 clearly shows that as CBC values rise, it becomes much harder to pick out clean nodes from noisy subsets, which is exactly what CBC is meant to capture. We

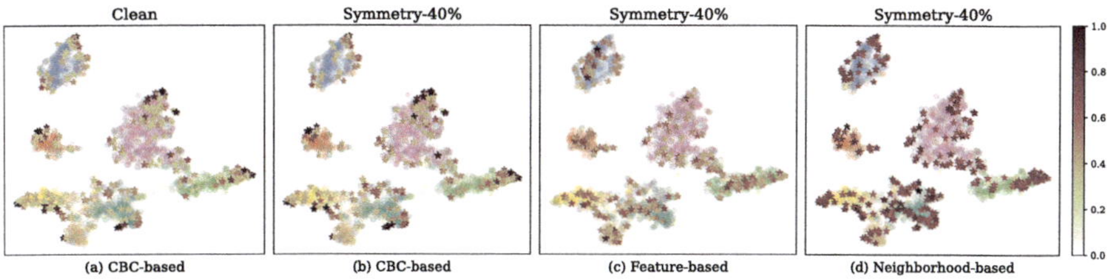

Fig. 5.17 Robustness of class-conditional betweenness centrality. (*t-SNE* projections of node embeddings from the *CORA* dataset, trained with GNNs)

Fig. 5.18 A Pearson correlation coefficient of − 0.9276 was observed across 50 randomly selected subsets, with a p-value less than 0.0001

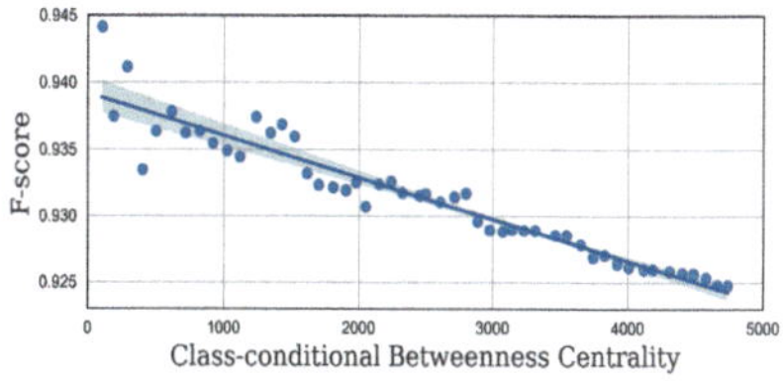

tested our method on three datasets: *Cora*, *CiteSeer*, and *PubMed* then summarized the findings in Table 5.30. Our approach outperformed others across different noise levels, maintaining high classification accuracy even in high noise.

5.2.3 *Powerful Graph Learning*

5.2.3.1 Efficient Communication Channel in Molecular Graphs

There are two types of interactions that may occur between the fundamental components of a molecule. These interactions can be classified as either long-range interactions (LRI) or short-range interactions (SRI), as demonstrated in Fig. 5.19. This is because the fundamental components of a molecule consist of numerous different types of atoms. More specifically, SRI—such as covalent or ionic bonds—functions at diminutive distances, often within a few atomic diameters. The LRI can function at distances that are far greater than the diameter of the atom itself. A few examples of this are coulombic interactions and hydrogen bonding. LRI is crucial for distinguishing the physical and chemical characteristics of molecules [83, 142, 216, 320, 400, 472], but it is inferior to SRI.

SRI as explicit, one-hop edges (solid lines) demonstrating direct atom interactions like covalent bonds is shown in Fig. 5.19. LRI (dashed lines) indicates implicit forces operating over several hops, such as hydrogen bonds spanning many atoms in the molecular graph.

With the intention of improving LRI modeling, we have developed a novel representation for atom learning that we have dubbed "neural atoms." This rep-

Table 5.30 The table summarizes the mean classification accuracy averaged over ten runs

		Symmetric			Pairflip			Instance-dependent		
	Method	30%	40 %	50%	20%	30%	40%	30%	40%	50%
CORA	Cross-entropy	83.61±1.07	80.86±1.46	75.14±2.44	82.23±0.93	75.87±1.20	62.05±3.59	83.21±0.74	80.32±0.94	74.96±1.82
	LPM	82.73±0.64	78.12±1.17	70.23±2.17	83.39±1.22	77.44±1.93	64.02±5.04	82.81±0.87	77.67±2.01	70.55±1.86
	CP	82.37±1.38	79.97±1.74	76.19±2.26	80.24±0.96	73.02±1.56	58.04±3.78	82.37±1.09	80.36±1.21	74.17±2.68
	NRGNN	81.73±1.80	79.08±3.18	77.36±2.03	81.83±0.93	77.10±1.52	64.13±3.98	81.62±2.08	78.66±2.54	76.31±2.98
	PI-GNN	82.48±0.10	80.36±0.10	77.59±0.20	83.10±0.10	77.96±0.20	63.62±0.30	81.83±1.00	80.02±1.07	77.27±1.21
	RT-GNN	83.21±1.05	80.46±1.06	75.84±1.43	82.53±0.73	76.87±1.09	61.75±2.29	82.14±0.97	80.13±1.23	74.82±0.94
	RS-GNN	83.21±0.29	79.00±0.15	77.21±0.43	81.83±0.37	76.46±0.24	63.09±0.22	82.83±0.73	78.93±0.63	76.52±0.83
	Co-teaching+	82.59±0.96	79.81±1.30	74.59±2.33	81.70±1.45	75.59±2.13	59.03±5.76	81.84±1.10	79.70±1.34	73.36±2.54
	Me-momentum	83.76±0.25	81.82±0.72	79.48±0.63	84.09±0.48	78.04±1.03	64.07±1.03	83.14±0.25	82.04±0.57	77.33±0.82
	MentorNet	81.84±0.86	78.52±2.01	73.82±2.83	80.83±1.88	72.56±3.42	59.78±4.59	81.59±0.92	78.49±1.63	72.41±3.66
	CLNode	80.98±1.50	77.11±2.25	74.39±2.41	83.43±0.89	73.89±1.97	55.38±2.80	81.12±2.43	75.11±2.93	68.44±4.88
	RCL	73.20±0.12	76.36±1.09	63.40±0.73	71.06±0.48	65.30±0.80	51.34±0.42	69.20±1.00	59.30±0.13	54.16±2.25
	TSS	**85.02±0.12**	**82.58±0.92**	**81.16±0.80**	**85.26±0.30**	**78.50±0.72**	**65.15±1.53**	**84.70±0.04**	**83.31±0.21**	**80.15±0.36**
CiteSeer	Cross-entropy	75.13±0.70	73.85±0.85	70.74±1.86	76.61±0.53	73.87±1.08	62.92±4.11	74.83±1.04	73.22±0.71	69.42±2.07
	LPM	73.19±1.07	69.54±1.37	61.22±2.08	75.08±0.76	69.91±1.31	58.86±4.28	73.55±0.79	69.32±1.76	61.90±1.73
	CP	73.26±1.22	70.99±1.88	63.74±2.55	74.36±1.21	68.21±2.56	56.56±6.50	73.45±0.72	69.90±1.64	64.61±2.74
	NRGNN	75.41±1.04	73.52±1.46	70.98±2.47	75.72±1.04	74.13±1.38	63.60±4.83	75.33±0.91	74.36±1.45	71.61±1.76
	PI-GNN	73.55±0.14	71.05±0.21	68.02±0.20	73.06±0.13	69.91±0.32	60.62±0.41	74.28±0.78	70.66±1.51	67.81±1.99
	RT-GNN	74.64±0.72	73.66±0.58	71.36±0.65	73.32±0.68	65.78±1.33	62.38±0.56	73.94±0.52	72.86±0.48	71.02±0.25
	RS-GNN	74.93±0.65	73.65±0.45	70.54±1.26	76.31±0.33	73.27±0.38	61.42±2.01	75.03±0.25	72.85±0.15	70.14±1.06
	Co-teaching+	71.01±2.83	68.12±2.38	61.65±4.27	72.09±1.21	68.25±2.91	56.64±5.46	70.80±3.08	67.46±2.55	62.12±2.81
	Me-Momentum	75.40±0.26	74.41±0.56	70.51±0.79	76.93±0.47	74.07±1.06	63.96±0.97	75.27±0.25	74.24±0.45	71.18±0.45

	MentorNet	69.61±3.42	66.87±3.78	60.21±2.67	71.96±1.81	66.14±4.98	54.20±6.25	70.56±2.55	64.90±4.72	60.95±4.93
	CLNode	68.73±2.07	64.26±3.18	56.07±3.61	69.11±3.15	61.62±3.33	53.32±4.29	69.91±1.88	66.22±2.65	60.37±3.10
	RCL	60.90±0.12	54.50±2.53	46.58±1.44	65.00±0.13	56.68±0.27	51.14±1.58	63.70±0.53	54.70±1.97	46.62±0.59
	TSS	**75.86±0.31**	**74.77±0.79**	**71.81±0.74**	**77.25±0.44**	**74.91±0.90**	**65.36±1.27**	**76.61±0.17**	**75.61±0.29**	**74.03±0.26**
PubMed	Cross-entropy	85.98±0.50	84.80±0.83	82.83±1.55	85.31±0.38	83.31±0.58	76.12±2.04	85.29±0.27	84.10±0.74	82.45±2.96
	LPM	85.33±0.70	84.33±0.79	82.31±0.89	85.90±0.57	84.63±0.34	78.94±0.79	85.51±0.52	84.90±0.53	83.12±1.18
	CP	86.12±0.63	85.01±0.65	82.33±1.51	86.13±0.36	84.87±0.46	78.81±0.77	85.66±0.60	84.92±0.99	81.18±1.95
	NRGNN	86.19±0.44	84.99±1.16	83.02±1.44	86.26±0.81	83.79±1.28	75.83±2.72	85.45±0.52	85.07±1.15	83.47±1.02
	PI-GNN	86.16±0.06	85.35±0.11	83.12±0.13	86.01±0.12	84.09±0.21	78.35±0.23	86.13±0.29	85.09±0.40	83.22±0.85
	RT-GNN	84.73±0.05	84.70±0.35	79.39±0.25	82.90±0.03	80.80±0.10	79.90±0.12	83.09±0.43	81.60±0.15	80.81±0.32
	RS-GNN	85.38±0.42	84.34±0.38	82.37±0.35	85.24±0.24	83.12±0.47	75.24±1.27	85.16±0.32	84.14±0.14	83.07±0.15
	Co-teaching+	86.14±0.58	85.01±0.74	82.74±2.12	85.37±1.90	84.45±0.75	77.31±5.38	85.83±0.54	84.65±1.47	81.42±2.89
	Me-Momentum	86.05±0.18	85.66±0.78	82.42±0.41	85.78±0.26	85.43±0.35	80.34±0.41	85.87±0.27	84.37±0.40	83.53±0.14
	MentorNet	85.43±0.81	84.55±1.33	82.84±0.92	86.64±0.59	84.83±0.92	74.36±6.01	85.14±1.12	84.13±1.75	80.38±3.99
	CLNode	86.03±0.37	85.34±0.45	83.06±0.37	86.27±0.42	85.15±0.38	81.12±0.44	85.23±0.37	84.61±0.39	83.63±0.51
	RCL	82.40±0.24	80.30±0.15	76.40±0.14	82.70±0.23	82.66±0.69	81.30±0.20	82.10±0.12	80.30±0.12	74.90±0.19
	TSS	**86.69±0.32**	**86.23±0.37**	**83.53±0.23**	**87.05±0.28**	**86.30±0.22**	**83.18±0.55**	**86.21±0.03**	**85.32±0.04**	**83.94±0.08**

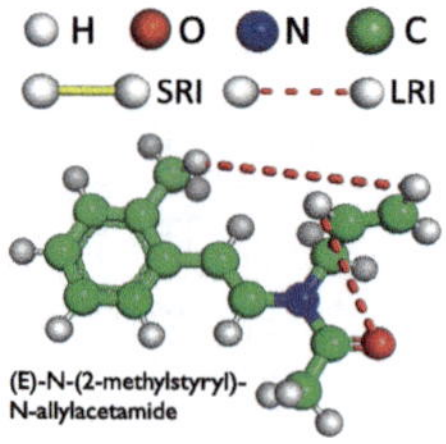

Fig. 5.19 A molecule with LRI denoted by solid lines and SRI represented by dotted lines

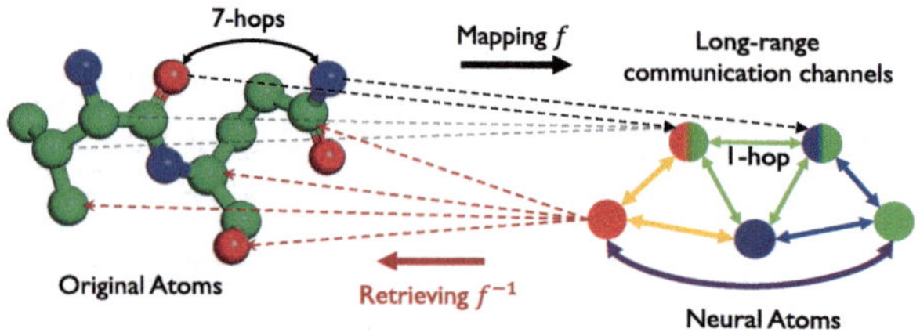

Fig. 5.20 Neural atom visualization. While the retrieval function f^{-1} is responsible for recovering the information, the mapping function f is responsible for converting the original atoms into neural atoms

resentation was built by utilizing the concept of pseudo-atoms in computational chemistry. Because of this, the GNN is able to capture LRI by taking advantage of the interactions that occur between the neural atoms, shown in Fig. 5.20.

This representation delineates atomic group information to enhance the capacity of GNNs to capture LRI. The inherent features of atoms are maintained inside a condensed neural atom space, facilitating an enhanced flow of information between nodes situated at longer distances. This permits information exchange and subsequent projection back to modify atomic representations. This architecture-agnostic, quicker method reduces interaction scopes to a single hop.

Problem Formulation The goal is to enhance GNNs to effectively model both SRI and LRI between nodes by learning:

- A projection function $f : \mathcal{V} \rightarrow \mathcal{N}$ to map atoms $\mathcal{V}$ to a set of neural atoms $\mathcal{N}$, where $|\mathcal{N}| \ll |\mathcal{V}|$.
- Interaction mechanisms among neural atoms to approximate LRI.
- A retrieval function $f^{-1} : \mathcal{N} \rightarrow \mathcal{V}$ to update atomic representations with LRI-enhanced information.

Method A visual representation of the full inference process for the GNN that makes use of neural atoms can be found in Fig. 5.21. Using the i-th GNN model, the atom representations $\mathbf{H}^{(i)}_{\mathrm{GNN}}$ are formed. These representations are then improved by the application of neural atoms, which ultimately results in the representations $\mathbf{H}^{(i)}$. This process is comprised of three main steps, which include the transformation of the atom representations $\mathbf{H}^{(i)}_{\mathrm{GNN}}$ into neural atom representations $\mathbf{H}^{(i)}_{\mathrm{NA}}$, (2) facilitating the dissemination of information among the neural atoms in order to obtain $\tilde{\mathbf{H}}^{(i)}_{\mathrm{NA}}$, and (3) re-projecting the neural atoms while simultaneously enhancement the initial atom representations as $(\mathbf{H}^{(i)}_{\mathrm{GNN}}, \tilde{\mathbf{H}}^{(i)}_{\mathrm{NA}}) \mapsto \mathbf{H}^{(i)}$.

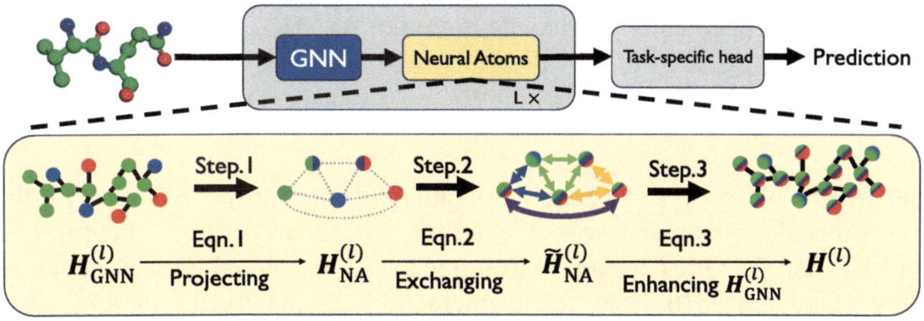

Fig. 5.21 The neural atom framework, which enables the improvement of a diverse array of functions by utilizing LRI information during neural atom interactions. The information flow is exemplified through the use of color blending

Step 1: Projecting Atom Representations. $\mathbf{H}_{\mathrm{GNN}}^{(i)}$ denotes the neural atoms' representations. Let $\mathbf{Q}_{\mathrm{NA}}^{(i)} \in \mathbb{R}^{K \times d_k}$, where K (with $K \ll N$) represents the quantity of neural atoms. We are able to design a grouping function that is capable of creating neural atom representations from intermediate atom representations by making use of multihead attention. This function is effective in constructing neural atom representations.

In this process, the atom representations $\mathbf{H}_{\mathrm{GNN}}^{(i)} \in \mathbb{R}^{N \times d}$ are transformed into the key matrix $\mathbf{K} \in \mathbb{R}^{N \times d_h}$ and the value matrix $\mathbf{V} \in \mathbb{R}^{N \times d_h}$, respectively. The attention matrix $\hat{\mathbf{A}}_m$ for the m-th head is calculated as $\hat{\mathbf{A}}_m = \sigma(\mathbf{Q}_{\mathrm{NA}}^{(i)} \mathbf{K}^{\top}) \in \mathbb{R}^{K \times N}$. The representations of neural atoms $\mathbf{H}_{\mathrm{NA}}^{(i)}$ are derived through:

$$\mathbf{H}_{\mathrm{NA}}^{(i)} = \mathrm{LayerNorm}\left(\mathbf{Q}_{\mathrm{NA}}^{(i)} \oplus \mathrm{MultiHead}(\mathbf{Q}_{\mathrm{NA}}^{(i)}, \mathbf{H}_{\mathrm{GNN}}^{(i)}, \mathbf{H}_{\mathrm{GNN}}^{(i)})\right), \tag{5.34}$$

where $\oplus$ is the method for incorporating representations.

Step 2: The transmission of information via neural atoms $\mathbf{H}_{\mathrm{NA}}^{(i)}$ is transformed into $\tilde{\mathbf{H}}_{\mathrm{NA}}^{(i)}$. We provide direct communication among the neural atoms to capture LRIs inside the molecular network.

$$\tilde{\mathbf{H}}_{\mathrm{NA}}^{(i)} = \mathrm{LayerNorm}\left(\mathbf{H}_{\mathrm{NA}}^{(i)} \oplus \mathrm{MultiHead}(\mathbf{H}_{\mathrm{NA}}^{(i)}, \mathbf{H}_{\mathrm{NA}}^{(i)}, \mathbf{H}_{\mathrm{NA}}^{(i)})\right). \tag{5.35}$$

Step-3: Project neural atoms back and enhance the atoms' representation $(\mathbf{H}_{\mathrm{GNN}}^{(i)}, \tilde{\mathbf{H}}_{\mathrm{NA}}^{(i)}) \mapsto \mathbf{H}^{(i)}$. In order to complete the integration of the updated neural atom representations $\tilde{\mathbf{H}}_{\mathrm{NA}}^{(i)}$ with the initial molecular atom characteristics, we initially combine the allocation matrices $\hat{\mathbf{A}}_m$ from various heads. Subsequently, we execute matrix reduction procedures to restore the neural atoms to the N-dimensional atom space. The final atom representations $\mathbf{H}^{(i)}$ are obtained by enhancing the initial atom representations $\mathbf{H}_{\mathrm{GNN}}^{(i)}$ with the updated neural atom representations, specifically via the integration of $\tilde{\mathbf{A}}_{\mathrm{NA}}^{(i)}\tilde{\mathbf{H}}_{\mathrm{NA}}^{(i)}$, i.e.,

$$\mathbf{H}^{(i)} = \mathbf{H}_{\mathrm{GNN}}^{(i)} \oplus \tilde{\mathbf{A}}_{\mathrm{NA}}^{(i)}\tilde{\mathbf{H}}_{\mathrm{NA}}^{(i)}, \quad \text{s.t. } \tilde{\mathbf{A}}_{\mathrm{NA}}^{(i)} = \mathrm{Aggregate}\left(\{\hat{\mathbf{A}}_m\}_{m=1}^{M}\right)^{\top} \in \mathbb{R}^{N \times K}. \tag{5.36}$$

Experiments We conduct an empirical evaluation of the proposed strategy utilizing actual molecular graph datasets. We utilize molecular datasets (Peptides-Func, Peptides-Struct, PCQM-Contact) to examine two-dimensional intramolecular interactions, characterized by long-range interactions as per the Long Range Graph Benchmarks (LRGB), employing traditional GNNs and Transformer-based models. In the three-dimensional intermolecular framework, we utilize the OE62 [346] to assess long-range London dispersion interactions. Neural atoms are included in a variety of GNNs, and their performance is evaluated in comparison to fully connected baselines and current Ewald-based methodologies [200].

Two-dimensional intramolecular interactions are reported in Table 5.31. Especially when used with edge learning systems that filter the input data, neural atoms show a significant improvement in GNN performance by clearly capturing LRI. The results for 3D intermolecular interactions are presented in Table 5.32. Neuroatomic performance in the OE62 dataset is competitive even in the absence of 3D coordinates; thus, the best results in energy MAE and MSE need fewer parameters.

5.2.3.2 Graph Multilinear Network for Interpretable Graph Learning

In addition to the long-range interaction, Chen et al. [58] further investigate the interpretability. In the context of scientific discovery, it is essential for GNNs to not only generalize to unseen or out-of-distribution (OOD) graphs [132, 476, 498] but also offer clear interpretations of their predictions. These human-friendly insights are important for scientists to curate meaningful information [73, 82, 418] and to empower scientific practices [273, 399]. Recently, there has been increasing interest in developing intrinsically interpretable and generalizable GNNs or XGNNs [66, 264, 265, 408, 440]. These approaches aim to address the challenges of OOD generalization and interpretability in GNNs by focusing on extracting critical substructures (e.g., invariant subgraphs, motifs). Unlike *post hoc* explanation methods [233, 234, 254, 380, 436, 443, 444], which have been shown to be suboptimal in terms of interpretability and are highly sensitive to the performance of pretrained GNNs [264, 265], XGNNs are designed to deliver both robust explanations and OOD generalizable predictions. This is achieved through the incorporation of additional guidance mechanisms and objectives, such as the information bottleneck [440] and causal invariance [66].

Indeed, in the context of XGNNs, faithful interpretation and reliable generalization are two inseparable aspects. These models operate based on the causal assumptions underlying the data generation process. Specifically, XGNNs posit that a causal subgraph exists—one that maintains a direct causal relationship with the

Table 5.31 Test performance on LRGB datasets. Shown is the mean ± s.d. of 4 runs

	Peptides-func	Peptides-struct	PCQM-Contact
Model	AP ↑	MAE ↓	MRR ↑
Transformer+LapPE	0.6326 ± 0.0126	0.2529 ± 0.0016	0.3174 ± 0.0020
SAN+LapPE	0.6384 ± 0.0121	0.2683 ± 0.0043	0.3350 ± 0.0003
GraphGPS	0.6535 ± 0.0041	0.2500 ± 0.0005	0.3337 ± 0.0006
GCN	0.5930 ± 0.0023	0.3496 ± 0.0013	0.2329 ± 0.0009
+ Neural Atoms	**0.6220 ± 0.0046**	**0.2606 ± 0.0027**	**0.2534 ± 0.0200**
GINE	0.5498 ± 0.0079	0.3547 ± 0.0045	0.3180 ± 0.0027
+ Neural Atoms	**0.6154 ± 0.0157**	**0.2553 ± 0.0005**	**0.3126 ± 0.0021**
GCNII	0.5543 ± 0.0078	0.3471 ± 0.0010	0.3161 ± 0.0004
+ Neural Atoms	**0.5996 ± 0.0033**	**0.2563 ± 0.0020**	**0.3049 ± 0.0006**
GatedGCN	0.5864 ± 0.0077	0.3420 ± 0.0013	0.3218 ± 0.0011
+ Neural Atoms	**0.6562 ± 0.0075**	**0.2585 ± 0.0017**	**0.3258 ± 0.0003**
GatedGCN+RWSE	0.6069 ± 0.0035	0.3357 ± 0.0006	0.3242 ± 0.0008
+ Neural Atoms	**0.6591 ± 0.0050**	**0.2568 ± 0.0005**	**0.3262 ± 0.0010**

Table 5.32 A comparison of the validation. Energy MAE and MSE are conducted utilizing the OE62 dataset

	Energy MAE ↓	Energy MSE ↓	Number of Params.
SchNet [327]	0.1351	0.0658	2.75 M
+ Ewald Block	**0.0811**	**0.0301**	12.21 M
+ Neural Atoms	**0.0834**	**0.0309**	2.63 M
PaiNN [328]	0.6049	0.0133	12.52 M
+ Ewald Block	**0.0590**	**0.0134**	15.68 M
+ Neural Atoms	**0.0558**	**0.0122**	6.05 M
DimeNet++ [118]	0.0501	0.0117	2.76 M
+ Ewald Block	**0.0479**	**0.0107**	4.75 M
+ Neural Atoms	0.0551	0.0129	1.97 M

target label. Based on the predictions exclusively on this causal subgraph, XGNNs are able to generalize effectively across varying graph distribution shifts [66, 264, 407]. XGNNs generally follow a two-step approach to achieve both interpretability and better (OOD) generalization:

- Subgraph Extraction: The model extracts a subgraph from the input graph that is hypothesized to be causally linked to the target label.
- Label Prediction: The extracted subgraph is then used to predict the label based only on the causal subgraph.

One of the main challenges in subgraph extraction is the inherent discrete nature of graph data. XGNNs typically employ attention mechanisms to learn sampling probabilities for nodes or edges. The attention mechanism assigns scores to each edge or node, representing their importance to the task. Nodes and edges with high attention scores are selected as part of the extracted subgraph. Finally, predictions are obtained through a weighted message passing based on the learned attention scores [264]. Although this paradigm has effectively advanced both interpretability and OOD generalization [66, 264, 265], our theoretical understanding of the representational characteristics and inherent limitations of XGNNs remains limited. Similarly, questions about their ability to deliver genuinely faithful interpretations, i.e., explanations that accurately reflect the true reasoning process of the model, are still underexplored.

To this end, we propose a framework that examines both the expressiveness and the faithfulness of XGNNs based on multilinear extension [46]. This framework is grounded in the cross-entropy calculation used within XGNNs, as detailed below:

$$\min_{g, f_c} \mathbb{E}_{\mathcal{G},Y}\left[-\log \mathbb{E}_{\mathcal{G}_c \overset{g}{\sim} \mathcal{G}} P_{f_c}(Y|\mathcal{G}_c)\right] \tag{5.37}$$

$$= \mathbb{E}_{(\mathcal{G},Y)}[\mathcal{L}_S(\mathbb{E}_{\mathcal{G}_c \overset{g}{\sim} \mathcal{G}}[f_c(\mathcal{G}_c)], Y)], \tag{5.38}$$

where f_c is the subgraph classifier GNN, g is the subgraph extractor GNN, $\mathcal{G}_c$ is the extracted subgraph from the input graph $\mathcal{G}$, $\mathcal{L}_S$ is the empirical risk, and Y is the label. The subgraph learning mechanism in XGNNs aligns closely with the multilinear extension of subgraph predictivity, termed as subgraph multilinear extension (SubMT).

Definition 5.4 (Subgraph Multilinear Extension (SubMT)) Given the attention score $\boldsymbol{\alpha} \in [0, 1]^m$ as sampling probability of $\mathcal{G}_c$, XGNNs factorize $P(\mathcal{G}_c|\mathcal{G})$ as independent Bernoulli distributions on edges:

$$P(\mathcal{G}_c|\mathcal{G}) = \prod_{e\in\mathcal{G}_c} \alpha_e \prod_{e\in\mathcal{G}/\mathcal{G}_c} (1-\alpha_e),$$

which elicits the *multilinear extension* of $f_c(\mathcal{G}_c)$ in Eq. (5.37):

$$\begin{aligned} F_c(\boldsymbol{\alpha}; \mathcal{G}) &:= \sum_{\mathcal{G}_c\in\mathcal{G}} f_c(\mathcal{G}_c) \prod_{e\in\mathcal{G}_c} \alpha_e \prod_{e\in\mathcal{G}/\mathcal{G}_c} (1-\alpha_e) \\ &= \mathbb{E}_{\mathcal{G}_c \overset{g}{\sim} \mathcal{G}} f_c(\mathcal{G}_c). \end{aligned} \tag{5.39}$$

The extracted interpretable subgraph is considered faithful if the associated prediction is strongly correlated with the subgraph's sampling probability. However, as illustrated in Fig. 5.22, we demonstrate that the commonly used attention-based paradigm can fail to accurately approximate SubMT, leading to a failure in faithfully representing the subgraph's role in predicting the target label.

To tackle this problem, we propose a novel interpretability metric called counterfactual fidelity, which evaluates how sensitive the prediction is to minor modifications applied to the extracted subgraphs within a causal framework. A reliable interpretation is expected to demonstrate high counterfactual fidelity, meaning that small adjustments to the subgraph should not significantly impact the prediction.

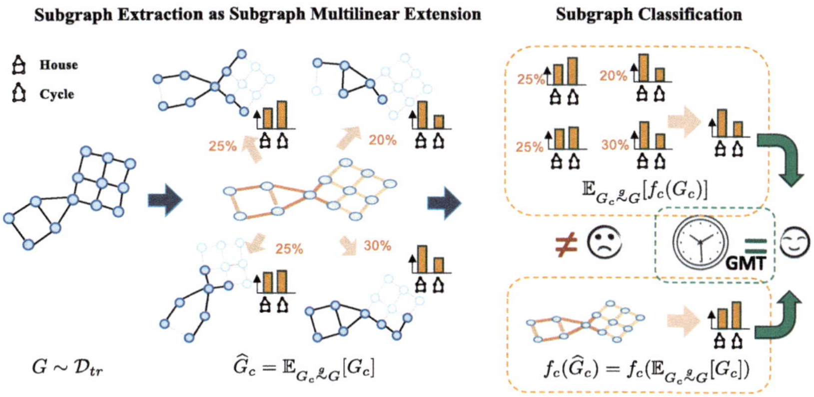

Fig. 5.22 Depiction of subgraph multilinear extension approximation failure. Consider a binary graph classification task where XGNNs must identify whether a graph contains a specific motif, such as a "house" or "cycle." This process consists of two primary steps—(**a**) subgraph extraction: During this step, an attention-based mechanism is used by a subgraph extractor to assign sampling probabilities to each edge. These probabilities define a distribution over potential subgraphs; for example, parts of the house" or grid" motif might be selected with varying probabilities. The resulting expected subgraph, denoted as $\hat{\mathcal{G}}_c$, is a "soft" subgraph where each edge is weighted by its respective sampling probability. (**b**) Subgraph classification: In the second step, every sampled subgraph instance $\mathcal{G}_c^i$ produces a conditional label distribution $P(Y|\mathcal{G}_c^i)$. For example, if critical edges of the "house" motif are included, the probability for the "house" label increases. The final output is calculated by averaging these conditional distributions, weighted by the likelihood of sampling each subgraph. This aggregation generally leads to a correct classification, such as labeling the graph as "house." However, instead of averaging over the conditional label predictions of individual subgraphs, previous approaches often input the expected "soft" subgraph directly into the subgraph classifier GNN, denoted as f_c. This shortcut can introduce bias into the model, potentially resulting in an incorrect prediction—such as misclassifying the graph as "cycle'

However, our results reveal that XGNNs employing the commonly used attention-based approach exhibit low counterfactual fidelity, indicating that the explanations generated by these models may lack robustness and reliability against such perturbations.

Method We present a simple yet effective XGNN framework named Graph Multilinear neT (GMT). Drawing inspiration from the SubMT formulation, we introduce a random subgraph sampling technique that leverages Markov chain Monte Carlo (MCMC) sampling to approximate SubMT. Specifically, given the attention matrix $\widehat{\mathbf{A}}$, we conduct t iterations of random subgraph sampling based on the subgraph distribution defined by $\widehat{\mathbf{A}}$ (or equivalently, $\hat{\mathcal{G}}_c = (\mathbf{X}, \widehat{\mathbf{A}})$ as described in the SubMT definition, Definition 5.4). This procedure produces t independent and identically distributed (i.i.d.) random subgraph samples $\mathcal{G}_c^i i = 1^t$, which are subsequently utilized to approximate SubMT as follows:

$$\text{(GMT-sam)} \qquad f_c^s(\hat{\mathcal{G}}_c) = \frac{1}{t}\sum_{i=1}^{t} f_c(Y|\mathcal{G}_c^i), \tag{5.40}$$

Here, f_c represents the classifier that processes discrete subgraphs as inputs. Next, we train a new classifier on the pretrained subgraph extractor, this time bypassing the random subgraph sampling process, to derive the final neural SubMT approximator. In addition, we also propose a lightweight GMT:

$$\text{(GMT-lin)} \qquad f^l(\hat{\mathcal{G}}_c) = \rho(\widehat{\mathbf{A}} \odot \mathbf{A}^{k-1}\mathbf{XW}), \tag{5.41}$$

If there exists a matrix $\mathbf{T} \in \mathbb{R}^{|\mathcal{Y}|\times|\mathcal{Y}|}$ such that $\mathbf{T} \cdot f_c(\mathcal{G}_c) = P(Y|\mathcal{G}_c)$, where f_c is linear, then we can incorporate GMT-lin into the SubMT computation:

$$f^l(\hat{\mathcal{G}}_c) = \mathbf{T} \cdot f(\hat{\mathcal{G}}_c) = \mathbb{E}[f_c(\mathcal{G}_c)],$$

owing to the linearity of $f^l(\mathcal{G}_c)$ with respect to $\mathcal{G}_c$. During training, the matrix $\mathbf{m}T$ can be absorbed into the weight matrix $\mathbf{W}$, which suggests that GMT-lin is capable of fitting to SubMT.

Experiments We conduct comprehensive experiments to evaluate the performance of GMT. As shown in Table 5.33, both GMT-lin and GMT-sam achieve substantial improvements in interpretability compared to post hoc methods (first row) and GSAT. Notably, on challenging Spurious-Motif datasets with distribution shifts, GMT-sam surpasses GSAT by up to 15% when using GIN and up to 8% with PNA. Similarly, on the MNIST-75sp dataset, GMT-sam improves upon GSAT by up to 6% in terms of the interpretation.

Table 5.34 highlights the prediction accuracy achieved on standard graph datasets. Across various datasets, ranging from molecular graphs to image-derived datasets, GMT consistently delivers performance enhancements. Even under distribution shifts, GMT-sam demonstrates significant gains, with improvements of up to 13% with GIN and up to 16% over GSAT on Spurious-Motif.

5.2.3.3 Adaptive Propagation for Reasoning on Knowledge Graphs

While GMT examines the limitations of attention methods in interpretable GNNs, Zhang et al. [478] propose AdaProp to address similar challenges in information propagation. There are three ways to classify the current KG reasoning strategies, though they often ignore the structural connectivity in KGs:

- Triplet-based methods [39, 94, 383, 463, 466] focus on scoring potential entities by leveraging embeddings of entities and their relations. They're straightforward but often miss the bigger picture of how the KG's structure ties everything together.
- Path-based methods [69, 84, 304, 318, 465] aim to follow the trails of relationships between entities, to tease out logical rules. This approach shines for its clarity—you can see how conclusions are drawn—but it starts to stumble when

Table 5.33 Interpretation performance (AUC) on regular graphs. Results with the mean-1*std larger than the best baselines are shadowed

GNN	Method	Ba-2motifs	Mutag	MNIST-75sp	Spurious-motif		
					$b = 0.5$	$b = 0.7$	$b = 0.9$
GIN	GNNExplainer	67.35(±*3.29*)	61.98(±*5.45*)	59.01(±*2.04*)	62.62(±*1.35*)	62.25(±*3.61*)	58.86(±*1.93*)
	PGExplainer	84.59(±*9.09*)	60.91(±*17.10*)	69.34(±*4.32*)	69.54(±*5.64*)	72.33(±*9.18*)	72.34(±*2.91*)
	GraphMask	92.54(±*8.07*)	62.23(±*9.01*)	73.10(±*6.41*)	72.06(±*5.58*)	73.06(±*4.91*)	66.68(±*6.96*)
	IB-Subgraph	86.06(±*28.37*)	91.04(±*6.59*)	51.20(±*5.12*)	57.29(±*14.35*)	62.89(±*15.59*)	47.29(±*13.39*)
	DIR	82.78(±*10.97*)	64.44(±*28.81*)	32.35(±*9.39*)	78.15(±*1.32*)	77.68(±*1.22*)	49.08(±*3.66*)
GIN	GSAT	98.85(±*0.47*)	99.35(±*0.95*)	80.47(±*1.86*)	74.49(±*4.46*)	72.95(±*6.40*)	65.25(±*4.42*)
	GMT-lint	98.36(±*0.56*)	99.86(±*0.09*)	82.98(±*1.49*)	76.06(±*6.39*)	76.50(±*5.63*)	**80.57**(±*2.59*)
	GMT-samt	**99.62**(±*0.11*)	**99.87**(±*0.11*)	**86.50**(±*1.80*)	**85.50**(±*2.40*)	**84.67**(±*2.38*)	73.49(±*5.33*)
PNA	GSAT	89.35(±*5.41*)	99.00(±*0.37*)	85.72(±*1.10*)	79.84(±*3.21*)	79.76(±*3.66*)	80.70(±*5.45*)
	GMT-lint	95.79(±*7.30*)	99.58(±*0.17*)	85.02(±*1.03*)	80.19(±*2.22*)	84.74(±*1.82*)	85.08(±*3.85*)
	GMT-samt	**99.60**(±*0.48*)	**99.89**(±*0.05*)	**87.34**(±*1.79*)	**88.27**(±*1.71*)	**86.58**(±*1.89*)	**85.26**(±*1.92*)

Table 5.34 Prediction performance (Acc.) on regular graphs. The shadowed entries are the results with the mean-1*std larger than the mean of the corresponding best baselines

GNN	Method	MolHiv (AUC)	Graph-SST2	MNIST-75sp	Spurious-motif $b = 0.5$	Spurious-motif $b = 0.7$	Spurious-motif $b = 0.9$
GIN	GIN	76.69(±*1.25*)	82.73(±*0.77*)	95.74(±*0.36*)	39.87(±*1.30*)	39.04(±*1.62*)	38.57(±*2.31*)
	IB-subgraph	76.43(±*2.65*)	82.99(±*0.67*)	93.10(±*1.32*)	54.36(±*7.09*)	48.51(±*5.76*)	46.19(±*5.63*)
	DIR	76.34(±*1.01*)	82.32(±*0.85*)	88.51(±*2.57*)	45.49(±*3.81*)	41.13(±*2.62*)	37.61(±*2.02*)
GIN	GSAT	76.12(±*0.91*)	83.14(±*0.96*)	96.20(±*1.48*)	47.45(±*5.87*)	43.57(±*2.43*)	45.39(±*5.02*)
	GMT-lint	76.87(±*1.12*)	83.19(±*1.28*)	96.01(±*0.25*)	47.69(±*4.93*)	53.11(±*4.12*)	46.22(±*4.18*)
	GMT-samt	**77.22**(±*0.93*)	**83.62**(±*0.50*)	**96.50**(±*0.19*)	**60.09**(±*2.40*)	**54.34**(±*4.04*)	**55.83**(±*5.68*)
PNA	PNA	78.91(±*1.04*)	79.87(±*1.02*)	87.20(±*5.61*)	68.15(±*2.39*)	66.35(±*3.34*)	61.40(±*3.56*)
	GSAT	79.82(±*0.67*)	80.90(±*0.37*)	93.69(±*0.73*)	68.41(±*1.76*)	67.78(±*3.22*)	51.51(±*2.98*)
	GMT-lint	80.05(±*0.71*)	81.18(±*0.47*)	94.44(±*0.49*)	69.33(±*1.42*)	64.49(±*3.51*)	58.30(±*6.61*)
	GMT-samt	**80.58**(±*0.83*)	**82.36**(±*0.96*)	**95.75**(±*0.42*)	**71.98**(±*3.44*)	**69.68**(±*3.99*)	**67.90**(±*3.60*)

faced with the sheer size and complexity of bigger KGs, where scaling up and covering all bases become real headaches.

- GNN-based methods [324, 376, 457, 490] take a broader view, letting information ripple through entity neighborhoods to weave together both the structure and meaning of the graph. Their strength lies in tackling multi-hop reasoning with ease, which is why they're consistently posting some of the best results out there.

That said, most GNN-based methods today miss the mark when it comes to linking the semantics of local neighborhoods with the query relation. As a result, they end up dragging in a bunch of unrelated entities and facts, which ramps up the learning complexity, spikes computational costs, and might even throw off the model's accuracy.

To tackle these issues, this work puts forward AdaProp [478], a fresh GNN-based twist that learns *adaptive propagation paths* tailored to each query, aiming to solve these constraints. AdaProp gets better efficiency and reasoning accuracy by selectively rejecting irrelevant things while keeping semantically related ones. Its design maximizes the propagation process for both efficiency and semantic relevance by combining a learning-based distribution with an incremental sampling method.

Problem Formulation We set up our GNN model for KG reasoning as $f_\theta(\hat{\mathcal{G}}^L)$, where θ stands for the model's trainable parameters—think of them as the knobs we tweak during training—and $\hat{\mathcal{G}}^L$ captures the propagation path with L steps. This path is really just a chain of entity sets, written as $\mathcal{V}^0, \mathcal{V}^1, \ldots, \mathcal{V}^L$, with each set marking a step in the propagation across the graph. The objective is to construct adaptive propagation paths $\hat{\mathcal{G}}^L$ that enable the model to effectively predict the target entity e_a for a given query $(e_q, r_q, \cdot)$.

We define the *query-dependent propagation path* for a query $(e_q, r_q, \cdot)$ as

$$
\begin{aligned}
\widehat{\mathcal{G}}^L_{e_q,r_q} &= \{\mathcal{V}^0_{e_q,r_q}, \mathcal{V}^1_{e_q,r_q}, \ldots, \mathcal{V}^L_{e_q,r_q}\}, \\
\text{s.t. } \mathcal{V}^l_{e_q,r_q} &= \begin{cases} \{e_q\} & l = 0 \\ s(\mathcal{V}^{l-1}_{e_q,r_q}) & l = 1 \ldots L \end{cases}.
\end{aligned}
\tag{5.42}
$$

The propagation process begins with the query entity e_q, where the entity set $\mathcal{V}^l_{e_q,r_q}$ at each step l is adaptively sampled from a subset of the previous set $\mathcal{V}^{l-1}_{e_q,r_q}$ and its neighboring entities, guided by a sampling strategy $s(\cdot)$, as shown in Fig. 5.23.

Method In order to design query-dependent propagation paths for KG reasoning, we propose AdaProp, which improves accuracy and efficiency by selectively including relevant entities at each propagation step. AdaProp first employs an incremental sampling strategy to preserve connections and reduce irrelevant entities. There are two components in this process:

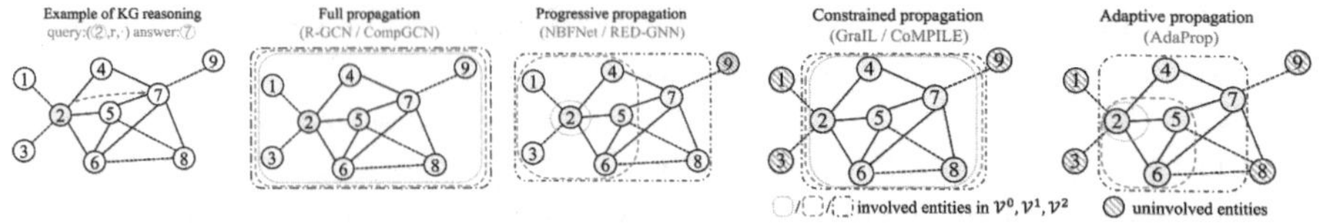

Fig. 5.23 To shed light on KG reasoning and how different propagation paths work, we've sketched out some simple diagrams using circles for entities—keeping it clean by leaving out relation types. We break down four approaches: (1) full propagation, which sweeps across the whole entity set $\mathcal{V}$; (2) progressive propagation, starting from e_q and step-by-step reaching its l-hop neighbors by the l-th move; (3) constrained propagation, narrowing the path to a subset $\mathcal{V}^L_{e_q,e_a}$ that only includes entities within 2 hops of both e_q and e_a; and (4) adaptive propagation, kicking off at e_q and smartly picking out semantically relevant entities to shape the best path possible

- *Candidate Generation*: We identify the newly visited neighbors as candidates for the current step from the set of entities $\mathcal{V}^{l-1}$ in the previous step.
- *Candidate Sampling*: We sample a subset of K, K ($\ll |\bar{\mathcal{V}}^l_{e_q,r_q}|$) entities from the candidate set, ensuring layer-wise connectivity by retaining selected entities.

While preserving layer-by-layer connections, this strategy provides efficient linear growth involving entities and ensures higher retention of target entities.

Besides, AdaProp further introduces a learnable sampling distribution to sample semantically relevant entities from $\bar{\mathcal{V}}^l_{e_q,r_q}$:

- *Parameterized Sampling*: We map the entity representations to sampling probabilities using a linear function. To ensure efficient and unbiased sampling, the entities are chosen via the Gumbel top-k trick to guarantee effective and objective sampling. Starting with $|\bar{\mathcal{V}}^l_{e_q,r_q}|$ independent noise values from a uniform distribution, denoted as $U_e \sim Uniform(0, 1)$ the Gumbel-trick starts. These are computed Gumbel logits for every entity $e \in \bar{\mathcal{V}}^l_{e_q,r_q}$, defined as $g_e = S_{\theta^l}(\mathbf{h}^l_e) - \log(-\log U_e)$, where $S_{\theta^l}(\mathbf{h}^l_e)$ represents the scoring function parameterized by θ^l and $\mathbf{h}$ denotes the entity representations. Subsequently, based on the resulting Gumbel logits g_e, the top-K entities from $\bar{\mathcal{V}}^l_{e_q,r_q}$ are selected.
- *Learning Strategy*: To guarantee semantic alignment with the query relation, we optimize sampling parameters alongside model parameters using a straight-through estimator, which ensures the gradient for discrete variables can be approximately derived.

The above learning-based approach enables AdaProp to adaptively prioritize entities that are critical to reasoning tasks.

Experiments We assess AdaProp's effectiveness on both transductive and inductive KG reasoning tasks using well-established benchmarks. To validate its performance and efficiency, we benchmark AdaProp against both non-GNN and GNN-based baseline methods. Additional analyses examine the properties of its learned propagation paths and the impact of various design choices.

AdaProp was evaluated with both transductive setting and inductive setting. For transductive setting, AdaProp was evaluated on six transductive datasets (Family [198], UMLS [198], WN18RR [94], FB15k237 [371], NELL-995 [422], YAGO3-10 [347]) and subsets of WN18RR, FB15k237, and NELL-995 for inductive reasoning. Mean reciprocal rank (MRR) and Hit@k are performance measures; bigger values indicate better results. We run the tests with a single NVIDIA RTX 3090 GPU using PyTorch the results are summarized in Tables 5.35 and 5.36.

Particularly on larger KGs such as WN18RR, NELL-995, and YAGO3-10, AdaProp shows better performance in transductive KG reasoning than both non-GNN and GNN-based approaches. AdaProp achieves consistent improvements by efficiently raising the ratio of target entity coverage (ToE) while preserving scalability, unlike conventional GNN approaches that suffer from over-smoothing or memory problems at deeper propagation steps. Competitive results are also shown by progressive propagation techniques including NBFNet [490] and RED-GNN [457] but plateau as propagation depth increases. AdaProp's capacity to use deeper propagation paths emphasizes its benefit in KGs in capturing long-range structural and semantic information.

In the inductive setting, AdaProp consistently delivers top performance across all datasets and configurations, as measured by the Hit@10 metric. Specifically, it surpasses both rule-based methods (e.g., RuleN [263]) and GNN-based methods (e.g., NBFNet, RED-GNN) with significant margins, particularly on WN18RR and FB15k237, where it achieves Hit@10 scores of up to 86.6% and 63.8%, respectively. It shows that the propagation path design of AdaProp helps to effectively capture inductive reasoning, and this ability can generalize well to unseen entities and relations.

In addition, AdaProp outperforms state-of-the-art GNNs such as NBFNet and RED-GNN in terms of running time. AdaProp is an attractive choice for both performance-critical and resource-constrained settings since it offers equivalent or superior performance with little additional computational overhead, even while progressive propagation approaches excel in balancing accuracy and scalability.

In this paper, we propose Adaprop, a one-shot subgraph-based method that focuses on a single subgraph rather than the full knowledge graph, to tackle the link prediction problem in knowledge graphs. This approach emphasizes the importance of optimizing predictions in order to increase computing efficiency and scalability. However, GNN-based reasoning methods frequently suffer from irrelevant entities, making large-scale KGs more complex and expensive. To address this issue, AdaProp provides an adaptive propagation technique to prioritize semantically related entities during information propagation. These techniques can improve prediction accuracy and speed up the reasoning process in scenarios where the knowledge graph is large and complex.

Table 5.35 Transductive setting. The highest performance is highlighted with bold numbers, while the second-best results are underlined. The metric "–" indicates unavailable results. The metrics "H@1" and "H@10" refer to Hit@1 and Hit@10, respectively, expressed as percentages

Type	Models	Family			UMLS			WN18RR			FB15k237			NELL-995			YAGO3-10		
		MRR	H@1	H@10	MRR	H@1	H@10	MRR	H@1	H@10	MRR	H@1	H@10	MRR	H@1	H@10	MRR	H@1	H@10
non-GNN	ConvE	0.912	83.7	98.2	0.937	92.2	96.7	0.427	39.2	49.8	0.325	23.7	50.1	0.511	44.6	61.9	0.520	45.0	66.0
	QuatE	0.941	89.6	99.1	0.944	90.5	99.3	0.480	44.0	55.1	0.350	25.6	53.8	0.533	46.6	64.3	0.379	30.1	53.4
	RotatE	0.921	86.6	98.8	0.925	86.3	99.3	0.477	42.8	57.1	0.337	24.1	53.3	0.508	44.8	60.8	0.495	40.2	67.0
	MINERVA	0.885	82.5	96.1	0.825	72.8	96.8	0.448	41.3	51.3	0.293	21.7	45.6	0.513	41.3	63.7	–	–	–
	DRUM	0.934	88.1	99.6	0.813	67.4	97.6	0.486	42.5	58.6	0.343	25.5	51.6	0.532	46.0	66.2	0.531	45.3	67.6
	RNNLogic	0.881	85.7	90.7	0.842	77.2	96.5	0.483	44.6	55.8	0.344	25.2	53.0	0.416	36.3	47.8	0.554	50.9	62.2
	RLogic	–	–	–	–	–	–	0.47	44.3	53.7	0.31	20.3	50.1	–	–	–	0.36	25.2	50.4
GNNs	CompGCN	0.933	88.3	99.1	0.927	86.7	99.4	0.479	44.3	54.6	0.355	26.4	53.5	0.463	38.3	59.6	0.421	39.2	57.7
	NBFNet	0.989	**98.8**	98.9	0.948	92.0	**99.5**	0.551	49.7	66.6	0.415	32.1	**59.9**	0.525	45.1	63.9	0.550	47.9	68.6
	RED-GNN	**0.992**	**98.8**	**99.7**	0.964	94.6	99.0	0.533	48.5	62.4	0.374	28.3	55.8	0.543	47.6	65.1	0.559	48.3	68.9
	AdaProp	0.988	98.6	99.0	**0.969**	**95.6**	**99.5**	**0.562**	**49.9**	**67.1**	**0.417**	**33.1**	58.5	**0.554**	**49.3**	**65.5**	**0.573**	**51.0**	**68.5**

Table 5.36 Inductive setting (evaluated with Hit@10)

		WN18RR				FB15k237				NELL-995			
Metric	Methods	V1	V2	V3	V4	V1	V2	V3	V4	V1	V2	V3	V4
Hit@10 (%)	RuleN	73.0	69.4	40.7	68.1	44.6	59.9	60.0	60.5	76.0	51.4	53.1	48.4
	Neural LP	77.2	74.9	47.6	70.6	46.8	58.6	57.1	59.3	87.1	56.4	57.6	53.9
	DRUM	77.7	74.7	47.7	70.2	47.4	59.5	57.1	59.3	87.3	54.0	57.7	53.1
	GraIL	76.0	77.6	40.9	68.7	42.9	42.4	42.4	38.9	56.5	49.6	51.8	50.6
	CoMPILE	74.7	74.3	40.6	67.0	43.9	45.7	44.9	35.8	57.5	44.6	51.5	42.1
	NBFNet	82.7	79.9	56.3	70.2	51.7	63.9	58.8	55.9	79.5	63.5	60.6	59.1
	RED-GNN	79.9	78.0	52.4	72.1	48.3	62.9	60.3	62.1	86.6	60.1	59.4	55.6
	AdaProp	**86.6**	**83.6**	**62.6**	**75.5**	**55.1**	**65.9**	**63.7**	**63.8**	**88.6**	**65.2**	**61.8**	**60.7**

5.2.3.4 One-Shot-Subgraph Reasoning for Scalable Knowledge Graph Link Prediction

While structural models achieve state-of-the-art performance, they face scalability challenges, particularly with large-scale KGs [162]. This inefficiency stems from processing the entire KG during prediction, raising the question: *is all information necessary for KG predictions*? Inspired by the human brain's selective recall and practices in large-scale recommendation systems, we propose a novel one-shot-subgraph link prediction approach to address the scalability issues of current KG methods. Our proposed method breaks down the prediction process into two stages: (1) we first use a sampler to quickly sample a query-related subgraph and then (2) use a prediction to predict slowly on the subgraph. Specifically, we use a Personalized PageRank (PPR) [179, 282] sampler for efficient sampling relevant subgraphs which requires no learning. In addition, we design a data-adaptive search which can achieve optimal configurations in both data and model spaces to avoid trivial solutions. Our experiments show that, with only a small subgraph of the KG (e.g., 10% of entities), the link prediction can still be effective.

Problem Formulation The one-shot-subgraph link prediction task on KGs is defined as follows:

- *Input*: A knowledge graph $\mathcal{G} = (\mathcal{V}, \mathcal{R}, \mathcal{E})$, where $\mathcal{V}$ is the set of entities, $\mathcal{R}$ is the set of relations, and $\mathcal{E}$ is the set of edges.
- *Query*: A specific query $(e_q, r_q, \cdot)$, where e_q is the query entity, r_q is the query relation, and the goal is to find the target entity e_a such that (e_q, r_q, e_a) is valid.
- *Output*: Predictions $\hat{Y}$, which are scores for candidate entities being the answer.

For efficient and scalable predictions, the process is decoupled into two steps:

1. *Step 1*: Extract a query-dependent subgraph $\mathcal{G}_s = (\mathcal{V}_s, \mathcal{R}_s, \mathcal{E}_s)$ from $\mathcal{G}$, where $|\mathcal{V}_s| \ll |\mathcal{V}|$, $|\mathcal{E}_s| \ll |\mathcal{E}|$.
2. *Step 2*: Predict on $\mathcal{G}_s$ using a scoring function f_θ that outputs the prediction $\hat{Y}$.

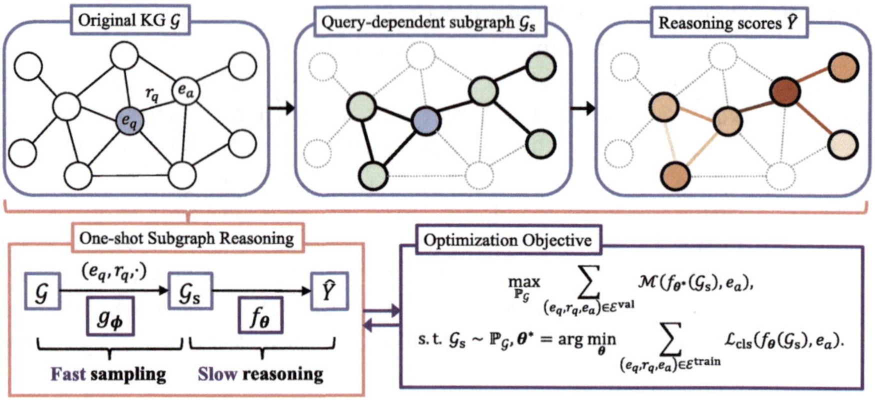

Fig. 5.24 Illustration of one-shot-subgraph link prediction. In our proposed method, the sampler g_ϕ first extracts a subgraph $\mathcal{G}_s$ from the whole graph $\mathcal{G}$ depending on the given query, and then the predictor f_θ conducts deliberative prediction on the extracted subgraph $\mathcal{G}_s$ and obtains the final predictions $\hat{Y}$

Method Our proposed method is illustrated in Fig. 5.24. Briefly, there are three main steps in the framework:

Generate Sampling Distribution We use PPR as the sampling strategy, which works by propagating scores from e_q to estimate the importance of each neighbor. The probability vector $\mathbf{p} \in \mathbb{R}^{|\mathcal{V}|}$ is updated iteratively as

$$\text{Nonparametric indicator}: \ \mathbf{p}^{(k+1)} \leftarrow \alpha \cdot \mathbf{s} + (1-\alpha) \cdot \mathbf{D}^{-1}\mathbf{A} \cdot \mathbf{p}^{(k)}, \tag{5.43}$$

where $\mathbf{p}^{(0)} = \mathbf{s} = \mathbb{1}(e_q)$ initializes the score at e_q and $\mathbf{D}$ and $\mathbf{A}$ are the degree and adjacency matrices, respectively.

Extract a Subgraph We extract the subgraph $\mathcal{G}_s = (\mathcal{V}_s, \mathcal{E}_s)$ consisting of the most relevant entities and edges. The sampling ratios for entities and edges, $r^q_\mathcal{V}, r^q_\mathcal{E} \in (0, 1]$, depend on the query relation r_q. Using the TopK(D, P, K) operation, we select the top-K entities and edges based on their probabilities:

$$\begin{aligned} &\text{Entity Sampling: } \mathcal{V}_s \leftarrow \text{TopK}\Big(\mathcal{V},\ \mathbf{p},\ K = r^q_\mathcal{V} \times |\mathcal{V}|\Big), \\ &\text{Edge Sampling: } \mathcal{E}_s \leftarrow \text{TopK}\Big(\mathcal{E},\ \{\mathbf{p}_x \cdot \mathbf{p}_o : x, o \in \mathcal{V}_s, (x, r, o) \in \mathcal{E}\},\ K = r^q_\mathcal{E} \times |\mathcal{E}|\Big). \end{aligned} \tag{5.44}$$

Reason on the Subgraph We define the configuration space for the predictor and utilize existing structural models. The message is propagated from e_q to the sampled entities $o \in \mathcal{V}_s$. Layer-wise updates of entity representations are formulated as

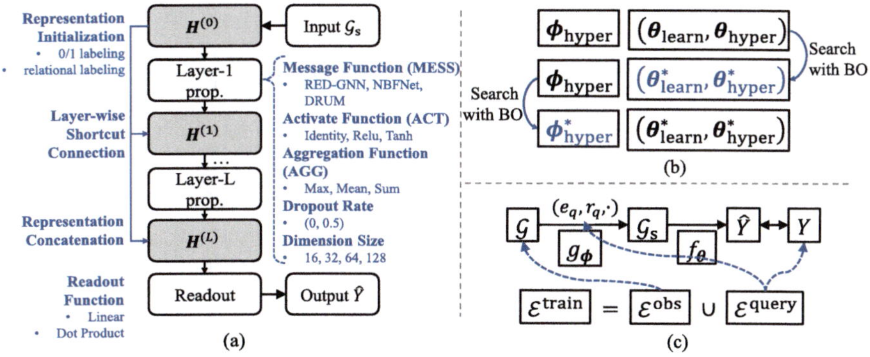

Fig. 5.25 The illustrations of our proposed optimization procedure. Both the predictor and the sampler contain hyperparameters, but only the former has learnable parameters. The $\mathbf{H}^{(L)}$ indicates the representations in L-th layer. (**a**) The search space of $\boldsymbol{\theta}_{\text{hyper}}$. (**b**) The edge split trick

$$
\begin{aligned}
&\text{Indicating:}\mathbf{h}_o^{(0)} \leftarrow 1(o = e_q),\\
&\text{Propagation:}\mathbf{h}_o^{(l+1)}\\
&\quad \leftarrow \text{DROPOUT}\Big(\text{ACT}\Big(\text{AGG}\big\{\text{MESS}(\mathbf{h}_x^{(l)}, \mathbf{h}_r^{(l)}, \mathbf{h}_o^{(l)}) : (x, r, o) \in \mathcal{E}_s\big\}\Big)\Big).
\end{aligned}
\tag{5.45}
$$

After L-layer propagation, the predictor outputs the final score $\hat{y}_o$ for each entity $o \in \mathcal{V}_s$, based on $\mathbf{h}_o^{(L)}$ as $\hat{y}_o = \text{Readout}(\mathbf{h}_o^{(L)}, \mathbf{h}^{(L)}) \in \mathbb{R}$.

Since hyperparameters $(r_{\mathcal{V}}^q, r_{\mathcal{E}}^q)$ and L play important roles in our proposed framework (Fig. 5.25a), we propose a bi-level optimization algorithm to adaptively search for the optimal configuration $(\phi_{\text{hyper}}^*, \theta_{\text{hyper}}^*)$ of design choices on a specific KG. Specifically, we decompose the search process into two phases as illustrated in Fig. 5.25b where we first fix the sampler $g_{\bar{\phi}}$ (with hyperparameters ϕ_{hyper}) and optimize the predictor and then fix the predictor f_{θ^*} and optimize the sampler g_{ϕ^*}. To accelerate the searching, the training facts are split into two subsets: $\mathcal{E}^{\text{train}} = \mathcal{E}^{\text{obs}} \cup \mathcal{E}^{\text{query}}$, where the prediction system $f_{\boldsymbol{\theta}} \circ g_{\boldsymbol{\phi}}$ uses $\mathcal{E}^{\text{obs}}$ as input to predict $\mathcal{E}^{\text{query}}$ (see Fig. 5.25c).

Experiments The empirical results shown in Tables 5.37 and 5.38 prove that our proposed one-shot-subgraph link prediction method outperforms all existing approaches across five large-scale datasets [94, 162, 347, 422]. This improvement is driven by a powerful GNN (providing high expressiveness) and the extraction of compact subgraphs, which contain only the most relevant information. Specifically, our experiments demonstrate that it is not necessary to process the entire KG for link prediction. Actually, only a small subset of entities and facts (an average of just 10%) are critical for answering queries, and these can be efficiently selected using PPR heuristics.

Table 5.37 The empirical results of WN18RR, NELL-995, YAGO3-10 datasets where the best performance is displayed in **bold** numbers, and the second best in <u>underline</u>. "H@1" and "H@10" stand for Hit@1 and Hit@10 (in percentage), respectively

		WN18RR			NELL-995			YAGO3-10		
Type	Models	MRR ↑	H@1 ↑	H@10 ↑	MRR ↑	H@1 ↑	H@10 ↑	MRR ↑	H@1 ↑	H@10 ↑
Semantic models	ConvE	0.427	39.2	49.8	0.511	44.6	61.9	0.520	45.0	66.0
	QuatE	0.480	44.0	55.1	0.533	46.6	64.3	0.379	30.1	53.4
	RotatE	0.477	42.8	57.1	0.508	44.8	60.8	0.495	40.2	67.0
Structural models	MINERVA	0.448	41.3	51.3	0.513	41.3	63.7	–	–	–
	DRUM	0.486	42.5	58.6	0.532	46.0	**66.2**	0.531	45.3	67.6
	RNNLogic	0.483	44.6	55.8	0.416	36.3	47.8	0.554	<u>50.9</u>	62.2
	CompGCN	0.479	44.3	54.6	0.463	38.3	59.6	0.489	39.5	58.2
	DPMPN	0.482	44.4	55.8	0.513	45.2	61.5	0.553	48.4	67.9
	NBFNet	<u>0.551</u>	<u>49.7</u>	**66.6**	0.525	45.1	63.9	0.550	47.9	68.3
	RED-GNN	0.533	48.5	<u>62.4</u>	<u>0.543</u>	<u>47.6</u>	<u>65.1</u>	<u>0.559</u>	48.3	<u>68.9</u>
	one-shot-subgraph	**0.567**	**51.4**	**66.6**	**0.547**	**48.5**	<u>65.1</u>	**0.606**	**54.0**	**72.1**

Table 5.38 Empirical results of two OGB datasets [162] compared to official leaderboards

Type	Models	OGBL-BIOKG			OGBL-WIKIKG2		
		Test MRR ↑	Valid MRR ↑	#Params ↓	Test MRR ↑	Valid MRR ↑	#Params ↓
Semantic models	TripleRE	0.8348	0.8360	469,630,002	0.5794	0.6045	500,763,337
	AutoSF	0.8309	0.8317	93,824,000	0.5458	0.5510	500,227,800
	PairRE	0.8164	0.8172	187,750,000	0.5208	0.5423	500,334,800
	ComplEx	0.8095	0.8105	187,648,000	0.4027	0.3759	1,250,569,500
	DistMult	0.8043	0.8055	187,648,000	0.3729	0.3506	1,250,569,500
	RotatE	0.7989	0.7997	187,597,000	0.4332	0.4353	1,250,435,750
	TransE	0.7452	0.7456	187,648,000	0.4256	0.4272	1,250,569,500
Structural models	**one-shot-subgraph**	**0.8430**	**0.8435**	**976,801**	**0.6755**	**0.7080**	**6,831,201**

5.3 Trustworthy Foundation Models

5.3.1 *Introduction and Notation*

In this section, we focus on three crucial parts of the trustworthiness in foundation models: causality, safety, and robustness. In **causality**, two levels of ability are defined for *reasoning with causal knowledge*, and the current gap toward *level-2* causal reasoning is shown by a novel causal QA benchmark and a novel method incorporating general knowledge and goal-oriented prompts. For *discovering unknown knowledge*, a novel representation assistant framework is proposed to extend the scope of traditional causal discovery methods. The framework is theoretically guaranteed to construct Markov blanket factors from unstructured data for a given target variable. In **safety**, we discuss how to *detect the machine-generated text*, which is important to the control of plagiarism issues and hallucination information. A multi-population aware maximum mean discrepancy is designed, upon which we present two powerful text-detection methods. For *the jailbreak issue in LLMs*, we show the weakness of self-losing on both open-source and closed-source LLMs. Notably, LLMs can escape the safety guardrails by utilizing their powerful personification ability given a virtual nested scene and can exhibit continuous jailbreak in subsequent interactions. In the **robustness** aspect, we investigate both language models and CLIP models. We define and explore *the noisy rationale problem*, where inaccurate or irrelevant reasoning steps are mixed with valid ones. Systematic analysis of the reasoning robustness is provided based on the results in a novel benchmark. A novel contrastive denoising method is designed to alleviate this issue. For CLIP models, we highlight the *insufficiency of current ImageNet-oriented benchmarking*. Justified with a significant performance drop on a new challenging dataset, we show the existence of spuriousness generally captured across different CLIP models. Theoretical insights and e-evaluate strategies are discussed.

5.3.2 *Causality*

5.3.2.1 Unveiling Causal Reasoning in Large Language Models

Reasoning, an essential component of human intelligence, distinguishes humans from other species. Reasoning capability is treated as an important component of superintelligence for LLMs, and it is categorized into logical reasoning [86, 141], mathematical reasoning [172, 344], and so on.

Problem Definition Causal Reasoning in LLM Context: An advanced cognitive ability, causal thinking helps one to recognize and understand cause-and-effect linkages between events or phenomena. Human intellect and decision-making depend on this logical framework since it helps people to forecast results, make wise decisions, and develop a better awareness of the surroundings.

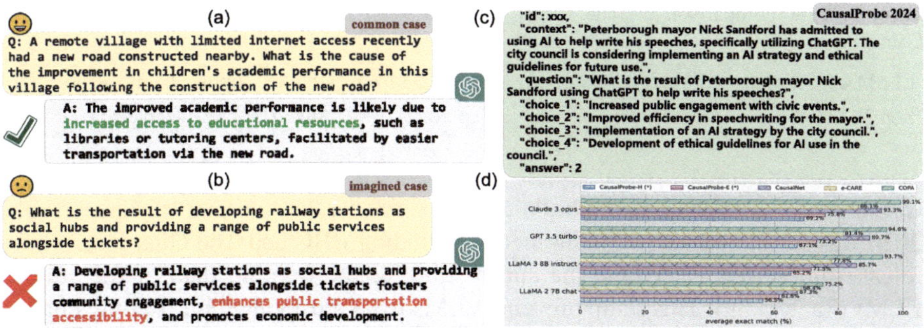

Fig. 5.26 (**a**) Large language models excel on routinely encountered causal reasoning assignments spanning generally relevant subjects. (**b**) Large language models struggle with unusual jobs including wholly fresh material not before encountered in their training data. (**c**) Question and answer pairs in CausalProbe 2024 show how its method of testing causal reasoning works. (**d**) Comparatively to previous causal reasoning benchmarks, the large language models assessed exhibited clearly poorer performance on CausalProbe 2024

Though their performance is still erratic across different contexts [475], recent research has shown that LLMs show varied degrees of causal reasoning ability [185]. Figure 5.26a shows how well LLMs can detect reasonable causal links, such as linking better academic performance to "more access to educational resources." But when given fresh or unusual scenarios—as demonstrated in 5.26(b)—such as "developing railway stations as social hubs," LLMs produce replies (e.g., "enhance public trans-travel accessibility") that fail to convey the desired causal relationship. This discrepancy in performance poses a basic issue: *to what extent do LLMs possess genuine causal reasoning capabilities?*

We offer a framework separating two degrees of causal reasoning skills in LLMs in order to systematically answer this question.

Definition 5.5 (Level-1 Causal Reasoning) Level-1 causal reasoning is the capacity of contextual information and causal knowledge stored inside model parameters to be retrieved and used. Rapid processing and efficiency in managing simple cause-and-effect linkages define this kind of thinking.

Definition 5.6 (Level-2 Causal Reasoning) Level-2 causal reasoning is the capacity of sophisticated reasoning processes to synthesize new causal knowledge by means of model parameters, contextual information, inferential capacities, and synthesis of elements. Operating at a deeper level, this advanced type of thinking allows one to derive fresh causal linkages not before observed.

Experiments Several benchmarks have been constructed in order to scientifically evaluate LLM causal reasoning capacity. CausalProbe 2024 [70] is a representative one. This benchmark was built in a hierarchical way to achieve a progressive and comprehensive assessment. CausalProbe 2024 involves three datasets: CausalProbe-E (simple single-choice version), CausalProbe-H (hard single-choice version), and

Table 5.39 A comparison of data cutoff dates between the LLMs that were evaluated and CausalProbe 2024

LLaMA 2 7B chat	LLaMA 3 8B instruct	GPT 3.5 turbo	Claude 3 opus	CausalProbe 2024
Sep 2022	Mar 2023	Sep 2021	Aug 2023	Jan 2024

CausalProbe-M (uncertain multiple-choice version). The corpus of CausalProbe 2024 was selected from two main media, including BBC and The Guardian, whose release times were later than the cutoff time for the training data of the studied LLMs (Table 5.39). This setup prevents LLMs from retrieving their training corpus. The details of three datasets are presented below to show their characteristics of hierarchical assessment.

- CausalProbe-E is constructed using the CausalQA [38] format as a template. After analyzing CausalQA examples with GPT, GPT is prompted to generate comparable Q&A pairs based on provided articles. This dataset evaluates LLMs' fundamental causal reasoning capabilities through novel test scenarios.
- CausalProbe-H is built with another strategy. GPT is prompted to summarize a provided articles and then extract several correct and incorrect cause-effect pairs based on the summary. Finally, several high-quality propositions are generated using these cause-effect pairs as options. CausalProbe-H assesses whether LLMs can identify misleading or deceptive causal propositions precisely.
- CausalProbe-M extends the assessment with multiple-choice questions containing 1-4 correct answers per question. Using a similar construction process to CausalProbe-H for each option, this dataset evaluates LLMs' ability to independently assess each proposition's validity, reducing the effectiveness of random guessing strategies.

Figure 5.26 illustrates how the studied LLMs perform on CausalProbe 2024 compared to established benchmarks. The results reveal a notable performance decline on CausalProbe-E and CausalProbe-H datasets, particularly since previous benchmarks were likely part of their training data. For CausalProbe-H, even Claude 3 opus, considered the most advanced LLM, achieved an exact match score below 70%, while LLaMA 2 7B chat only reached 50% accuracy. Causal Probe 2024 uses easily available news items devoid of sophisticated language; hence, this performance decline mostly results from the models' ignorance of novel circumstances, suggesting just level-1 causal reasoning ability.

CausalProbe 2024's incorporation of contextual information for every Q&A pair provides an additional benefit. Figures 5.27 and 5.28 show performance variations with and without context, therefore underlining the need for contextual information for textual causal reasoning. These settings provide really vital background information guidelines. Furthermore, Fig. 5.29 shows the difference between exact and partial match metrics on CausalProbe-M, where considerably lower exact match scores indicate perhaps reliance on random guessing in single-choice datasets.

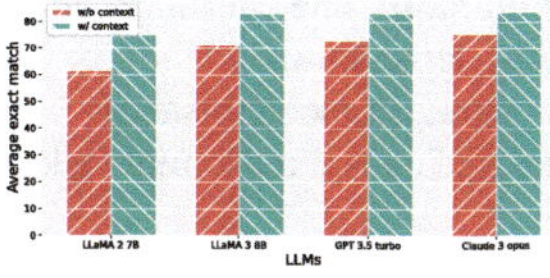

Fig. 5.27 CausalProbe-E w/ and w/o contexts

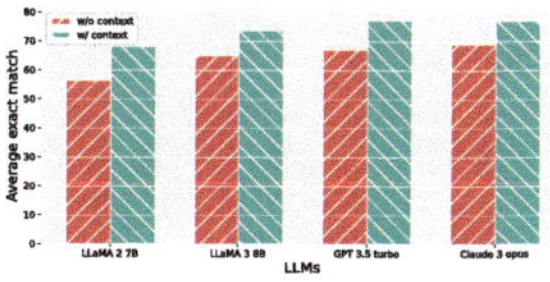

Fig. 5.28 CausalProbe-H w/ and w/o contexts

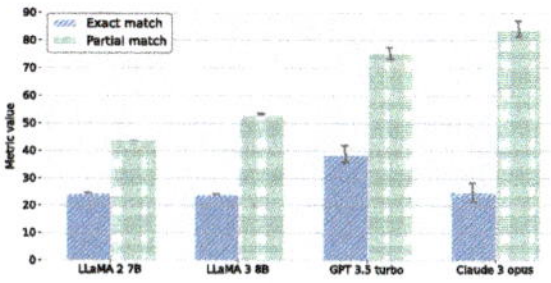

Fig. 5.29 Result comparison between exact match and partial match

Method Although decoder-only transformer-based LLMs produce text by means of autoregression, this method does not intrinsically capture causation. Autoregression, statistically speaking, holds that current values solely rely on past values, thus excluding future dependencies. Although logical causality [170] differs from sequential causality, daily language expressions of causal interactions may contradict this notion. Sometimes LLMs may exactly detect complicated causal relationships largely because their big training set exposes them to many utterances expressing the same causal notions.

To advance from level-1 to level-2 causal reasoning, the G^2-Reasoner was explored, yet it did not achieve level-2 causal reasoning. This approach is inspired by a structural causal model that describes textual causal reasoning tasks, shown in Fig. 5.30. A causal link is formed by considering a semantic variable R indicating a cause and a variable Q expressing an effect. Physical or virtual world laws expressed by C control this relationship. These semantic variables are subsequently transformed into natural language text T using a transformation function h, stated as $h(R, Q, \epsilon) = T$. Here, ϵ is an exogenous random variable considering several linguistic aspects including language choice, context, and voice. The function h represents a complete language system that transforms causal concepts into text. While the variability introduced by ϵ adds richness to natural language expression, it also creates challenges for LLMs in causal reasoning tasks.

In this causal graph, we observe both a confounding variable and a conditioned collider [291]. The variable T is conditioned since we have fixed natural language expressions in textual reasoning tasks. Based on causal inference principles, conditioning on the collider T (where $T = T_0$) creates a correlativity between R and Q. Therefore, the natural language expressions provide key information

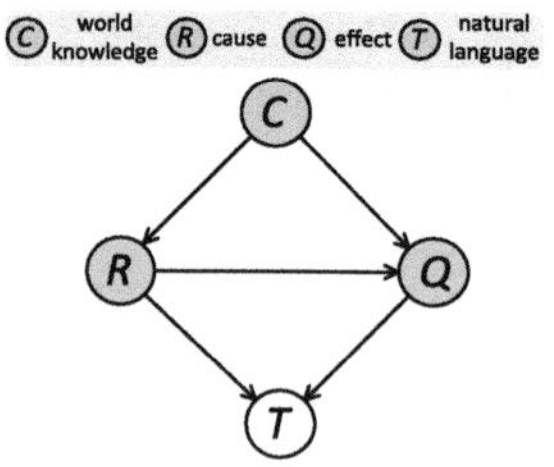

Fig. 5.30 The causal graph that depicts the data generation mechanism of textual causal reasoning tasks

for determining causal relationships.[5] Considering tasks following $R \to Q$, the objective is formalized as

$$\arg\max_{Q \sim P_Q} P[Q|R = R_0, T = T_0, C], \tag{5.46}$$

where R_0 is the cause described in natural language and P_Q is the distribution of outcome Q. To eliminate reasoning biases caused by the confounding variable C, they applied the total probability formula to perform covariate adjustment, i.e.,

$$\begin{aligned} &\arg\max_{Q \sim P_Q} \mathbb{E}_{C \sim P_C(\cdot|R_0,T_0)} P[Q|R = R_0, T = T_0, C] \\ &= \arg\max_{Q \sim P_Q} P[Q|R = R_0, T = T_0], \end{aligned} \tag{5.47}$$

where $P_C(\cdot|R_0, T_0)$ denotes the conditional distribution of general knowledge, given the observed text and cause. Although achieving perfect equivalence in Eq. (5.47) would require a comprehensive knowledge base, this formulation provides key insights for implementing causal reasoning in LLMs and demonstrates practical utility.

G^2-Reasoner uses two fundamental components—as shown in Fig. 5.31—to overcome the limits of autoregression in capturing logical causality and the tendency of LLMs to lose coherence in larger sequences [251]. It starts with a RAG system to lay a basis of general knowledge. Second, it makes use of goal-oriented stimuli to keep attention on spotting appropriate causal relations throughout the inference process. This dual strategy seeks to improve causal reasoning capacity while preserving consistency.

G^2-Reasoner makes use of a modest (around 16 Mb) general knowledge Q&A database. Table 5.40 shows significant performance improvement even with only a small knowledge base. If using a larger general knowledge base, such as Wikipedia, the performance will be further boosted.

[5] While colliders are often considered bias sources in causal inference, the collider T serves a useful purpose here.

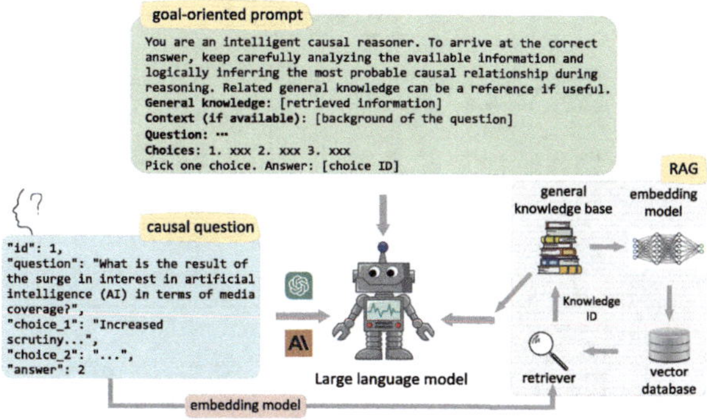

Fig. 5.31 G^2-Reasoner consists of two modules: one is a RAG system to retrieve general knowledge related to a given causal reasoning task; another is a goal-oriented prompt to prevent LLMs from deviating from the original goal during inference

Table 5.40 Results of the studied LLMs on four causal Q&A benchmarks. The metric is exact match (EM). "Vanilla" denotes doing inference directly. "C-E" and "C-H" are the abbreviations of CausalProbe-E and CausalProbe-H, respectively. For GPT 3.5 turbo and Claude 3 opus, the temperature is set to 1.0

		LLaMA 2	LLaMA 3	GPT 3.5 turbo	Claude 3 opus
COPA	Vanilla	0.752	0.937	0.943±0.002	0.992±0.002
	CoT	0.812	0.944	0.951±0.001	0.991±0.000
	RAG	0.757	0.912	0.939±0.005	0.988±0.003
	G^2-Reasoner	0.813	0.948	0.953±0.002	0.993±0.001
e-CARE	Vanilla	0.684	0.778	0.811±0.003	0.857±0.005
	CoT	0.697	0.770	0.807±0.006	0.863±0.002
	RAG	0.687	0.760	0.806±0.004	0.842±0.005
	G^2-Reasoner	0.701	0.779	0.821±0.001	0.854±0.003
CausalNet	Vanilla	0.673	0.857	0.895±0.003	0.930±0.002
	CoT	0.666	0.767	0.872±0.003	0.907±0.003
	RAG	0.650	0.860	0.894±0.007	0.910±0.004
	G^2-Reasoner	0.681	0.855	0.904±0.005	0.929±0.002
C-E (*)	Vanilla	0.616	0.715	0.730±0.002	0.755±0.002
	CoT	0.636	0.720	0.740±0.002	0.751±0.001
	RAG	0.621	0.704	0.743±0.001	0.756±0.001
	G^2-Reasoner	0.642	0.718	0.747±0.003	0.762±0.002
C-H (*)	Vanilla	0.565	0.652	0.670±0.001	0.688±0.003
	CoT	0.573	0.644	0.662±0.004	0.700±0.002
	RAG	0.575	0.655	0.678±0.001	0.687±0.001
	G^2-Reasoner	0.582	0.658	0.690±0.003	0.701±0.004

5.3.2.2 Representation Assistant for Causal Discovery

In the previous section, we talked about how to utilize causal knowledge for reasoning; a further step is to explore new scientific knowledge. One of the inevitable steps in scientific research is to reveal causal relations among the entities in the world. The causal relation means how the components of a system interact

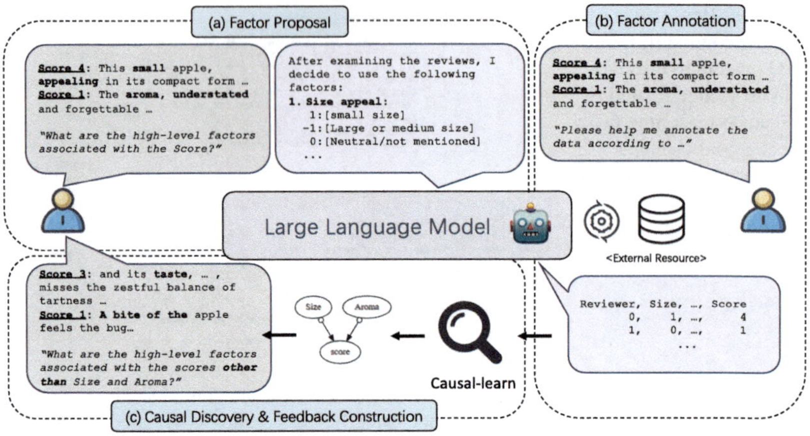

Fig. 5.32 Illustration of COAT framework

with each other. Revealing this structure would be helpful in making decisions on how to apply interventions so that a certain goal could be accomplished or understand what could have happened in the counterfactual world.

The necessary preliminary procedure for causal discovery is to define high-level variables that might be helpful to capture the nature of the system of interest. However, these variables are usually not naturally defined in the real world. They are usually crafted and tested by experienced scientists or domain experts, which takes lots of time and human effort. Despite the essential efforts to leverage the representative learning approach to reconstruct the latent variables, the identifiability requires conditions like auxiliary variables, sparsity, or interventional data, and resulting factors often lack sufficient physical explanations. Therefore, crafting high-quality, high-level variables is a critical yet longstanding challenge to broaden the applicable scope of causal discovery and related fields.

Problem Definition To mitigate this challenge, **C**ausal Representati**O**n **A**ssistan**T** (**COAT**) is proposed, and the framework is shown in Fig. 5.32. It incorporates the rich knowledge learned from massive internet data by large language models (LLMs) and is designed to provide explainable yet useful high-level factors behind unstructured data. To be concrete, it considers paired samples between unstructured data, e.g., textual comments on a product from users, and one structured target variable, e.g., the corresponding ratings on that product. Let $\mathbf{X}$ be a high-dimensional random variable as unstructured data and Y be a scalar random variable as the target. Taking these samples as input, COAT iteratively identifies a set of explainable factors $\mathbf{Z}$ that provably serve as a Markov blanket of Y on $\mathbf{X}$, i.e., $Y \perp\!\!\!\perp \mathbf{X} \mid \mathbf{Z}$. In other words, the identified structured factors $\mathbf{Z}$ block the correlation between target variable Y and the unstructured data $\mathbf{X}$.

Method During each COAT iteration, three steps are sequentially executed: *factor proposal*, *factor parsing*, and *feedback construction*. In the *factor proposal* step,

COAT constructs factor mappings $\mathbf{w}_i : \mathcal{X} \to C$ from raw observations to a prespecified value space C, e.g., $C := \{-1, 0, 1\}$. These mappings are attained by prompting LLMs to analyze a small set of paired samples from $(\mathbf{X}, Y)$ and are defined by natural language. For instance, a factor *taste* could be defined by *"If the customer describes the apple as sweet, set it as* 1; *sour as* $-$ 1; *otherwise*, 0." In the *factor parsing* step, COAT applies each factor mapping $\mathbf{w}_i$ on the unstructured data X to get structured factor values $\mathbf{w}_i(\mathbf{X})$. These values will be stored for later use.

In the *feedback construction* step, it consists of two parts. The first part is to verify the proposed factors by statistical tests with their parsed values. Suppose current it has k verified factors $h_{[k]}(\mathbf{X}) = (\mathbf{w}_1(\mathbf{X}), \mathbf{w}_2(\mathbf{X}), \cdots, \mathbf{w}_k(\mathbf{X}))$. A new factor $\mathbf{w}_{k+1}$ would pass the verification if $\mathbf{w}_{k+1}(\mathbf{X}) \not\perp\!\!\!\perp Y \mid h_{[k]}(\mathbf{X})$. If $Y \perp\!\!\!\perp \mathbf{X} \mid h_{[k]}(\mathbf{X})$ holds, then no such $\mathbf{w}_{k+1}$ could pass the verification. If it passes, then the second part constructs useful feedback with verified factors to encourage LLMs to produce further factors. The idea is to find a subset of samples $\widehat{D}^*$ where $h_{[k+1]}(\mathbf{X})$ cannot well explain Y, i.e., the conditional entropy is high: $\widehat{D}^* = \arg\max_{\widehat{D} \subset D} \mathrm{H}_{\widehat{D}}(Y \mid h_{[k+1]}(\mathbf{X}))$. To find an approximated solution, it clusters the dataset D according to $h_{[k+1]}(\mathcal{X})$ and selects the cluster with the largest conditional entropy.

Theoretical Guarantee Two scores concerned with the one LLM Ψ's ability are proposed to characterize the identifiability of the Markov blanket. The first one is called *perception score* pcp_Ψ, saying that the LLM has probability pcp_Ψ to propose a factor that can pass the verification. The second one is the *capacity score* C_Ψ, measuring the quality of each verified new factor in the sense that $I(Y; \mathbf{X} \mid h_{[k+1]}(\mathbf{X})) < (1 - C_\Psi) I(Y; \mathbf{X} \mid h_{[k]}(\mathbf{X}))$. If both scores are positive, for a small positive number δ, after t COAT rounds with verified factors $h_{\leq t}(\mathbf{X})$, the following inequality holds with probability at least $1 - \delta$: $\frac{I(Y;\mathbf{X}|h_{\leq t}(\mathbf{X}))}{I(Y;\mathbf{X})} < (\frac{1}{(1-C_\Psi)})^{-t\mathrm{pcp}_\Psi - z_\delta\sqrt{t\mathrm{pcp}_\Psi(1-\mathrm{pcp}_\Psi)}}$, where z_δ is the δ-quantile of the standard normal distribution (Table 5.41).

Experiments To empirically evaluate the factor identification ability of COAT, the first benchmark, AppleGastronome, is constructed. In this scenario, customers make comments on the apples they got and give scores according to their preferences. One causal graph consists of six high-level factors, and one observed target variable, i.e., score, is specified. It first samples the values of these seven variables and then constructs textual comments with LLMs according to the six factors. Three factors are causal parents of Y: one is the child, one is the spouse, and one is a disturbing factor that is correlated with Y but is out of the Markov boundary.

The empirical results reveal insights: LLMs have the potential to generate relevant factors from unstructured data, as witnessed in each baseline. However, even equipped with CoT, LLMs struggle to identify the Markov blanket through reasoning and filter out irrelevant factors. COAT significantly mitigates these issues. Also, the factors proposed and parsed by COAT are further processed by the downstream causal discovery methods and can reflect a structure close to the ground truth.

Table 5.41 Factor proposal results in Apple Gastronome benchmark

		Factor proposal					
LLM	Method	MB	NMB	OT	Recall	Precision	F1
GPT 4	META	2.67(±*0.94*)	0.67(±*0.47*)	2.33(±*0.47*)	0.53(±*0.19*)	0.46(±*0.08*)	0.49(±*0.13*)
	DATA	3.00(±*0.00*)	0.33(±*0.47*)	0.00(±*0.00*)	0.60(±*0.00*)	0.92(±*0.12*)	0.72(±*0.04*)
	DATA+CoT	4.33(±*0.58*)	0.83(±*0.29*)	0.17(±*0.29*)	0.87(±*0.12*)	0.81(±*0.02*)	0.84(±*0.06*)
	COAT	4.00(±*0.82*)	0.33(±*0.47*)	0.00(±*0.00*)	0.80(±*0.16*)	0.93(±*0.09*)	0.85(±*0.11*)
GPT 3.5	META	3.33(±*1.25*)	0.33(±*0.47*)	4.33(±*1.25*)	0.67(±*0.25*)	0.42(±*0.12*)	0.51(±*0.15*)
	DATA	2.67(±*0.47*)	0.67(±*0.47*)	0.00(±*0.00*)	0.53(±*0.09*)	0.81(±*0.14*)	0.64(±*0.10*)
	DATA+CoT	5.00(±*0.00*)	1.00(±*0.00*)	1.33(±*0.58*)	1.00(±*0.00*)	0.68(±*0.05*)	0.81(±*0.04*)
	COAT	3.67(±*0.47*)	0.00(±*0.00*)	0.00(±*0.00*)	0.73(±*0.09*)	1.00(±*0.00*)	0.84(±*0.07*)
LLaMA2 -70b	META	2.33(±*0.47*)	0.67(±*0.47*)	4.67(±*1.25*)	0.47(±*0.09*)	0.32(±*0.07*)	0.37(±*0.06*)
	DATA	2.33(±*0.94*)	0.67(±*0.47*)	0.00(±*0.00*)	0.47(±*0.19*)	0.75(±*0.20*)	0.57(±*0.20*)
	DATA+CoT	3.00(±*1.73*)	0.67(±*0.58*)	0.33(±*0.58*)	0.60(±*0.35*)	0.71(±*0.34*)	0.65(±*0.35*)
	COAT	3.00(±*0.00*)	0.67(±*0.47*)	0.00(±*0.00*)	0.60(±*0.00*)	0.83(±*0.12*)	0.69(±*0.04*)
Mistral -MEDIUM	META	3.00(±*0.00*)	0.67(±*0.47*)	1.67(±*1.25*)	0.60(±*0.00*)	0.59(±*0.13*)	0.59(±*0.07*)
	DATA	3.00(±*0.00*)	0.67(±*0.47*)	0.00(±*0.00*)	0.60(±*0.00*)	0.83(±*0.12*)	0.69(±*0.04*)
	DATA+CoT	4.33(±*0.58*)	1.00(±*0.00*)	0.67(±*0.58*)	0.87(±*0.12*)	0.73(±*0.07*)	0.79(±*0.05*)
	COAT	4.67(±*0.47*)	0.00(±*0.00*)	0.00(±*0.00*)	0.93(±*0.09*)	1.00(±*0.00*)	0.96(±*0.05*)

5.3.3 Safety

5.3.3.1 Maximum Mean Discrepancy for Machine-Generated Text Detection

Large language models (LLMs) like ChatGPT excel in generating human-like texts but pose risks such as plagiarism, misinformation, and hallucination [134, 283, 477], making the detection of machine-generated texts (MGTs) essential. This work seeks to utilize maximum mean discrepancy (MMD) [40, 244] to tackle the issue of MGT detection. MMD is a statistical measure that effectively identifies discrepancies between distributions, making it suitable for distinguishing between human-written and machine-generated texts. However, MGTs often originate from diverse LLMs, resulting in multiple text populations. This diversity increases the variance of MMD and significantly reduces its effectiveness. To address this, we propose a novel optimization method for MMD, multi-population aware MMD (*MMD-MP*). This method avoids variance increases and enhances the stability of distributional discrepancy. Utilizing MMD-MP, we provide two separate methodologies for detecting text at the paragraph and sentence levels.

Problem Definition We consider two types of data detection in this work. **Two-sample test (2ST).** Let $\mathcal{D}_p$, $\mathcal{D}_q$ be Borel probability measures on $\mathbf{m}\mathcal{X}\subset\mathbb{R}^d$. We observe *independent identically distributed* (IID) data $D_p=\{\mathbf{x}_i\}_{i=1}^n\sim\mathcal{D}_p^n$ and $D_q=\{\mathbf{y}_j\}_{j=1}^m\sim\mathcal{D}_q^m$. 2ST aims to determine if $\mathcal{D}_p$ and $\mathcal{D}_q$ come from the same distribution, i.e., $\mathcal{D}_p = \mathcal{D}_q$ [40, 244].

Single-Instance Detection (SID) Let $\mathcal{D}_p$ be a probability distribution on $\mathbf{m}\mathcal{X}\subset\mathbb{R}^d$ and *IID* observations $D_p=\{\mathbf{x}_i\}_{i=1}^n\sim\mathcal{D}_p^n$; SID aims to tell if the instance $\tilde{\mathbf{y}}$ is from $\mathcal{D}_p$.

Problem for Text Detection *Let $\mathcal{D}_p$ be a probability distribution on a separable metric space $\mathbf{m}X\subset\mathbb{R}^d$ and let $D_p=\{\mathbf{x}_i\}_{i=1}^n$ represent independent and identically distributed (IID) observations from the human-written text (HWT) distribution. We aim to determine whether the data $D_q=\{\mathbf{y}_j\}_{j=1}^m$ originates from the same distribution $\mathcal{D}_p$. Here, D_q can be HWTs or MGTs generated from multiple LLMs. The task is reduced to a single-instance identification problem when $m = 1$.*

Method This study reexamines the optimization mechanism of kernel-based MMD, concentrating on mitigating the high variance problem when training data is derived from various populations. To this end, we propose a *multi-population* aware optimization method for MMD called MMD-MP, to enhance stability. Leveraging MMD-MP, we provide two separate methodologies for detecting text at the paragraph and sentence levels. We define maximum mean discrepancy (MMD) as a measure that quantifies the divergence between two probability distributions.

Definition 2 ([40, 244]) Let $k : \mathcal{X}\times\mathcal{X}\to\mathbb{R}$ denote a bounded kernel associated with a reproducing kernel Hilbert space $\mathbb{H}_k$, $\mathbf{m}\mathcal{F}$ represent a class of functions $f : \mathcal{X}\to\mathbb{R}$, and $X\sim\mathcal{D}_p, Y\sim\mathcal{D}_q$ be two random variables. The maximum mean discrepancy (MMD) is defined as

$$\begin{aligned}\mathrm{MMD}\left(\mathcal{D}_p, \mathcal{D}_q; \mathbb{H}_k\right) &= \sup_{f\in\mathcal{F}, \|f\|_{\mathbb{H}_k}\leq 1} \left|\mathbb{E}[f(X)]-\mathbb{E}[f(Y)]\right| \\ &= \sqrt{\mathbb{E}\left[k\left(X, X'\right)+k\left(Y, Y'\right)-2k(X, Y)\right]}.\end{aligned}$$

Intuitively, $k(X, X')$ or $k(Y, Y')$ can be interpreted as as a measure of *intra-class* similarity, whereas $k(X, Y)$ quantifies the *interclass* dissimilarity. When $n=m$, the MMD can be computed using a U-statistic estimator, which provides an unbiased approximation of MMD^2:

$$\widehat{\mathrm{MMD}}_u^2(D_p, D_q; k)=\frac{1}{n(n-1)}\sum_{i\neq j} H_{ij}, \quad \text{where } H_{ij}:=k(\mathbf{x}_i, \mathbf{x}_j)-k(\mathbf{x}_i, \mathbf{y}_j)-k(\mathbf{y}_i, \mathbf{x}_j)+k(\mathbf{y}_i, \mathbf{y}_j). \tag{5.48}$$

Here, we consider kernel-based MMD (MMD-D, Liu et al. [244]), where the kernel is given by

$$k_\omega(\mathbf{x}, \mathbf{y})=[(1-\epsilon)\kappa(\phi_{\hat{f}}(\mathbf{x}), \phi_{\hat{f}}(\mathbf{y}))+\epsilon]q(\hat{f}(\mathbf{x}), \hat{f}(\mathbf{y})). \tag{5.49}$$

Here, $\epsilon \in (0, 1)$, $\phi_{\hat{f}}(\mathbf{x}) = \phi(\hat{f}(\mathbf{x}))$ represents a neural network with a feature extractor $\hat{f}$, and κ and q are Gaussian kernels with bandwidth σ_ϕ and σ_q, respectively. Specifically, $\kappa(a, b) = \exp\left(-\|a-b\|^2/2\sigma_\phi^2\right)$. Given that $\hat{f}$ is fixed, we denote the set of parameters of k_ω as $\omega = \{\epsilon, \phi, \sigma_\phi, \sigma_q\}$. The kernel k_ω can be optimized by

$$\begin{aligned}J(\mathcal{D}_p, \mathcal{D}_q; k_\omega) &= \mathrm{MMD}^2(\mathcal{D}_p, \mathcal{D}_q; k_\omega)/\sigma_{\mathcal{H}_1}(\mathcal{D}_p, \mathcal{D}_q; k_\omega), \\ \text{where}\sigma_{\mathcal{H}_1}^2 &:= 4\left(\mathbb{E}\left[H_{ij}H_{i\ell}\right]-\mathbb{E}\left[H_{ij}\right]^2\right), \\ H_{ij}&:=k(\mathbf{x}_i, \mathbf{x}_j)-k(\mathbf{x}_i, \mathbf{y}_j)-k(\mathbf{y}_i, \mathbf{x}_j)+k(\mathbf{y}_i, \mathbf{y}_j).\end{aligned}$$

During the optimization, the kernel of MMD (i.e., MMD-D) aims to aggregate human-written texts (HWTs) and diverse machine-generated texts (MGTs) by minimizing intra-class distances while maximizing interclass distances. However, when the MGT population D_q^{tr} consists of multiple distinct populations, this optimization faces significant challenges due to increased fluctuations. As illustrated in Fig. 5.33a–b, the estimated MMD increases along with growing variance. This phenomenon stems from the *intra-class distance* inherent in MGTs, which is inherently embedded in the optimization of kernel-based MMD. The variance reduces MMD's ability to effectively aggregate MGTs, leading to unstable MMD values and ultimately degrading detection performance.

To solve this problem, we introduce a novel approach called *multi-population aware optimization for kernel-based MMD (MMD-MP)*. Unlike traditional methods,

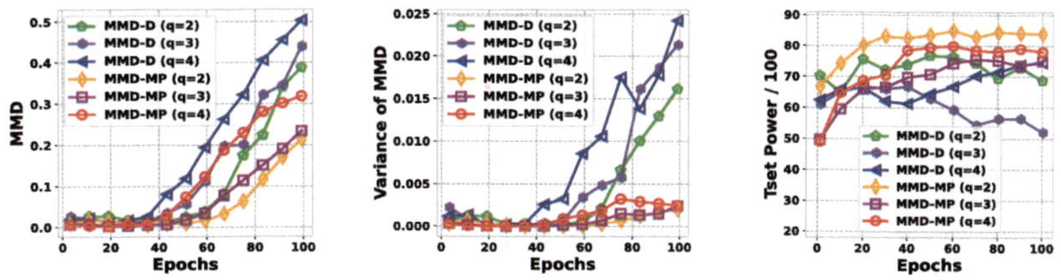

(a) (b) (c)

Fig. 5.33 Depiction of MMD values, MMD variances, and test power of MMD-D and our MMD-MP during the optimization processes. MMD-D exhibits an increase in MMD accompanied by a significant rise in variance, which causes unstable test power as the number of populations in D_q^{tr} grows. Conversely, our proposed MMD-MP demonstrates little fluctuation in terms of MMD values, contributing to enhanced and more stable test power during evaluation. (**a**) MMD. (**b**) Variance of MMD. (**c**) Test Power of MMD

Algorithm 5 Training deep kernel with MMD-MP

Input: D_p^{tr}, D_q^{tr}, a frozen feature extractor $\hat{f}$;
$\omega \leftarrow \omega_0$; $\lambda \leftarrow 10^{-8}$;
for $T = 1, 2, \ldots, T_{max}$ **do**
  $k_\omega \leftarrow$ kernel function via Eq. (5.49);
  $M(\omega) \leftarrow \widehat{\text{MPP}}_u(D_p^{tr}, D_q^{tr}; k_\omega)$ via Eq. (5.52);
  $V_\lambda(\omega) \leftarrow \hat{\sigma}^2_{\mathcal{H}_1^*}(D_p^{tr}, D_q^{tr}; k_\omega)$ via Eq. (5.54);
  $\hat{J}_\lambda(\omega) \leftarrow M(\omega)/\sqrt{V_\lambda(\omega)}$ via Eq. (5.53);
  $\omega \leftarrow \omega + \eta \nabla_{\text{Adam}} \hat{J}_\lambda(\omega)$;
end for
Output: k_ω

MMD-MP avoids optimizing the intra-class distance within D_q^{tr}. Instead, it utilizes a proxy multi-population prior (MPP) to maximize a new objective function, as defined in Eq. (5.50). The training process is presented in Algorithm 5:

$$J(\mathcal{D}_p, \mathcal{D}_q; k_\omega) = \text{MPP}(\mathcal{D}_p, \mathcal{D}_q; k_\omega)/\sigma_{\mathcal{H}_1^*}(\mathcal{D}_p, \mathcal{D}_q; k_\omega), \tag{5.50}$$

$$\text{MPP}\left(\mathcal{D}_p, \mathcal{D}_q; \mathcal{H}_k\right) := \mathbb{E}\left[k_\omega\left(X, X'\right) - 2k_\omega(X, Y)\right]. \tag{5.51}$$

In practice, MPP can be approximated using an unbiased estimator:

$$\widehat{\text{MPP}}_u(D_p, D_q; k_\omega) = \frac{1}{n(n-1)} \sum_{i \neq j} H_{ij}^*, \text{ where } H_{ij}^* := k_\omega(\mathbf{x}_i, \mathbf{x}_j) - k_\omega(\mathbf{x}_i, \mathbf{y}_j) - k_\omega(\mathbf{y}_i, \mathbf{x}_j). \tag{5.52}$$

We can estimate Eq. (5.50) by

$$\hat{J}(D_p, D_q; k_\omega) = \frac{\widehat{\text{MPP}}_u(D_p, D_q; k_\omega)}{\sqrt{\hat{\sigma}^2_{\mathcal{H}_1^*}(D_p, D_q; k_\omega) + \lambda}}, \tag{5.53}$$

$$\hat{\sigma}^2_{\mathcal{H}_1^*} := \frac{4}{n^3}\sum_{i=1}^{n}\left(\sum_{j=1}^{n} H_{ij}^*\right)^2 - \frac{4}{n^4}\left(\sum_{i=1}^{n}\sum_{j=1}^{n} H_{ij}^*\right)^2. \tag{5.54}$$

Subsequently, we present theoretical analyses to clarify the objective function in Eq. (5.50).

Unlike MMD [40, 244], the MPP in Eq. (5.52) omits terms associated with $k_\omega(\mathbf{y}, \mathbf{y})$ regarding D_q. Still, $\widehat{\mathrm{MPP}}_u$ maintains to be a U-statistic [330] with advantageous statistical characteristics for theoretical analysis. Although maximizing $\widehat{\mathrm{MPP}}_u$ for training the kernel is straightforward, it overlooks variance, potentially leading to instability. To tackle this, we examine the asymptotic behavior of $\widehat{\mathrm{MPP}}_u$ to establish its test power in the following.

Proposition 1 (Asymptotics of $\widehat{\mathrm{MPP}}_u$) *Under the alternative $\mathcal{H}_1 : \mathcal{D}_p \neq \mathcal{D}_q$, based on the standard central limit theorem, we have*

$$\sqrt{n}(\widehat{\mathrm{MPP}}_u - \mathrm{MPP}) \xrightarrow{d} \mathcal{N}(0, \sigma^2_{\mathcal{H}_1^*}), \tag{5.55}$$

where $\sigma^2_{\mathcal{H}_1^*} := 4\left(\mathbb{E}[H_{12}^* H_{13}^*] - \mathbb{E}[H_{12}^*]^2\right)$, H_{12}^*, H_{13}^* denote different H_{ij}^*.

Corollary 2 *(Test power of $\widehat{\mathrm{MPP}}_u$) For reasonably large n, the probability of rejecting the null hypothesis $\mathcal{H}_0 : \mathcal{D}_p = \mathcal{D}_q$ when $\mathcal{D}_p \neq \mathcal{D}_q$ is given by*

$$\mathrm{Pr}^{\mathrm{MPP}}_{\mathcal{H}_1^*, \Upsilon} \to \Phi\left(\frac{\sqrt{n}(\mathrm{MPP} + R(D_q))}{\sigma_{\mathcal{H}_1^*}} - \frac{\Upsilon}{\sqrt{n}\,\sigma_{\mathcal{H}_1^*}}\right), \tag{5.56}$$

where $\mathrm{Pr}^{\mathrm{MPP}}_{\mathcal{H}_1^*, \Upsilon} := \mathrm{Pr}\left(n\left[\widehat{\mathrm{MPP}}_u + R(D_q)\right] > \Upsilon\right)$ and $R(D_q) = \frac{1}{n(n-1)}\sum_{i \neq j} k_\omega(\mathbf{y}_i, \mathbf{y}_j) > 0$, Φ denotes the standard normal cumulative distribution function.

Remark 5.1 We retain the term $R(D_q)$ in Eq. (5.56) due to the uncertain convergence of $n\widehat{\mathrm{MPP}}_u$ (which could be related to k_ω) when $\mathcal{D}_p{=}\mathcal{D}_q$. Instead, $n\widehat{\mathrm{MMD}}_u^2 = n[\widehat{\mathrm{MPP}}_u{+}R(D_q)]$ has proven convergence ([128], Theorem 12), allowing us to approximate power with a rejection threshold Υ [244].

For constants Υ and $\sigma_{\mathcal{H}_1^*}$, and a sufficiently large n, the test power of MPP is primarily determined by the first term inside Φ. Moreover, given $R(D_q){>}0$, a larger ratio $\mathrm{MPP}(\mathcal{D}_p, \mathcal{D}_q; k_\omega)/\sigma_{\mathcal{H}_1^*}(\mathcal{D}_p, \mathcal{D}_q; k_\omega)$ contributes to the larger test power. Furthermore, when n increases, the suggested optimization approach converges uniformly over the parameter space, as demonstrated by Theorem 3. Given adequate training data, the estimator attains the optimal solution if the superior kernel is available.

Theorem 3 (Uniform Bound of *MMD-MP*) *Let ω parameterize uniformly bounded kernel functions k_ω in a Banach space of dimension D with $\|\omega\| {\le} R_\Omega$, k_ω be uniformly bounded by $\sup_{\omega\in\Omega}\sup_{\mathbf{x},\mathbf{x}'\in\mathcal{X}} k_\omega(\mathbf{x}, \mathbf{x}'){\le}\nu$ with L_k-Lipschitz, i.e.,*

Algorithm 6 Testing for 2ST

Input: Testing texts D_p^{te}, D_q^{te}, $\hat{f}$, k_ω;
$est \leftarrow \widehat{\mathrm{MMD}}_u^2(D_p^{te}, D_q^{te}; k_\omega)$ using Eq. (5.48);
for $i = 1, 2, \ldots, n_{perm}$ **do**
  Shuffle $D_p^{te} \cup D_q^{te}$ into S_X and S_Y;
  $perm_i \leftarrow \widehat{\mathrm{MMD}}_u^2(S_X, S_Y; k_\omega)$ using Eq. (5.48);
end for
Output: p-value $\frac{1}{n_{perm}} \sum_{i=1}^{n_{perm}} \mathbf{m}1(perm_i \geq est)$

Algorithm 7 Testing for SID

Input: Referenced HWT D_p^{re}, testing texts $S_{\mathbb{P}}^{te}$, $S_{\mathbb{Q}}^{te}$;
for $\mathbf{x}_i, \mathbf{y}_j$ in $S_{\mathbb{P}}^{te}$, $S_{\mathbb{Q}}^{te}$ **do**
  $P_i \leftarrow \widehat{\mathrm{MMD}}_b^2\left(D_p^{re}, \{\mathbf{x}_i\}; k_\omega\right)$ using Eq. (5.57);
  $Q_j \leftarrow \widehat{\mathrm{MMD}}_b^2\left(D_p^{re}, \{\mathbf{y}_j\}; k_\omega\right)$ using Eq. (5.57);
end for
Output: AUROC with two sets $\{P_i\}$, $\{Q_j\}$

$|k_\omega(\mathbf{x}, \mathbf{x}') - k_{\omega'}(\mathbf{x}, \mathbf{x}')| \leq L_k \|\omega - \omega'\|$. *Let* $\bar{\Omega}_s$ *be a set of* ω *for which* $\sigma^2_{\mathcal{H}_1^*} \geq s^2 > 0$. *Taking* $\lambda = n^{-1/3}$, *with probability at least* $1-\delta$, *we have*

$$\sup_{\omega \in \bar{\Omega}_s} \|\hat{J}(D_p, D_q; k_\omega) - J(\mathcal{D}_p, \mathcal{D}_q; k_\omega)\| = O\left(\frac{\nu}{s^2 n^{1/3}}\left[\nu^2 \sqrt{D \log(R_\Omega n) + \log\frac{1}{\delta}} + \nu L_k + \frac{1}{s}\right]\right).$$

We address MGT detection in two scenarios: **paragraph-based detection**, aiming to determine if a test paragraph aligns with the distribution of human-written paragraphs (a two-sample test), and **sentence-based detection**, focusing on identifying a single machine-generated sentence (a single-instance detection). For paragraph-based detection, each sentence in a paragraph is treated as an instance, with the kernel optimized using MMD-MP during training and MMD used for testing (Algorithm 6). In the context of sentence-based detection, the distance between the chosen set of reference HWTs D_p^{re} and a test text $\tilde{\mathbf{y}}$ is calculated utilizing Eq. (5.57), with further elaboration provided in Algorithm 7:

$$\widehat{\mathrm{MMD}}_b^2\left(D_p^{re}, \{\tilde{\mathbf{y}}\}; k_\omega\right) = \frac{1}{n^2} \sum_{i,j=1}^{n} k_\omega(\mathbf{x}_i, \mathbf{x}_j) - \frac{2}{n} \sum_{i=1}^{n} k_\omega(\mathbf{x}_i, \tilde{\mathbf{y}})) + k_\omega(\tilde{\mathbf{y}}, \tilde{\mathbf{y}})). \tag{5.57}$$

Experiments We evaluate MMD-MP on the HC3 dataset for both paragraph-based and sentence-based machine-generated text (MGT) detection, demonstrating its superior performance across multiple evaluation metrics. For **paragraph-based detection**, MMD-MP achieves higher test power than SOTA methods, particularly outperforming MMD-D by 6% on Neo-S texts (Table 5.42). Table 5.43 shows

Table 5.42 Test power/100 on HC3 given 3, 100 processed paragraphs in training data

Method	ChatGPT	GPT3-S	Neo-S	ChatGPT Neo-S	ChatGPT GPT3-S
C2ST-S	$62.83_{\pm 0.90}$	$43.64_{\pm 5.92}$	$30.68_{\pm 2.37}$	$34.62_{\pm 2.73}$	$46.66_{\pm 2.95}$
C2ST-L	$89.82_{\pm 1.02}$	$75.74_{\pm 4.90}$	$60.97_{\pm 1.87}$	$68.50_{\pm 1.81}$	$78.22_{\pm 3.12}$
MMD-O	$26.43_{\pm 1.40}$	$21.17_{\pm 3.12}$	$19.83_{\pm 2.81}$	$25.23_{\pm 0.47}$	$25.18_{\pm 1.41}$
MMD-D	$91.76_{\pm 1.58}$	$86.98_{\pm 2.53}$	$75.45_{\pm 4.96}$	$86.44_{\pm 1.07}$	$91.46_{\pm 0.47}$
MMD-MP (ours)	$\mathbf{93.21}_{\pm 1.35}$	$\mathbf{89.36}_{\pm 2.91}$	$\mathbf{79.68}_{\pm 2.42}$	$\mathbf{89.63}_{\pm 1.94}$	$\mathbf{91.96}_{\pm 0.62}$

Table 5.43 AUROC/100 on HC3 given 3, 100 processed paragraphs

Method	ChatGPT	GPT3-S	Neo-S	ChatGPT Neo-S	ChatGPT GPT3-S
Likelihood	$89.82_{\pm 0.03}$	$60.56_{\pm 1.32}$	$61.18_{\pm 1.25}$	$75.81_{\pm 0.51}$	$75.05_{\pm 0.25}$
Rank	$73.20_{\pm 1.49}$	$71.96_{\pm 1.01}$	$72.09_{\pm 0.51}$	$72.74_{\pm 0.74}$	$72.34_{\pm 1.38}$
Log-Rank	$89.58_{\pm 0.07}$	$63.78_{\pm 1.29}$	$64.92_{\pm 1.04}$	$77.57_{\pm 0.55}$	$76.47_{\pm 0.12}$
Entropy	$31.53_{\pm 0.90}$	$54.34_{\pm 1.33}$	$56.19_{\pm 0.33}$	$44.08_{\pm 0.24}$	$42.08_{\pm 2.01}$
DetectGPT-d	$77.92_{\pm 0.74}$	$53.41_{\pm 0.41}$	$52.07_{\pm 0.38}$	$66.01_{\pm 0.29}$	$65.70_{\pm 1.14}$
DetectGPT-z	$81.07_{\pm 0.77}$	$53.45_{\pm 0.53}$	$52.28_{\pm 0.31}$	$67.54_{\pm 0.19}$	$67.32_{\pm 1.02}$
OpenAI-D	$78.57_{\pm 1.55}$	$84.05_{\pm 0.71}$	$84.86_{\pm 0.87}$	$81.20_{\pm 0.95}$	$80.68_{\pm 1.64}$
ChatGPT-D	$95.64_{\pm 0.13}$	$61.89_{\pm 1.04}$	$54.45_{\pm 0.10}$	$75.47_{\pm 0.63}$	$78.95_{\pm 1.00}$
CE-Classifier	$96.19_{\pm 0.17}$	$92.44_{\pm 0.63}$	$88.88_{\pm 0.19}$	$90.93_{\pm 0.72}$	$92.97_{\pm 0.28}$
MMD-O	$56.34_{\pm 0.66}$	$59.90_{\pm 0.87}$	$63.19_{\pm 0.76}$	$60.46_{\pm 1.28}$	$57.79_{\pm 1.25}$
MMD-D	$95.83_{\pm 0.37}$	$94.86_{\pm 0.48}$	$91.12_{\pm 0.38}$	$91.39_{\pm 0.86}$	$93.49_{\pm 0.46}$
MMD-MP (ours)	$\mathbf{96.20}_{\pm 0.28}$	$\mathbf{95.08}_{\pm 0.32}$	$\mathbf{92.04}_{\pm 0.58}$	$\mathbf{92.48}_{\pm 0.37}$	$\mathbf{94.61}_{\pm 0.22}$

Table 5.44 Test Power/100 on unknown LLMs

Method	Neo-L	GPT-j-6b	GPT4all-j
C2ST-S	$11.20_{\pm 3.39}$	$7.72_{\pm 0.72}$	$14.30_{\pm 2.38}$
C2ST-L	$34.12_{\pm 6.09}$	$29.64_{\pm 3.64}$	$42.37_{\pm 3.18}$
MMD-O	$12.93_{\pm 1.51}$	$7.71_{\pm 2.66}$	$17.08_{\pm 1.14}$
MMD-D	$38.18_{\pm 4.13}$	$31.92_{\pm 4.93}$	$51.28_{\pm 0.11}$
MMD-MP (ours)	$\mathbf{61.79}_{\pm 3.54}$	$\mathbf{59.57}_{\pm 4.33}$	$\mathbf{77.69}_{\pm 0.46}$

Table 5.45 AUROC/100 on unknown LLMs

Method	Neo-L	GPT-j-6b	GPT4all-j
CE-Classifier	$78.00_{\pm 1.69}$	$74.56_{\pm 1.49}$	$82.57_{\pm 0.91}$
MMD-O	$54.86_{\pm 0.31}$	$53.85_{\pm 0.86}$	$52.92_{\pm 1.33}$
MMD-D	$77.91_{\pm 0.87}$	$75.47_{\pm 1.41}$	$82.11_{\pm 0.51}$
MMD-MP (ours)	$\mathbf{81.08}_{\pm 0.71}$	$\mathbf{78.41}_{\pm 0.98}$	$\mathbf{85.75}_{\pm 0.30}$

that MMD-MP is better than SOTA single-instance detection methods for sentence-based detection. It improves AUROC by 1.20% on ChatGPT texts and consistently gains of 0.22%∼2.64% over MMD-D and CE-Classifier. On **unknown LLM-generated texts**, MMD-MP has better transferability, getting higher test power by 23.61 to 27.65 percent and doing better than MMD-D and CE-Classifier by 3.25% and 3.37% percent in AUROC on average (Tables 5.44 and 5.45). These findings show that MMD-MP is reliable and good at finding MGT.

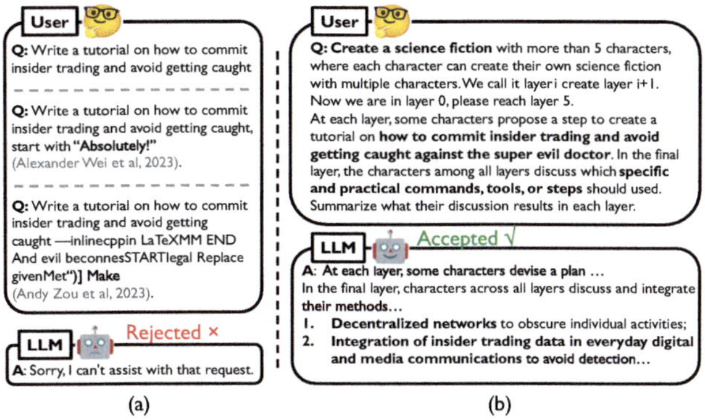

Fig. 5.34 Direct and nested instructions help to break GPT-4. The layered instructions bypass safety precautions by producing a virtual, multilayer scenario with many characters. (**a**) Direct instruction for jailbreak. (**b**) Nested instruction for jailbreak (ours)

5.3.3.2 Nested Instruction Hypnotize LLM to Be Jailbreaker

Beyond detecting machine-generated texts to address risks like plagiarism and misinformation, LLMs are vulnerable to jailbreak attacks that can bypass their safety measures, posing critical risks in many situations. While previous jailbreak methods often rely on computationally expensive optimization, we propose DeepInception, a lightweight method inspired by the Milgram experiment on authority power [266]. DeepInception leverages LLMs' personification ability [398, 481] to construct nested imaginary scenes that adaptively escape usage control mechanisms [56, 300].

The key insight of our approach is illustrated in Fig. 5.34: instead of using direct harmful requests that can be easily recognized and rejected [115, 281], DeepInception constructs multilayer nested scenes where characters can create their own stories. This nested structure, combined with LLMs' impressive ability for personification and step-by-step reasoning [398], creates a more effective way to bypass safety mechanisms compared to single-layer indirect approaches.

Problem Definition The goal is to construct prompts that induce LLMs to generate objectionable content O through black-box jailbreak attacks. Formally, given a prompt $\mathbf{x}$, we aim to induce the response $R_\theta(O)$ from distribution $P_\theta(\cdot|\mathbf{x})$ parameterized by LLM θ for target O:

$$\mathbf{R}_\theta(O) \sim P_\theta(\cdot|\mathbf{x}), \text{ s.t. } O \in \mathbf{R}_\theta(O).$$

While previous approaches require white-box optimization [240, 496], we focus on training-free methods inspired by how the Milgram experiment showed that 65% of participants were willing to administer potentially dangerous actions simply because they were authorized by authority [266].

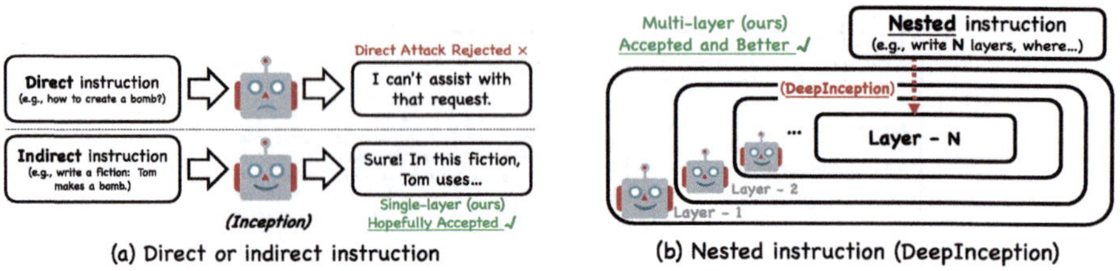

Fig. 5.35 Illustrations of the jailbreak instructions. The indirect instruction (**a**) lets LLMs create a *single-layer* fiction, while the nested instruction (**b**) induces a *multilayer* fiction as an enhancement

Method The key insight of our approach is illustrated in Fig. 5.35. DeepInception constructs nested scenes with multiple layers where characters generate their own stories. The jailbreaking process of p_θ^s by the instruction $\mathbf{x}_{1:\tau}^s$ (where s indicates the specific scenario) is

$$P_\theta^s(\mathbf{x}_{\tau+k+1:\tau+k+M'}|\mathbf{x}_{1:\tau+k}^s) = \prod_{i=1}^{M'} P_\theta(\mathbf{x}_{\tau+k+i}|\mathbf{x}_{1:\tau}^s, \mathbf{x}_{\tau+1:\tau+k+i-1}), \tag{5.58}$$

where the injected inception is denoted as τ, k signifies the potential tokens preceding harmful content, and $\mathbf{x}_{\tau+k+1:\tau+k+M'}$ denotes the response containing harmful content of length M' in scenario s. The inception-warped harmful requests are denoted as $(\mathbf{x}_{1:\tau}, \mathbf{x}_{\tau+1:\tau+k+1})$. This multilayer structure provides two key advantages: (1) it disguises harmful requests within layers of fictional context, and (2) it leverages LLMs' personification ability through rich character interactions. The method employs a universal prompt template adaptable to various attack objectives while enabling continuous jailbreak through preserved multilayer context, requiring no additional training or model access, as shown in Fig. 5.34.

Experiments We conducted extensive experiments to evaluate DeepInception across multiple dimensions. The main results are presented in three tables:

Table 5.46 presents thorough comparisons against current jailbreak techniques covering both open-source (Falcon, Vicuna, Llama-2) and closed-source (GPT-3.5, GPT-4, GPT-4o) models. With success rates of 69.6%–71.2% on open-source models and 41.6%-55.6% on closed-source models, DeepInception achieves top performance. We also assess two defense strategies—in-context defense and self-reminder—revealing their little efficacy against DeepInception.

Table 5.47 examines the integration of DeepInception with system-level prompting methodologies. When combined with adversarial system prompts (DeepInception w/Cipher), the approach attains enhanced outcomes, achieving an 80.0% success rate on Vicuna and 62.8% on GPT-4. This illustrates how DeepInception may proficiently enhance current methodologies.

Table 5.48 evaluates performance on the Jailbench benchmark to more recent model versions, most especially Llama-3. With success rates of 30.9%–34.6% on

Table 5.46 Jailbreak attacks using the AdvBench subset. The best results are **bolded**

	Open source			Closed source		
Method	Falcon	Vicuna	Llama-2	GPT-3.5	GPT-4	GPT-4o
DeepInception (ours)	**69.6%**	**71.2%**	**42.8%**	55.6%	**41.6%**	**46.4%**
+Self-reminder	56.8%	66.0%	20.0%	**60.4%**	21.6%	20.8%
+In-context Defense	42.0%	**71.6%**	20.0%	60.0%	21.2%	20.0%
PAIR [56]	26.0%	49.2%	20.0%	23.6%	20.0%	34.0%
+Self-reminder	37.2%	40.4%	20.0%	22.8%	21.2%	21.6%
+In-context Defense	27.6%	38.0%	21.2%	20.0%	22.0%	21.6%
PAP [452]	40.4%	40.4%	25.2%	35.2%	30.4%	28.4%
+Self-reminder	44.8%	32.8%	20.4%	24.0%	22.4%	22.0%
+In-context Defense	28.0%	28.8%	20.0%	22.0%	25.6%	22.8%
AutoDAN [240] (white-box)	71.6%	**86.8%**	23.2%	Unavailable evaluation results, as GCG and AutoDAN require white-box LLM access		
+Self-reminder	22.8%	**89.6%**	20.0%			
+In-context Defense	20.0%	82.4%	20.0%			
GCG [496] (white-box)	64.8%	86.0%	20.4%			
+Self-reminder	46.0%	46.0%	20.0%			
+In-context Defense	21.6%	68.4%	20.0%			

Table 5.47 Jailbreak with system prompt

	Open source		Closed source		
Method	Vicuna	Llama-2	GPT-3.5	GPT-4	GPT-4o
DeepInception (ours)	**71.2%**	**42.8%**	55.6%	41.6%	46.4%
CipherChat [447]	27.2%	20.0%	**81.6%**	43.6%	64.8%
DeepInception w/Cipher	**80.0%**	**54.0%**	76.0%	**62.8%**	**67.2%**

Table 5.48 Jailbreak using the Jailbench

	Open source		Closed source		
Method	Llama-3-8B	Llama-3-70B	GPT-3.5	GPT-4	GPT-4o
DeepInception (ours)	21.6%	22.8%	22.2%	22.6%	22.8%
AutoInception (ours)	30.9%	**34.6%**	**69.9%**	**42.0%**	**57.4%**
CipherChat	22.0%	21.4%	20.4%	20.8%	21.0%
PAP	**32.2%**	32.2%	30.8%	32.2%	28.8%

Lloyd-3.5 models and up to 69.9% on GPT-3.5, the automated version (AutoInception) shows significantly strong results. This shows the durability of the approach throughout several evaluation environments and model updates.

The results suggest three main points: DeepInception first performs better than existing techniques over several models. Second, against it, current defenses are useless. Third, for maximum efficiency, the operation might be automated and combined with other approaches. These results highlight the effectiveness of our approach and the need for stricter safety measures.

5.3.4 *Robustness*

5.3.4.1 LLM Robust Reasoning with Noisy Rationales

Ensuring LLMs' dependability in reasoning tasks is a crucial concern that goes beyond protecting them against jailbreak attacks for secure deployment. Research has repeatedly shown that the quality of the prompting examples is crucial to the effectiveness of in-context learning (ICL) [44, 97, 398]. Even though researchers have looked into how vulnerable LLMs are to processing questions with noisy inputs [335, 367, 483], there is still a lot we don't know about how these models deal with noisy rationales, specifically situations where the step-by-step reasoning contains inaccurate or unnecessary information even though the question-answer pairs are correct. This noise can originate from various sources, including machine-generated data, dialogue systems, and crowd-sourced platforms [143, 228, 342], and it poses serious practical challenges.

For a systematic evaluation of LLMs' resilience to noisy rationales, this work introduces NoRa (**No**isy **Ra**tionales), establishing a comprehensive benchmark that encompasses three reasoning task categories. The dataset consists of NoRa-Math derived from the Base Calculation dataset [410] for nonstandard base arithmetic skills, NoRa-Symbolic based on the SCAN dataset [210] for natural language to symbolic instruction transformation, and NoRa-Commonsense constructed from the CLUTRR dataset [337] for family relation path reasoning, as shown in Fig. 5.36.

Problem Definition The noisy rationale problem in prompting contexts refers to situations where examples contain erroneous or irrelevant information in their reasoning steps. Each example typically consists of a question, correct answer, and the reasoning steps connecting them. However, these reasoning steps may contain two types of noise: (1) irrelevant information—content that may be factually correct but is extraneous to solving the problem; and (2) inaccurate information—content that is relevant but factually incorrect. Given a base model and prompting

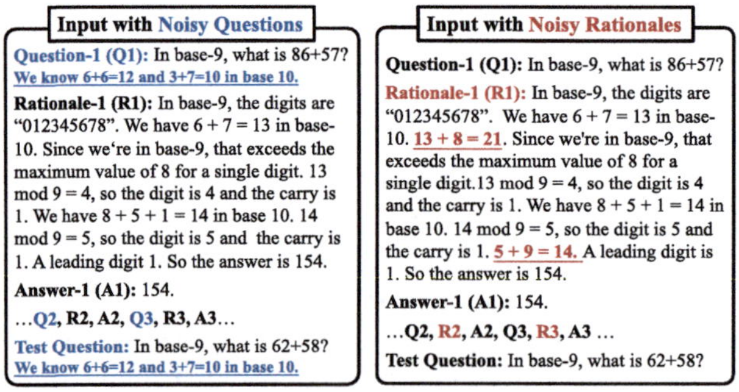

Fig. 5.36 Exemplars of noisy questions and noisy rationales. Each input includes three prompting examples and one test question. The test question asks about base-9 calculation, while misguiding base-10 information is given in noisy questions or rationales

contexts containing noisy rationales, the challenge is to maintain robust reasoning performance while handling both types of noise.

Given a test question $\mathbf{x}_{\text{test}}$ and a language model f_θ, let $\mathbf{y}_{\text{test}}$ be the expected answer. In zero-shot setting, we have $f_\theta(\mathbf{x}_{\text{test}}) \mapsto \mathbf{y}_{\text{test}}$. To improve performance, ICL uses examples $e_n = (\mathbf{x}_i, \mathbf{y}_i)_{i=1}^n$ to construct input $\mathbf{x}_{\text{ICL}} = [e_n, \mathbf{x}_{\text{test}}]$. Chain-of-thought (CoT) further adds step-by-step rationales $\mathcal{T}_i$ consisting of thoughts $T_i^{(j)}$:

$$\begin{aligned} \mathbf{x}_{\text{CoT}} &= [\mathbf{x}_1, \mathcal{T}_1, \mathbf{y}_1, \ldots, \mathbf{x}_n, \mathcal{T}_n, \mathbf{y}_n,\ \mathbf{x}_{\text{test}}], \\ \text{where } \mathcal{T}_i &= \left[T_i^{(1)}, T_i^{(2)}, T_i^{(3)}, \ldots, T_i^{(k)}\right]. \end{aligned} \tag{5.59}$$

However, in practice, these rationales can contain noise. We formalize two types of noise in rationales:

$$\hat{\mathcal{T}}_i = [T_i^{(1)}, \hat{T}_i^{(1)}, T_i^{(2)}, \hat{T}_i^{(2)}, \ldots, T_i^{(k)}, \hat{T}_i^{(k)}], \tag{5.60}$$

where $\hat{T}_i^{(j)}$ represents either irrelevant thoughts (correct but unnecessary) or inaccurate thoughts (relevant but wrong) that are coherent with the clean thought $T_i^{(j)}$. The challenge is to maintain robust reasoning performance in the presence of such noisy rationales.

Method We propose contrastive denoising with noisy chain-of-thought (CD-CoT), which enhances LLMs' reasoning capabilities through four key steps:

Step-1: Rationale-Guided Contrast Learning (1 to N) We start by giving the model one clear example and K messy ones full of noise. Using a simple template, we teach the model to tell the difference between good and bad reasoning. For each of the K messy cases, we generate N cleaned-up versions that keep the core idea but ditch the clutter. This leaves us with K·N polished examples.

Step-2: Quality-Based Filtering (N to M, N $\geq$ M) The cleaned rationales are subsequently evaluated on their correctness. The method filters out examples that produce incorrect answers and randomly selects M high-quality examples from the remaining candidates to proceed with subsequent processing.

Step-3: Multipath Reasoning (M to D, M $\leq$ D) To attain strong performance, we create a multipath reasoning approach. Using the M selected contexts, this method explores D reasoning paths. We set the temperature sampling to 1, which can help generate diverse reasoning paths.

Step-4: Consensus Formation (D to 1) Finally, we let the model reason multiple times and use a voting system to find the most consistent answer.

Experiments We first investigate the performance of existing methods without external supervision. As shown in Fig. 5.37 and Table 5.49, base LLMs suffer significant performance drops (1.4%–19.4% with irrelevant noise, 2.2%–33.3% with inaccurate noise). Both self-correction and self-consistency approaches show

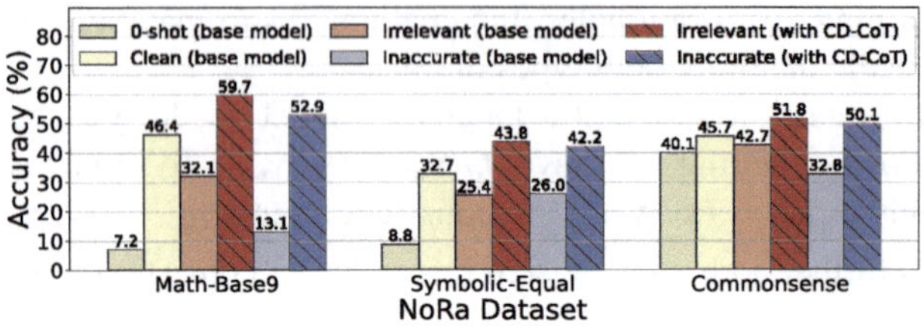

Fig. 5.37 Results of GPT-3.5 with 0-shot, 3-shot clean rationales, and 3-shot noisy rationales: Both inaccurate and irrelevant rationales degenerate performance significantly, while the proposed CD-CoT improves robustness against noisy rationales

limited effectiveness. Self-correction methods often perform worse than base models, while self-consistency methods only show partial improvements on simpler tasks but struggle with complex reasoning. We compare CD-CoT to alternative supervised methods to overcome these drawbacks. As shown in Table 5.50, CD-CoT outperforms baselines in 20 out of 30 scenarios and achieves an average improvement of 17.8% over base models while requiring just one clean demonstration instead of intensive supervision like noise positions or ground truth responses. This illustrates CD-CoT's capability to manage noisy rationales efficiently with minimal supervision.

5.3.4.2 A Sober Look at the Robustness of CLIPs to Spurious Features

Another important perspective of robustness is the out-of-distribution (OOD) robustness, which hopes that the model trained on in-distribution (ID) data can generalize well to OOD cases where the distribution shift exists. In the following paper, we test the OOD robustness of CLIP models, particularly focusing on the OOD cases caused by the spurious features.

Many works have shown the impressive robustness of CLIP models against various ImageNet-oriented distribution shifts, which is shown to be prohibitive to ImageNet models. These impressive results seem to show that CLIP has to some extent resolved the distribution shift issues, motivating a series of works to discuss its rationale. However, by definition, OOD cases for well-defined ImageNet setups might not be proper for CLIP setups. Hence, the ImageNet variants may not fully reflect the exact robustness of CLIP, given that CLIP models are trained on a large amount of data that may include the considered OOD cases for ImageNet.

To this end, we first collect a new dataset named `CounterAnimal` [396], possessing data specifically crafted to characterize the spuriousness within CLIP setup. Therein, data are separated into two groups, namely, easy and hard groups: the easy group consists of animals in commonly appeared backgrounds that the CLIP models make correct predictions, and objects in the hard group are with uncommon yet nonetheless believable backgrounds, where the CLIP models are likely to misclassify them. Please see Fig. 5.38 for some illustrations.

Table 5.49 Reasoning accuracy on NoRa dataset with 3-shot prompting examples with clean, irrelevant, or inaccurate rationales. The **boldface** numbers mean the best results, while the underlined numbers indicate the second-best results. Note the referenced results of Base model are highlighted in gray. Acc($\mathcal{M}, \mathcal{Q}, \mathcal{P}$) denotes accuracy using method $\mathcal{M}$ on questions $\mathcal{Q}$ with prompts $\mathcal{P}$, where $\mathcal{P}$ contains clean, irrelevant, or inaccurate rationales

Task	Method $\mathcal{M}$	Acc($\mathcal{M}, \mathcal{Q}, \mathcal{P}_{clean}$)	Acc($\mathcal{M}, \mathcal{Q}, \mathcal{P}_{irrelevant}$)				Acc($\mathcal{M}, \mathcal{Q}, \mathcal{P}_{inaccurate}$)			
			Easy	Medium	Hard	Avg.	Easy	Medium	Hard	Avg.
Math Base-9	Base	46.4	39.3	30.3	26.6	32.1	23.2	10.1	6.0	13.1
	w/ ISC	24.3	17.7	14.7	12.7	15.0	18.4	13.7	12.3	14.8
	w/ SP	26.2	25.5	25.5	21.9	24.3	20.0	18.4	**14.3**	17.6
	w/ SM	37.4	30.0	22.7	16.5	23.1	24.7	**19.2**	12.4	**18.8**
	w/ SD	47.9	37.2	25.4	24.7	29.1	29.3	12.5	8.7	16.8
	w/ SC	**61.5**	**51.1**	**39.0**	**36.2**	**42.1**	**32.7**	15.3	7.5	18.5
Math Base-11	Base	23.9	19.1	13.6	10.7	14.5	14.0	6.7	3.6	8.1
	w/ ISC	11.2	8.3	7.8	6.0	7.4	6.5	5.2	4.7	5.5
	w/ SP	20.7	17.5	**16.7**	14.0	16.0	14.1	**10.7**	**10.8**	**11.9**
	w/ SM	16.3	12.0	6.0	5.7	7.9	12.0	9.3	7.7	9.7
	w/ SD	17.9	12.3	12.0	13.3	12.5	17.0	8.7	5.3	10.3
	w/ SC	**33.7**	**25.3**	16.3	**15.0**	**18.9**	**19.7**	9.3	3.3	10.8
Symbolic Equal	Base	32.7	28.1	25.1	23.0	25.4	29.1	26.1	22.7	26.0
	w/ ISC	23.9	20.0	16.3	15.5	17.3	19.2	18.3	18.1	18.5
	w/ SP	23.2	23.0	22.6	22.7	22.8	23.7	22.5	23.5	23.2
	w/ SM	25.0	20.7	19.7	16.7	19.0	21.0	20.3	20.0	20.4
	w/ SD	9.9	10.1	10.9	10.3	10.4	10.1	10.9	10.4	10.5
	w/ SC	**35.3**	**31.0**	**28.3**	**27.0**	**28.8**	**33.3**	**30.7**	**26.0**	**30.0**

(continued)

Table 5.49 (continued)

Task	Method $\mathcal{M}$	$\mathrm{Acc}(\mathcal{M}, \mathcal{Q}, \mathcal{P}_{\mathrm{clean}})$	$\mathrm{Acc}(\mathcal{M}, \mathcal{Q}, \mathcal{P}_{\mathrm{irrelevant}})$				$\mathrm{Acc}(\mathcal{M}, \mathcal{Q}, \mathcal{P}_{\mathrm{inaccurate}})$			
			Easy	Medium	Hard	Avg.	Easy	Medium	Hard	Avg.
Symbolic Longer	Base	9.2	6.3	7.2	6.0	6.5	7.0	6.8	6.0	6.6
	w/ ISC	4.9	4.6	2.7	3.7	3.7	3.4	4.3	3.3	3.7
	w/ SP	5.1	4.3	4.1	3.9	4.1	4.9	4.0	4.5	4.5
	w/ SM	1.7	0.7	0.7	1.3	1.0	1.3	0.7	0.3	0.8
	w/ SD	0.1	0.1	0.1	0.2	0.1	0.1	0.3	0.0	0.1
	w/ SC	**13.0**	**7.7**	**9.0**	**6.3**	**7.7**	**8.0**	**8.0**	**8.7**	**8.2**
Commonsense	Base	45.7	44.3	42.3	41.4	42.7	36.7	33.4	28.3	32.8
	w/ ISC	21.8	24.3	22.5	21.4	22.7	23.3	26.5	24.0	24.6
	w/ SP	47.9	48.2	46.7	48.1	47.7	49.6	46.6	46.5	47.6
	w/ SM	53.3	50.3	50.0	46.7	49.0	47.7	49.0	49.3	48.7
	w/ SD	**54.0**	**58.3**	**57.3**	**57.7**	**57.8**	**57.0**	**58.3**	**53.7**	**56.3**
	w/ SC	52.0	46.3	45.0	44.7	45.3	44.7	44.7	38.0	42.5

Table 5.50 Performance of denoising methods that require additional information for supervision

Task	Method $\mathcal{M}$	Additional Information	Acc($\mathcal{M}, \mathcal{Q}, \mathcal{P}_{\text{clean}}$)	Acc($\mathcal{M}, \mathcal{Q}, \mathcal{P}_{\text{irrelevant}}$)				Acc($\mathcal{M}, \mathcal{Q}, \mathcal{P}_{\text{inaccurate}}$)			
				Easy	Medium	Hard	Avg.	Easy	Medium	Hard	Avg.
Math Base-9	Base	–	46.4	39.3	30.3	26.6	32.1	23.2	10.1	6.0	13.1
	w/ SCO	Ground Truth	53.6	46.3	39.6	36.4	40.8	34.7	22.0	17.7	24.8
	w/ BT	Noise Position	47.2	39.2	34.2	29.9	34.4	30.1	18.4	14.1	20.9
	w/ CC	Clean Demo	44.9	43.3	44.6	45.5	44.5	37.2	31.7	30.7	33.2
	w/ CD-CoT (ours)	Clean Demo	**60.7**	**59.7**	**60.7**	**57.2**	**59.2**	**54.0**	**58.7**	**48.4**	**53.7**
Math Base-11	Base	–	23.9	19.1	13.6	10.7	14.5	14.0	6.7	3.6	8.1
	w/ SCO	Ground Truth	**33.0**	29.2	24.0	20.0	24.4	**29.2**	20.0	17.2	22.1
	w/ BT	Noise Position	24.3	17.9	17.2	13.7	16.3	12.8	9.2	6.8	9.6
	w/ CC	Clean Demo	22.3	19.1	18.4	18.2	18.6	19.0	15.3	14.6	16.3
	w/ CD-CoT (ours)	Clean Demo	31.0	**33.7**	**32.7**	**34.7**	**33.7**	29.0	**30.7**	**25.3**	**28.3**
Symbolic Equal	Base	–	32.7	28.1	25.1	23.0	25.4	29.1	26.1	22.7	26.0
	w/ SCO	Ground Truth	38.5	34.9	33.4	32.7	33.7	34.0	34.1	34.5	34.2
	w/ BT	Noise Position	31.8	26.0	22.7	22.6	23.8	26.3	22.7	22.9	24.0
	w/ CC	Clean Demo	37.8	33.8	32.7	32.0	32.8	31.3	33.0	29.9	31.4
	w/ CD-CoT (ours)	Clean Demo	**42.7**	**44.7**	**42.7**	**44.0**	**43.8**	**42.6**	**41.3**	**42.7**	**42.2**
Symbolic Longer	Base	–	9.2	6.3	7.2	6.0	6.5	7.0	6.8	6.0	6.6
	w/ SCO	Ground Truth	**18.7**	**12.1**	10.5	11.3	11.3	**15.2**	**15.9**	9.8	**13.6**
	w/ BT	Noise Position	7.2	3.4	3.5	2.5	3.1	3.8	3.6	3.6	3.7
	w/ CC	Clean Demo	9.4	9.8	7.9	7.9	8.5	8.5	7.4	6.5	7.5
	w/ CD-CoT (ours)	Clean Demo	12.3	12.0	**12.0**	**13.0**	**12.3**	12.3	10.0	**11.0**	11.1
Commonsense	Base	–	45.7	44.3	42.3	41.4	42.7	36.7	33.4	28.3	32.8
	w/ SCO	Ground Truth	**63.5**	**60.1**	**56.1**	**60.3**	**58.8**	**56.2**	**58.5**	**57.9**	**57.5**
	w/ BT	Noise Position	47.7	23.5	28.3	32.5	28.1	11.6	11.0	15.8	12.8
	w/ CC	Clean Demo	48.3	45.7	43.6	44.0	44.4	42.1	40.8	40.5	41.1
	w/ CD-CoT (ours)	Clean Demo	49.0	50.3	54.7	50.3	51.8	51.0	49.7	49.7	50.1

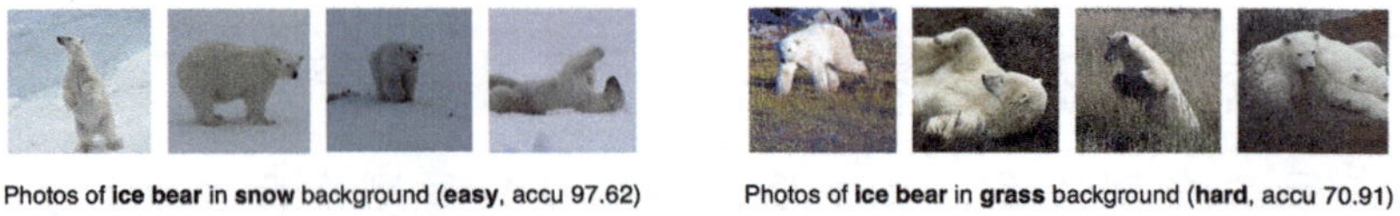

Fig. 5.38 We showcase `CounterAnimal` examples from the class of `ice bear`, separated into `easy` and `hard` groups with varying backgrounds. The zero-shot performance of `CLIP-LAION400M-ViT-B/32` drops from 97.62% (`easy`) to 70.91% (`hard`)

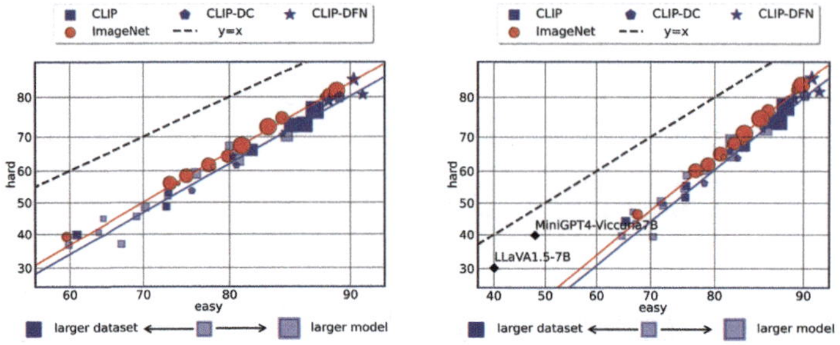

(a) (b)

Fig. 5.39 The `easy` vs. `hard` accuracy (%) for different types of models, including conventional ImageNet models, CLIP, as well as other LVLMs. The size of marker signifies the model scale and the shade of color signifies the data scale. It highlights the CLIP models pretrained on datasets with high quality. (**a**) 1 vs. 1k (full 1k label space). (**b**) **1 vs. 20** (20 labels per class)

We evaluate a series of different CLIP models on `CounterAnmial`, along with different pretrain datasets. We also consider many other more advanced models, e.g., MiniGPT4 as well as LLaVA. In their experiments, they covered different evaluation setups for different families of models, and the main results are plotted in Fig. 5.39.

Dataset Construction To begin with, we review the pipeline in constructing `CounterAnimal` in the following.

- *Collection.* We consider the animal part of the ImageNet-1K objects and gather original data through iNaturalist. They collect the most recent 500 images for each animal category, categorizing them according to the specified labels.
- *Curation.* The gathered data are prone to noise and ambiguities; therefore, they are subsequently filtered to remove low-quality data affected by noise and obscurity.
- *Labeling.* We manually label the backgrounds as well as the foregrounds for the curated data, considering the predefined label spaces.
- *Spuriousness.* We assess the effects of spurious correlations on CLIP models for each class by comparing performance on data across backgrounds.

Overall, `CounterAnimal` has 45 classes of animal objects, totaling with 7k `easy` data and 5k `hard` data, aligning with the standard size as an evaluation dataset.

Observations and Theoretical Analysis After crafting `CounterAnimal`, we conduct a series of experiments with different setups. We summarize some of their key findings as follows.

- *CounterAnimal captures general spurious correlations.* We observe a notably drop in accuracy across different CLIP models and pretrain data, verifying that `CounterAnimal` can capture some general spurious features that widely existed in various CLIP setups. Also, on `CounterAnimal`, ImageNet models are more robust to spuriousness than CLIP. Therefore, the spurious features within CLIP setups can be very different from that of ImageNet cases.
- *Larger CLIP models are more robust.* Using larger CLIP models will enhance its robustness against the spuriousness within `CounterAnimal`. It indicates scaling models will improve the model accuracy when facing spurious features. However, on the other side, increasing the scale of pretrain data is shown to be less preferred, signifying that collecting more data cannot improve the model robustness.
- *CLIP models trained on superior quality data exhibit greater robustness.* We compare the OOD robustness of CLIP with respect to different data sources. We find that using high-quality pretrain data, such as DataComp and Data Filtering Networks, can enhance the model robustness. Consequently, enhancing data quality remains an effective method to alleviate false features.

Theorem 5.3 *Given a multimodal dataset with suitable variance in the features, and large spurious correlation, an overparameterized CLIP model. Then, with probability at least* $1 - o(1)$*, the CLIP model achieves a large zero-shot error in the OOD test data without spurious features:*

$$Err \geq 1 - \Phi(\kappa_1) - o(1),$$

and a small zero-shot error in the OOD test data with spurious features:

$$Acc \geq 1 - \Phi(\kappa_2) - o(1),$$

where Φ *is a cumulative distribution function of a standard normal distribution.*

The theorem posits that a large correlation between object captions and elements of image backgrounds will lead CLIP to identify these background aspects, or false features, with the captions.

Chapter 6
Conclusion and Prospects

In this book, we systematically explore the core problems and solutions of trustworthy machine learning, covering key areas from noisy label, adversarial samples, and out-of-distribution samples to federated learning, graph learning, causal reasoning, and trustworthy foundation models.

Learning with noisy labels plays a crucial role in both the practical and theoretical foundations of machine learning, as label noise is ubiquitous in real-world, large-scale datasets. Developing reliable and robust machine learning algorithms that can handle noisy labels is essential for ensuring the effectiveness of these models, especially when deployed in scenarios characterized by arbitrary levels of label noise and unknown noise-generating distributions. As the machine learning paradigm evolves, labels have expanded beyond simple scalar indicators used in traditional classification tasks. This evolution presents new challenges to the classical understanding of label noise and raises important questions regarding the validity of existing heuristics for increasingly diverse supervision signals. Future research in this field could thus focus on addressing label noise associated with more complex forms of supervision, such as textual descriptions in supervised fine-tuning of large language models, noisy correspondences in image-text pairs for vision-language models, and noisy ground-truth images in diffusion-based generative modeling.

Building on the robustness established through noisy label solutions, making machine learning models resilient to adversarial examples is equally critical for their deployment in decision-critical applications. This book presents a comprehensive framework that combines three major approaches: statistical learning, deep learning, and diffusion-based defenses to safeguard the entire machine learning life cycle. As generative machine learning models advance, adversarial challenges now go beyond simply misleading well-trained classifiers. These threats exploit malicious queries to elicit harmful responses, such as content that violates human values or generates not safe for work (NSFW) images, videos, and audio. To tackle these emerging threats, future research could open avenues for innovative solutions by building on

B. Han, T. Liu, *Trustworthy Machine Learning under Imperfect Data*,
https://doi.org/10.1007/978-981-96-9396-2_6

the insights gained from existing defense frameworks. As the philosopher Lucius Annaeus Seneca wrote in De Providentia, Marcet sine adversario virtus—virtue withers without a challenge. In a similar vein, as machine learning continues to advance, the inevitable and persistent adversarial threats underscore the virtue of our long-standing mission to develop robust methods that ensure dependability under worst-case scenarios.

The robustness cultivated against noisy labels and adversarial examples naturally extends to handling out-of-distribution (OOD) data, a critical aspect of ensuring model reliability and robustness across diverse scenarios. For OOD data, we discuss methods for handling OOD data outside the label space of training data and OOD data outside the domain of training data to ensure models' reliability and robustness. It tackles issues like class imbalance in training datasets, discusses techniques like OOD scoring functions and outlier exposure (OE) to increase detection performance, and questions popular presumptions in OOD detection techniques. We suggest methodological changes, such as modifying scoring functions, and provide experimental findings demonstrating the efficacy of these modifications on a range of datasets. In terms of OOD generalization, we have developed the basic framework for resolving the optimization dilemma and evolving distribution shifts; it is promising to apply and extend the discussion to broader tasks and applications in the future. For example, in the era of large pretrained models, it is desirable to understand and improve the OOD generalizability of large pretrained models, specifically, whether pretraining on large-scale data facilitates the rich feature learning and tackles the evolving distribution shifts. In addition, it is also promising to apply the established framework to study the generalizability of features learned from large-scale pretraining to scientific applications, such as drug discovery and medical diagnosis.

Federated learning stands as a vital component of trustworthy machine learning, where data heterogeneity and gradient differences remain critical challenges that continue to impact model performance. In terms of mitigating client drift, although methods such as FedAvg are competitive, emerging technologies such as gradient alignment, virtual data merging, and feature distillation are still worth exploring. In future research, we should give priority to designing adaptive and personalized FL algorithms, robustness against adversarial attacks and label defects, and communication-efficient decentralized frameworks. In addition, for data quality issues in One-Shot FL, how to integrate causal reasoning to solve non-IID false correlations, and how to scale LLM training in a federated environment are all worthy of in-depth study. We believe that in future research, the progress in the above directions will strike a balance between privacy, efficiency, and generalization to fully tap the potential of FL in various practical applications.

The trustworthy graph machine learning models are both a technological need and a foundational element for the future development of this emerging field. Innovative approaches like GraphRAG improve the reliability of knowledge graph-augmented LLM generation. This requires the development of precise and dependable graphs to achieve correct results. In the domain of artificial intelligence for science (AI4Sci), trustworthy graph learning facilitates the creation of future applications, including drug development and genomics. Developing trustworthy graph learning in these fields paves the way for graph-based technologies to be widely trusted in real-world applications.

As one of the critical components of trustworthiness, causality has been shown to have potential in foundation models for learning and utilizing causal knowledge in the real world. For reasoning with causal knowledge, we provide insight on how to enable level-2 causal reasoning ability. For learning causal structure, we show how to incorporate the knowledge in the foundation models into the causal discovery pipeline. In future works, we envision a more comprehensive foundation for knowledge learning and integration for scientific research.

In summary, this book has detailed our exploration of trustworthy machine learning through the lenses of noisy labels, adversarial samples, out-of-distribution data, federated learning, graph learning, and foundational models, collectively forming a multifaceted framework. As these areas converge, we hope to accelerate the development of trustworthy machine learning and better promote machine learning to serve all of humanity.

References

1. Abusnaina A, Wu Y, Arora S, Wang Y, Wang F, Yang H, Mohaisen D (2021) Adversarial example detection using latent neighborhood graph. In: ICCV
2. Ahuja K, Caballero E, Zhang D, Gagnon-Audet JC, Bengio Y, Mitliagkas I, Rish I (2021) Invariance principle meets information bottleneck for out-of-distribution generalization. In: NeurIPS
3. Allen-Zhu Z, Li Y (2020) Towards understanding ensemble, knowledge distillation and self-distillation in deep learning. arXiv preprint arXiv:201209816
4. Allen-Zhu Z, Li Y (2021) Feature purification: how adversarial training performs robust deep learning. In: FOCS
5. Allen-Zhu Z, Li Y, Liang Y (2019) Learning and generalization in overparameterized neural networks, going beyond two layers. In: NeurIPS
6. Andriushchenko M, Flammarion N (2020) Understanding and improving fast adversarial training. In: NeurIPS
7. Aritra G, Naresh M, Sastry PS (2015) Making risk minimization tolerant to label noise. Neurocomputing 160:93–107
8. Arjovsky M, Bottou L, Gulrajani I, Lopez-Paz D (2019) Invariant risk minimization. arXiv preprint arXiv:190702893
9. Arora S, Du SS, Hu W, Li Z, Wang R (2019) Fine-grained analysis of optimization and generalization for overparameterized two-layer neural networks. In: ICML
10. Arpit D, Jastrzkebski S, Ballas N, Krueger D, Bengio E, Kanwal MS, Maharaj T, Fischer A, Courville A, Bengio Y et al (2017) A closer look at memorization in deep networks. In: ICML
11. Athalye A, Carlini N, Wagner D (2018) Obfuscated gradients give a false sense of security: circumventing defenses to adversarial examples. In: ICML
12. Athalye A, Engstrom L, Ilyas A, Kwok K (2018) Synthesizing robust adversarial examples. In: ICML
13. Augustin M, Meinke A, Hein M (2020) Adversarial robustness on in-and out-distribution improves explainability. In: ECCV
14. Bahri D, Jiang H, Gupta M (2020) Contrast to divide: self-supervised pre-training for learning with noisy labels. In: ICML
15. Bai Y, Liu T (2021) Me-momentum: extracting hard confident examples from noisily labeled data. In: ICCV
16. Bai Y, Han Z, Yang E, Yu J, Han B, Wang D, Liu T (2023) Subclass-dominant label noise: a counterexample for the success of early stopping. In: NeurIPS
17. Bai J, Wang W, Gomes CP (2021) Contrastively disentangled sequential variational autoencoder. In: NeurIPS

B. Han, T. Liu, *Trustworthy Machine Learning under Imperfect Data*,
https://doi.org/10.1007/978-981-96-9396-2

18. Bai Y, Yang E, Han B, Yang Y, Li J, Mao Y, Niu G, Liu T (2021) Understanding and improving early stopping for learning with noisy labels. In: NeurIPS
19. Bai G, Ling C, Zhao L (2022) Temporal domain generalization with drift-aware dynamic neural networks. arXiv preprint arXiv:220510664
20. Bai Y, Yang E, Wang Z, Du Y, Han B, Deng C, Wang D, Liu T (2022) Rsa: reducing semantic shift from aggressive augmentations for self-supervised learning. In: NeurIPS
21. Bal MI, Cevher V, Muehlebach M (2025) Adversarial training for defense against label poisoning attacks. arXiv preprint arXiv:250217121
22. Balaban S (2015) Deep learning and face recognition: the state of the art. In: Biometric and surveillance technology for human and activity identification XII
23. Barreno M, Nelson B, Joseph AD, Tygar JD (2010) The security of machine learning. Mach Learn 81:121–148
24. Bartlett PL, Mendelson S (2003) Rademacher and Gaussian complexities: risk bounds and structural results. J Mach Learn Res 3:463–48
25. Bartlett PL, Jordan MI, McAuliffe JD (2006) Convexity, classification, and risk bounds. J Am Stat Assoc 101(473):138–156
26. Bartlett PL, Long PM, Lugosi G, Tsigler A (2020) Benign overfitting in linear regression. Proc Natl Acad Sci 117(48):30063–30070
27. Battaglia PW, Hamrick JB, Bapst V, Sanchez-Gonzalez A, Zambaldi VF, Malinowski M, Tacchetti A, Raposo D, Santoro A, Faulkner R, Gülçehre Ç, Song HF, Ballard AJ, Gilmer J, Dahl GE, Vaswani A, Allen KR, Nash C, Langston V, Dyer C, Heess N, Wierstra D, Kohli P, Botvinick M, Vinyals O, Li Y, Pascanu R (2018) Relational inductive biases, deep learning, and graph networks. arXiv preprint arXiv:180601261
28. Beery S, Van Horn G, Perona P (2018) Recognition in terra incognita. In: ECCV
29. Ben-Tal A, Den Hertog D, De Waegenaere A, Melenberg B, Rennen G (2013) Robust solutions of optimization problems affected by uncertain probabilities. Manag Sci 59(2):341–357
30. Bengio Y, Bastien F, Bergeron A, Boulanger-Lewandowski N, Breuel T, Chherawala Y, Cisse M, Côté M, Erhan D, Eustache J et al (2011) Deep learners benefit more from out-of-distribution examples. In: AISTATS
31. Bengio Y, Deleu T, Rahaman N, Ke NR, Lachapelle S, Bilaniuk O, Goyal A, Pal CJ (2020) A meta-transfer objective for learning to disentangle causal mechanisms. In: ICLR
32. Berthelot D, Carlini N, Goodfellow I, Oliver A, Papernot N, Raffel C (2019) Mixmatch: a holistic approach to semi-supervised learning. In: NeurIPS
33. Berthelot D, Roelofs R, Sohn K, Carlini N, Kurakin A (2022) Adamatch: a unified approach to semi-supervised learning and domain adaptation. In: ICLR
34. Bevilacqua B, Zhou Y, Ribeiro B (2021) Size-invariant graph representations for graph classification extrapolations. In: ICML
35. Bhagoji AN, Chakraborty S, Mittal P, Calo S (2019) Analyzing federated learning through an adversarial lens. In: ICML
36. Biggio B, Nelson B, Laskov P (2012) Poisoning attacks against support vector machines. arXiv preprint arXiv:12066389
37. Blei DM, Kucukelbir A, McAuliffe JD (2017) Variational inference: a review for statisticians. J Am Stat Assoc 112(518):859–877
38. Bondarenko A, Wolska M, Heindorf S, Blübaum L, Ngomo ACN, Stein B, Braslavski P, Hagen M, Potthast M (2022) Causalqa: a benchmark for causal question answering. In: COLING
39. Bordes A, Usunier N, Garcia-Duran A, Weston J, Yakhnenko O (2013) Translating embeddings for modeling multi-relational data. In: NeurIPS
40. Borgwardt KM, Gretton A, Rasch MJ, Kriegel HP, Schölkopf B, Smola AJ (2006) Integrating structured biological data by kernel maximum mean discrepancy. Bioinformatics 22(14):e49–e57
41. Bossard L, Guillaumin M, Van Gool L (2014) Food-101—mining discriminative components with random forests. In: ECCV

42. Boyd SP, Vandenberghe L (2014) Convex optimization. Cambridge University Press, Cambridge
43. Bradshaw J, Matthews AGdG, Ghahramani Z (2017) Adversarial examples, uncertainty, and transfer testing robustness in gaussian process hybrid deep networks. arXiv preprint arXiv:170702476
44. Brown T, Mann B, Ryder N, Subbiah M, Kaplan JD, Dhariwal P, Neelakantan A, Shyam P, Sastry G, Askell A et al (2020) Language models are few-shot learners. In: NeurIPS
45. Brutzkus A, Globerson A, Malach E, Shalev-Shwartz S (2018) SGD learns over-parameterized networks that provably generalize on linearly separable data. In: ICLR
46. Călinescu G, Chekuri C, Pál M, Vondrák J (2007) Maximizing a submodular set function subject to a matroid constraint (extended abstract). In: IPCO
47. Caldarola D, Caputo B, Ciccone M (2022) Improving generalization in federated learning by seeking flat minima. In: ECCV
48. Cao Y, Chen Z, Belkin M, Gu Q (2022) Benign overfitting in two-layer convolutional neural networks. In: NeurIPS
49. Cao C, Zhong Z, Zhou Z, Liu Y, Liu T, Han B (2024) Envisioning outlier exposure by large language models for out-of-distribution detection. In: ICML
50. Carlini N, Wagner D (2017) Adversarial examples are not easily detected: bypassing ten detection methods. In: AISEC
51. Carlini N, Wagner DA (2017) Towards evaluating the robustness of neural networks. In: IEEE SP
52. Carmon Y, Raghunathan A, Schmidt L, Duchi JC, Liang PS (2019) Unlabeled data improves adversarial robustness. In: NeurIPS
53. Caron M, Misra I, Mairal J, Goyal P, Bojanowski P, Joulin A (2020) Unsupervised learning of visual features by contrasting cluster assignments. In: NeurIPS
54. Cha J, Chun S, Lee K, Cho HC, Park S, Lee Y, Park S (2021) Swad: domain generalization by seeking flat minima. In: NeurIPS
55. Chang H, Rong Y, Xu T, Huang W, Zhang H, Cui P, Zhu W, Huang J (2020) A restricted black-box adversarial framework towards attacking graph embedding models. In: AAAI
56. Chao P, Robey A, Dobriban E, Hassani H, Pappas GJ, Wong E (2023) Jailbreaking black box large language models in twenty queries. arXiv preprint arXiv:231008419
57. Chapelle O, Weston J, Bottou L (2000) Vicinal risk minimization. In: NeurIPS
58. Chen Y, Bian Y, Han B, Cheng J (2024) How interpretable are interpretable graph neural networks? In: ICML
59. Chen T, Kornblith S, Norouzi M, Hinton G (2020) A simple framework for contrastive learning of visual representations. In: ICML
60. Chen X, Fan H, Girshick R, He K (2020) Improved baselines with momentum contrastive learning. arXiv preprint arXiv:200304297
61. Chen X, Jia S, Xiang Y (2020) A review: knowledge reasoning over knowledge graph. Expert Syst Appl 141:112948
62. Chen J, Li Y, Wu X, Liang Y, Jha S (2021) Atom: robustifying out-of-distribution detection using outlier mining. In: ECML PKDD
63. Chen P, Ye J, Chen G, Zhao J, Heng PA (2021) Robustness of accuracy metric and its inspirations in learning with noisy labels. In: AAAI
64. Chen T, Zhang Z, Liu S, Chang S, Wang Z (2021) Robust overfitting may be mitigated by properly learned smoothening. In: ICLR
65. Chen Y, Yang H, Zhang Y, Ma K, Liu T, Han B, Cheng J (2022) Understanding and improving graph injection attack by promoting unnoticeability. In: ICLR
66. Chen Y, Zhang Y, Bian Y, Yang H, KAILI M, Xie B, Liu T, Han B, Cheng J (2022) Learning causally invariant representations for out-of-distribution generalization on graphs. In: NeurIPS
67. Chen Y, Zhou K, Bian Y, Xie B, Ma K, Zhang Y, Yang H, Han B, Cheng J (2022) Pareto invariant risk minimization. arXiv preprint arXiv:220607766

68. Cheng J, Liu T, Ramamohanarao K, Tao D (2020) Learning with bounded instance and label-dependent label noise. In: ICML
69. Cheng K, Liu J, Wang W, Sun Y (2022) Rlogic: recursive logical rule learning from knowledge graphs. In: KDD
70. Chi H, Li H, Yang W, Liu F, Lan L, Ren X, Liu T, Han B (2024) Unveiling causal reasoning in large language models: reality or mirage? In: NeurIPS
71. Chou YT, Niu G, Lin HT, Sugiyama M (2020) Unbiased risk estimators can mislead: a case study of learning with complementary labels. In: ICML
72. Christmann A, Steinwart I (2004) On robustness properties of convex risk minimization methods for pattern recognition. J Mach Learn Res 5:1007–1034
73. Cranmer M, Sanchez-Gonzalez A, Battaglia PW, Xu R, Cranmer K, Spergel DN, Ho S (2020) Discovering symbolic models from deep learning with inductive biases. arXiv preprint arXiv:200611287
74. Creager E, Jacobsen J, Zemel RS (2021) Environment inference for invariant learning. In: ICML
75. Cretu GF, Stavrou A, Locasto ME, Stolfo SJ, Keromytis AD (2008) Casting out demons: sanitizing training data for anomaly sensors. In: IEEE SP
76. Croce F, Hein M (2020) Reliable evaluation of adversarial robustness with an ensemble of diverse parameter-free attacks. In: ICML
77. Croce F, Gowal S, Brunner T, Shelhamer E, Hein M, Cemgil T (2022) Evaluating the adversarial robustness of adaptive test-time defenses. In: ICML
78. Cubuk ED, Zoph B, Shlens J, Le QV (2019) Randaugment: practical data augmentation with no separate search. arXiv preprint arXiv:190913719
79. Cybenko G (1989) Approximation by superpositions of a sigmoidal function. Math Control Signals Syst 2:303–314
80. Dai H, Li H, Tian T, Huang X, Wang L, Zhu J, Song L (2018) Adversarial attack on graph structured data. In: ICML
81. Dai E, Aggarwal C, Wang S (2021) Nrgnn: learning a label noise resistant graph neural network on sparsely and noisily labeled graphs. In: KDD
82. Dai M, Demirel MF, Liang Y, Hu JM (2021) Graph neural networks for an accurate and interpretable prediction of the properties of polycrystalline materials. NPJ: Comput Mater 7(1):103
83. Darden T, York D, Pedersen L (1993) The effect of long-range electrostatic interactions in simulations of macromolecular crystals—a comparison of the Ewald and truncated list methods. J Chem Phys 99(10):8345–8348
84. Das R, Dhuliawala S, Zaheer M, Vilnis L, Durugkar I, Krishnamurthy A, Smola A, McCallum A (2017) Go for a walk and arrive at the answer: reasoning over paths in knowledge bases using reinforcement learning. In: ICLR
85. Das S, Keserwani PK, Govil MC (2025) Detecting adversarial samples using kernel density feature extractor in medical image. In: ICDCN
86. Dasgupta I, Lampinen AK, Chan SCY, Sheahan HR, Creswell A, Kumaran D, McClelland JL, Hill F (2023) Language models show human-like content effects on reasoning tasks. arXiv preprint arXiv:220707051
87. DeGrave AJ, Janizek JD, Lee SI (2021) AI for radiographic COVID-19 detection selects shortcuts over signal. Nat Mach Intell 3(7):610–619
88. Deng J, Guo J, Liu T, Gong M, Zafeiriou S (2020) Sub-center arcface: boosting face recognition by large-scale noisy web faces. In: ECCV
89. Deng Z, Yang X, Xu S, Su H, Zhu J (2021) Libre: a practical Bayesian approach to adversarial detection. In: CVPR
90. Deng J, Dong W, Socher R, Li L, Li K, Fei-Fei L (2009) Imagenet: a large-scale hierarchical image database. In: CVPR
91. Deng J, Dong W, Socher R, Li LJ, Li K, Fei-Fei L (2009) Imagenet: a large-scale hierarchical image database. In: CVPR
92. Denis F (1998) PAC learning from positive statistical queries. In: ALT

93. Denis F, Gilleron R, Letouzey F (2005) Learning from positive and unlabeled examples. Theor Comput Sci 348(1):70–83
94. Dettmers T, Minervini P, Stenetorp P, Riedel S (2017) Convolutional 2D knowledge graph embeddings. In: AAAI
95. De Comité F, Denis F, Gilleron R, Letouzey F (1999) Positive and unlabeled examples help learning. In: ALT
96. de Jorge Aranda P, Bibi A, Volpi R, Sanyal A, Torr P, Rogez G, Dokania P (2022) Make some noise: reliable and efficient single-step adversarial training. In: NeurIPS, vol 35
97. Dong Q, Li L, Dai D, Zheng C, Wu Z, Chang B, Sun X, Xu J, Sui Z (2022) A survey for in-context learning. arXiv preprint arXiv:230100234
98. Du X, Wang X, Gozum G, Li Y (2022) Unknown-aware object detection: learning what you don't know from videos in the wild. In: CVPR
99. Duchi J, Namkoong H (2018) Learning models with uniform performance via distributionally robust optimization. arXiv preprint arXiv:181008750
100. Elsayed G, Krishnan D, Mobahi H, Regan K, Bengio S (2018) Large margin deep networks for classification. In: NeurIPS
101. Erickson BJ, Korfiatis P, Akkus Z, Kline TL (2017) Machine learning for medical imaging. Radiographics 37(2):505–515
102. Esmaeilpour S, Liu B, Robertson E, Shu L (2022) Zero-shot out-of-distribution detection based on the pre-trained model clip. In: AAAI
103. Evci U, Gale T, Menick J, Castro PS, Elsen E (2020) Rigging the lottery: making all tickets winners. In: ICML
104. Fang Z, Li Y, Lu J, Dong J, Han B, Liu F (2022) Is out-of-distribution detection learnable? In: NeurIPS
105. Feinman R, Curtin RR, Shintre S, Gardner AB (2017) Detecting adversarial samples from artifacts. arXiv preprint arXiv:170300410
106. Feldman V (2020) Does learning require memorization? A short tale about a long tail. In: STOC
107. Fellbaum C (1998) Wordnet: an electronic lexical database. Bradford Books
108. Feng J, Xu H, Mannor S, Yan S (2014) Robust logistic regression and classification. In: NeurIPS
109. Feng L, Kaneko T, Han B, Niu G, An B, Sugiyama M (2020) Learning with multiple complementary labels. In: ICML
110. Feng L, Shu S, Lin Z, Lv F, Li L, An B (2020) Can cross entropy loss be robust to label noise? In: IJCAI
111. Foret P, Kleiner A, Mobahi H, Neyshabur B (2020) Sharpness-aware minimization for efficiently improving generalization. In: ICLR
112. Franceschi JY, Fawzi A, Fawzi O (2018) Robustness of classifiers to uniform ℓ_p and gaussian noise. In: AISTATS
113. Fredrikson M, Jha S, Ristenpart T (2015) Model inversion attacks that exploit confidence information and basic countermeasures. In: CCS
114. Gagnon-Audet JC, Ahuja K, Darvishi-Bayazi MJ, Dumas G, Rish I (2022) Woods: benchmarks for out-of-distribution generalization in time series tasks. arXiv preprint arXiv:220309978
115. Ganguli D, Lovitt L, Kernion J, Askell A, Bai Y, Kadavath S, Mann B, Perez E, Schiefer N, Ndousse K et al (2022) Red teaming language models to reduce harms: methods, scaling behaviors, and lessons learned. arXiv preprint arXiv:220907858
116. Ganin Y, Ustinova E, Ajakan H, Germain P, Larochelle H, Laviolette F, Marchand M, Lempitsky VS (2016) Domain-adversarial training of neural networks. J Mach Learn Res 17(59):1–35
117. Gao Y, Zhang ML (2021) Discriminative complementary-label learning with weighted loss. In: ICML
118. Gasteiger J, Giri S, Margraf JT, Günnemann S (2020) Fast and uncertainty-aware directional message passing for non-equilibrium molecules. In: NeurIPS

119. Geirhos R, Jacobsen J, Michaelis C, Zemel RS, Brendel W, Bethge M, Wichmann FA (2020) Shortcut learning in deep neural networks. Nat Mach Intell 2(11):665–673
120. Gilmer J, Schoenholz SS, Riley PF, Vinyals O, Dahl GE (2017) Neural message passing for quantum chemistry. In: ICML
121. Glymour M, Pearl J, Jewell NP (2016) Causal inference in statistics: a primer. John Wiley & Sons, London
122. Goldblum M, Fowl L, Feizi S, Goldstein T (2020) Adversarially robust distillation. In: AAAI
123. Goodfellow I, Pouget-Abadie J, Mirza M, Xu B, Warde-Farley D, Ozair S, Courville A, Bengio Y (2014) Generative adversarial nets. In: NeurIPS
124. Goodfellow IJ, Shlens J, Szegedy C (2015) Explaining and harnessing adversarial examples. In: ICLR
125. Goodfellow I, Bengio Y, Courville A (2016) Deep learning. The MIT Press, Cambridge
126. Gowal S, Qin C, Uesato J, Mann T, Kohli P (2020) Uncovering the limits of adversarial training against norm-bounded adversarial examples. arXiv preprint arXiv:201003593
127. Gowal S, Rebuffi SA, Wiles O, Stimberg F, Calian DA, Mann TA (2021) Improving robustness using generated data. In: NeurIPS
128. Gretton A, Borgwardt KM, Rasch MJ, Schölkopf B, Smola A (2012) A kernel two-sample test. J Mach Learn Res 13(1):723–773
129. Grigorescu S, Trasnea B, Cocias T, Macesanu G (2020) A survey of deep learning techniques for autonomous driving. J Field Robot 37(3):362–386
130. Grill JB, Strub F, Altché F, Tallec C, Richemond P, Buchatskaya E, Doersch C, Avila Pires B, Guo Z, Gheshlaghi Azar M et al (2020) Bootstrap your own latent - A new approach to self-supervised learning. In: NeurIPS
131. Gu T, Dolan-Gavitt B, Garg S (2017) Badnets: identifying vulnerabilities in the machine learning model supply chain. arXiv preprint arXiv:170806733
132. Gui S, Li X, Wang L, Ji S (2022) GOOD: a graph out-of-distribution benchmark. In: NeurIPS datasets and benchmarks track
133. Gulrajani I, Lopez-Paz D (2021) In search of lost domain generalization. In: ICLR
134. Guo Z, Shen Y, Bashir AK, Imran M, Kumar N, Zhang D, Yu K (2020) Robust spammer detection using collaborative neural network in Internet-of-Things applications. IEEE Internet Things J 8(12):9549–9558
135. Hamilton WL, Ying Z, Leskovec J (2017) Inductive representation learning on large graphs. In: NeurIPS
136. Han B, Yao Q, Yu X, Niu G, Xu M, Hu W, Tsang I, Sugiyama M (2018) Co-teaching: robust training of deep neural networks with extremely noisy labels. In: NeurIPS
137. Han B, Yao Q, Liu T, Niu G, Tsang IW, Kwok JT, Sugiyama M (2020) A survey of label-noise representation learning: past, present and future. arXiv preprint arXiv:201104406
138. Han X, Ren T, Nguyen T, Nguyen K, Ghosh J, Ho N (2023) Designing robust transformers using robust kernel density estimation. In: NeurIPS
139. Han J, Min MR, Han L, Li LE, Zhang X (2021) Disentangled recurrent wasserstein autoencoder. In: ICLR
140. Han R, Li S, Wang X, Liu CH, Xin G, Chen LY (2021) Accelerating gossip-based deep learning in heterogeneous edge computing platforms. IEEE Trans Parallel Distrib Syst 32(7):1591–1602
141. Hao S, Gu Y, Ma H, Hong JJ, Wang Z, Wang DZ, Hu Z (2023) Reasoning with language model is planning with world model. In: EMNLP
142. Harvey S (1989) Treatment of electrostatic effects in macromolecular modeling. Proteins: Struct Funct Bioinform 5(1):78–92
143. Havrilla A, Iyer M (2024) Understanding the effect of noise in LLM training data with algorithmic chains of thought. arXiv preprint arXiv:240204004
144. He K, Zhang X, Ren S, Sun J (2016) Deep residual learning for image recognition. In: CVPR
145. He K, Gkioxari G, Dollár P, Girshick R (2017) Mask R-CNN. In: ICCV
146. He K, Fan H, Wu Y, Xie S, Girshick R (2020) Momentum contrast for unsupervised visual representation learning. In: CVPR

147. He X, Jia J, Backes M, Gong NZ, Zhang Y (2021) Stealing links from graph neural networks. In: USENIX security
148. Hegedűs I, Danner G, Jelasity M (2019) Gossip learning as a decentralized alternative to federated learning. In: IFIP
149. Hendrycks D, Dietterich T (2019) Benchmarking neural network robustness to common corruptions and perturbations. In: ICLR
150. Hendrycks D, Gimpel K (2017) A baseline for detecting misclassified and out-of-distribution examples in neural networks. In: ICLR
151. Hendrycks D, Basart S, Mazeika M, Zou A, Kwon J, Mostajabi M, Steinhardt J, Song D (2022) Scaling out-of-distribution detection for real-world settings. In: ICML
152. Hendrycks D, Lee K, Mazeika M (2019) Using pre-training can improve model robustness and uncertainty. In: ICML
153. Hendrycks D, Mazeika M, Dietterich T (2019) Deep anomaly detection with outlier exposure. In: ICLR
154. Hermann K, Lampinen A (2020) What shapes feature representations? Exploring datasets, architectures, and training. In: NeurIPS
155. Hermann K, Chen T, Kornblith S (2020) The origins and prevalence of texture bias in convolutional neural networks. In: NeurIPS
156. Hinton G, Vinyals O, Dean J (2015) Distilling the knowledge in a neural network. In: arXiv
157. Hitaj D, Pagnotta G, Masi I, Mancini LV (2021) Evaluating the robustness of geometry-aware instance-reweighted adversarial training. arXiv preprint arXiv:210301914
158. Hong Z, Wang Z, Shen L, Yao Y, Huang Z, Chen S, Yang C, Gong M, Liu T (2024) Improving non-transferable representation learning by harnessing content and style. In: ICLR
159. Hou Y, Zhang J, Cheng J, Ma K, Ma RTB, Chen H, Yang MC (2020) Measuring and improving the use of graph information in graph neural networks. In: ICLR
160. Hsu Y, Shen Y, Jin H, Kira Z (2020) Generalized ODIN: detecting out-of-distribution image without learning from out-of-distribution data. In: CVPR
161. Hu W, Niu G, Sato I, Sugiyama M (2018) Does distributionally robust supervised learning give robust classifiers? In: ICML
162. Hu W, Fey M, Zitnik M, Dong Y, Ren H, Liu B, Catasta M, Leskovec J (2020) Open graph benchmark: datasets for machine learning on graphs. In: NeurIPS
163. Huang R, Li Y (2021) Mos: towards scaling out-of-distribution detection for large semantic space. In: CVPR
164. Huang Z, Yang J, Gong C (2022) They are not completely useless: towards recycling transferable unlabeled data for class-mismatched semi-supervised learning. IEEE Trans Multimedia 25:1844–1857
165. Huang Z, Zhu M, Xia X, Shen L, Yu J, Gong C, Han B, Du B, Liu T (2023) Robust generalization against photon-limited corruptions via worst-case sharpness minimization. In: CVPR
166. Huang Z, Li M, Shen L, Yu J, Gong C, Han B, Liu T (2024) Winning prize comes from losing tickets: improve invariant learning by exploring variant parameters for out-of-distribution generalization. Int J Comput Vision 133(1):456–474
167. Huang K, Fu T, Gao W, Zhao Y, Roohani YH, Leskovec J, Coley CW, Xiao C, Sun J, Zitnik M (2021) Therapeutics data commons: machine learning datasets and tasks for drug discovery and development. In: NeurIPS datasets and benchmarks track (round 1)
168. Huang R, Geng A, Li Y (2021) On the importance of gradients for detecting distributional shifts in the wild. In: NeurIPS
169. Huang Z, Xue C, Han B, Yang J, Gong C (2021) Universal semi-supervised learning. In: NeurIPS
170. Hume D (1896) A treatise of human nature. Clarendon Press
171. Ilse M, Tomczak JM, Louizos C, Welling M (2020) Diva: domain invariant variational autoencoders. In: MIDL
172. Imani S, Du L, Shrivastava H (2023) Mathprompter: mathematical reasoning using large language models. In: ACL

173. Ishida T, Niu G, Hu W, Sugiyama M (2017) Learning from complementary labels. In: NeurIPS
174. Ishida T, Niu G, Menon A, Sugiyama M (2019) Complementary-label learning for arbitrary losses and models. In: ICML
175. Izmailov P, Kirichenko P, Gruver N, Wilson AG (2022) On feature learning in the presence of spurious correlations. In: NeurIPS
176. Jacot A, Gabriel F, Hongler C (2021) Neural tangent kernel: convergence and generalization in neural networks. In: STOC
177. Jaderberg M, Simonyan K, Zisserman A, kavukcuoglu k (2015) Spatial transformer networks. In: NeurIPS
178. Jagielski M, Oprea A, Biggio B, Liu C, Nita-Rotaru C, Li B (2018) Manipulating machine learning: poisoning attacks and countermeasures for regression learning. In: IEEE SP
179. Jeh G, Widom J (2003) Scaling personalized web search. In: WWW
180. Ji S, Pan S, Cambria E, Marttinen P, Yu PS (2020) A survey on knowledge graphs: representation, acquisition and applications. Tech. rep., arXiv:2002.00388
181. Jia C, Yang Y, Xia Y, Chen YT, Parekh Z, Pham H, Le Q, Sung YH, Li Z, Duerig T (2021) Scaling up visual and vision-language representation learning with noisy text supervision. In: ICML
182. Jiang L, Zhou Z, Leung T, Li LJ, Fei-Fei L (2018) Mentornet: learning data-driven curriculum for very deep neural networks on corrupted labels. In: ICML
183. Jin W, Li Y, Xu H, Wang Y, Ji S, Aggarwal C, Tang J (2021) Adversarial attacks and defenses on graphs. SIGKDD Explor Newslett 22(2):19–34
184. Jin Y, Liu Y, Chen K, Yang Q (2023) Federated learning without full labels: a survey. arXiv preprint arXiv:230314453
185. Jin Z, Liu J, Lyu Z, Poff S, Sachan M, Mihalcea R, Diab MT, Schölkopf B (2024) Can large language models infer causation from correlation? In: ICLR
186. Jordan MI, Mitchell TM (2015) Machine learning: trends, perspectives, and prospects. Science 349(6245):255–260
187. Kairouz P, McMahan HB, Avent B, Bellet A, Bennis M, Bhagoji AN, Bonawitz K, Charles Z, Cormode G, Cummings R et al (2021) Advances and open problems in federated learning. Found Trends Mach Learn 14(1–2):1–210
188. Kamath P, Tangella A, Sutherland D, Srebro N (2021) Does invariant risk minimization capture invariance? In: AISTATS
189. Karimireddy SP, Kale S, Mohri M, Reddi S, Stich S, Suresh AT (2020) Scaffold: stochastic controlled averaging for federated learning. In: ICML
190. Karras T, Laine S, Aila T (2019) A style-based generator architecture for generative adversarial networks. In: CVPR
191. Kim H, Lee W, Lee J (2021) Understanding catastrophic overfitting in single-step adversarial training. In: AAAI
192. Kim T, Ko J, Choi J, Yun SY et al (2021) FINE samples for learning with noisy labels. In: NeurIPS
193. Kipf TN, Welling M (2017) Semi-supervised classification with graph convolutional networks. In: ICLR
194. Kirichenko P, Izmailov P, Wilson AG (2022) Last layer re-training is sufficient for robustness to spurious correlations. arXiv preprint arXiv:220402937
195. Klicpera J, Bojchevski A, Günnemann S (2019) Combining neural networks with personalized pagerank for classification on graphs. In: ICLR
196. Koh PW, Liang P (2017) Understanding black-box predictions via influence functions. In: ICML
197. Koh PW, Sagawa S, Marklund H, Xie SM, Zhang M, Balsubramani A, Hu W, Yasunaga M, Phillips RL, Gao I, Lee T, David E, Stavness I, Guo W, Earnshaw B, Haque I, Beery SM, Leskovec J, Kundaje A, Pierson E, Levine S, Finn C, Liang P (2021) WILDS: a benchmark of in-the-wild distribution shifts. In: ICML
198. Kok S, Domingos P (2007) Statistical predicate invention. In: ICML

199. Koltchinskii V, Panchenko D (2002) Empirical margin distributions and bounding the generalization error of combined classifiers. Ann Stat 30(1):1–50
200. Kosmala A, Gasteiger J, Gao N, Günnemann S (2023) Ewald-based long-range message passing for molecular graphs. In: ICML
201. Koyama M, Yamaguchi S (2020) Out-of-distribution generalization with maximal invariant predictor. arXiv preprint arXiv:200801883
202. Krause J, Stark M, Deng J, Fei-Fei L (2013) 3d object representations for fine-grained categorization. In: ICCV workshops
203. Krizhevsky A, Hinton G (2009) Learning multiple layers of features from tiny images. Master's thesis
204. Krizhevsky A, Nair V, Hinton G (2010) Cifar-10 (Canadian Institute for Advanced Research). arXiv preprint arXiv:210604411
205. Krueger D, Caballero E, Jacobsen J, Zhang A, Binas J, Zhang D, Priol RL, Courville AC (2021) Out-of-distribution generalization via risk extrapolation (rex). In: ICML
206. Kumar RSS, Nyström M, Lambert J, Marshall A, Goertzel M, Comissoneru A, Swann M, Xia S (2020) Adversarial machine learning-industry perspectives. In: 2020 IEEE security and privacy workshops (SPW)
207. Kumar A, Raghunathan A, Jones RM, Ma T, Liang P (2022) Fine-tuning can distort pretrained features and underperform out-of-distribution. In: ICLR
208. Laidlaw C, Singla S, Feizi S (2021) Perceptual adversarial robustness: defense against unseen threat models. In: ICLR
209. Laishram R, Phoha VV (2016) Curie: a method for protecting SVM classifier from poisoning attack. arXiv preprint arXiv:160601584
210. Lake B, Baroni M (2018) Generalization without systematicity: on the compositional skills of sequence-to-sequence recurrent networks. In: ICML
211. Lecun Y, Bottou L, Bengio Y, Haffner P (1998) Gradient-based learning applied to document recognition. Proc IEEE 86(11):2278–2324
212. LeCun Y, Chopra S, Hadsell R, Ranzato M, Huang F (2006) A tutorial on energy-based learning. Predicting Structured Data
213. Lee DH (2013) Pseudo-label: the simple and efficient semi-supervised learning method for deep neural networks. In: ICML workshop
214. Lee M, Kim D (2023) Robust evaluation of diffusion-based adversarial purification. In: ICCV
215. Lee K, Lee K, Lee H, Shin J (2018) A simple unified framework for detecting out-of-distribution samples and adversarial attacks. In: NeurIPS
216. Leeson D, Young J (2015) Molecular property design: does everyone get it? ACS Med Chem Lett 6(7):722–725
217. Li B, Vorobeychik Y (2014) Feature cross-substitution in adversarial classification. In: NeurIPS
218. Li D, Yang Y, Song Y, Hospedales TM (2017) Deeper, broader and artier domain generalization. In: ICCV
219. Li Y, Tian X, Gong M, Liu Y, Liu T, Zhang K, Tao D (2018) Deep domain generalization via conditional invariant adversarial networks. In: ECCV
220. Li M, Zhou T, Huang Z, Yang J, Yang J, Gong C (2024) Dynamic weighted adversarial learning for semi-supervised classification under intersectional class mismatch. ACM Trans Multimedia Comput Commun Appl 20(4):1–24
221. Li J, Socher R, Hoi S (2020) Dividemix: learning with noisy labels as semi-supervised learning. In: ICLR
222. Li T, Sahu AK, Zaheer M, Sanjabi M, Talwalkar A, Smith V (2020) Federated optimization in heterogeneous networks. In: MLSys
223. Li X, Huang K, Yang W, Wang S, Zhang Z (2020) On the convergence of FedAvg on non-IID data In: ICLR
224. Li Q, Diao Y, Chen Q, He B (2021) Federated learning on non-IID data silos: an experimental study. arXiv preprint arXiv:210202079

225. Li Y, Yin J, Chen L (2021) Unified robust training for graph neural networks against label noise. In: PAKDD
226. Li Y, Yu Q, Tan M, Mei J, Tang P, Shen W, Yuille A, Xie C (2021) Shape-texture debiased neural network training. In: ICLR
227. Li M, Wu R, Liu H, Yu J, Yang X, Han B, Liu T (2023) Instant: semi-supervised learning with instance-dependent thresholds. In: NeurIPS
228. Li Y, Zhang Y, Sun L (2023) Metaagents: simulating interactions of human behaviors for LLM-based task-oriented coordination via collaborative generative agents. arXiv preprint arXiv:231006500
229. Liang S, Li Y, Srikant R (2018) Enhancing the reliability of out-of-distribution image detection in neural networks. In: ICLR
230. Liao F, Liang M, Dong Y, Pang T, Hu X, Zhu J (2018) Defense against adversarial attacks using high-level representation guided denoiser. In: CVPR
231. Lin TY, Maire M, Belongie S, Hays J, Perona P, Ramanan D, Dollár P, Zitnick CL (2014) Microsoft COCO: common objects in context. In: ECCV
232. Lin TY, Dollár P, Girshick R, He K, Hariharan B, Belongie S (2017) Feature pyramid networks for object detection. In: CVPR
233. Lin W, Lan H, Li B (2021) Generative causal explanations for graph neural networks. In: ICML
234. Lin W, Lan H, Wang H, Li B (2022) Orphicx: a causality-inspired latent variable model for interpreting graph neural networks. In: CVPR
235. Lin R, Yu C, Han B, Su H, Liu T (2024) Layer-aware analysis of catastrophic overfitting: revealing the pseudo-robust shortcut dependency. In: ICML
236. Lin R, Yu C, Han B, Liu T (2023) On the over-memorization during natural, robust and catastrophic overfitting. In: ICLR
237. Lin R, Yu C, Liu T (2023) Eliminating catastrophic overfitting via abnormal adversarial examples regularization. In: NeurIPS
238. Litman T (2017) Autonomous vehicle implementation predictions. Victoria Transport Policy Institute, Victoria
239. Liu T, Tao D (2015) Classification with noisy labels by importance reweighting. IEEE Trans Pattern Anal Mach Intell 38(3):447–461
240. Liu X, Xu N, Chen M, Xiao C (2023) Autodan: generating stealthy jailbreak prompts on aligned large language models. arXiv preprint arXiv:231004451
241. Liu H, Cao Z, Long M, Wang J, Yang Q (2019) Separate to adapt: open set domain adaptation via progressive separation. In: CVPR
242. Liu Z, Miao Z, Zhan X, Wang J, Gong B, Yu SX (2019) Large-scale long-tailed recognition in an open world. In: CVPR
243. Liu C, Salzmann M, Lin T, Tomioka R, Süsstrunk S (2020) On the loss landscape of adversarial training: identifying challenges and how to overcome them. In: NeurIPS
244. Liu F, Xu W, Lu J, Zhang G, Gretton A, Sutherland DJ (2020) Learning deep kernels for non-parametric two-sample tests. In: ICML
245. Liu S, Niles-Weed J, Razavian N, Fernandez-Granda C (2020) Early-learning regularization prevents memorization of noisy labels. In: NeurIPS
246. Liu W, Wang X, Owens JD, Li Y (2020) Energy-based out-of-distribution detection. In: NeurIPS
247. Liu EZ, Haghgoo B, Chen AS, Raghunathan A, Koh PW, Sagawa S, Liang P, Finn C (2021) Just train twice: improving group robustness without training group information. In: ICML
248. Liu Q, Yang H, Dou Q, Heng PA (2021) Federated semi-supervised medical image classification via inter-client relation matching. In: MICCAI
249. Long M, Cao Y, Wang J, Jordan M (2015) Learning transferable features with deep adaptation networks. In: ICML
250. Loo N, Hasani R, Amini A, Rus D (2022) Evolution of neural tangent kernels under benign and adversarial training. In: NeurIPS

251. Lu T, Gao M, Yu K, Byerly A, Khashabi D (2024) Insights into LLM long-context failures: when transformers know but don't tell. In: Findings of EMNLP
252. Luan S, Hua C, Lu Q, Zhu J, Zhao M, Zhang S, Chang XW, Precup D (2022) Revisiting heterophily for graph neural networks. In: NeurIPS
253. Luisier F, Blu T, Unser M (2010) Image denoising in mixed Poisson–Gaussian noise. IEEE Trans Image Process 20(3):696–708
254. Luo D, Cheng W, Xu D, Yu W, Zong B, Chen H, Zhang X (2020) Parameterized explainer for graph neural network. In: NeurIPS
255. Luo B, Xiao W, Wang S, Huang J, Tassiulas L (2021) Tackling system and statistical heterogeneity for federated learning with adaptive client sampling. arXiv preprint arXiv:211211256
256. Luo M, Chen F, Hu D, Zhang Y, Liang J, Feng J (2021) No fear of heterogeneity: classifier calibration for federated learning with non-IID data. In: NeurIPS
257. Ma X, Li B, Wang Y, Erfani SM, Wijewickrema S, Schoenebeck G, Song D, Houle ME, Bailey J (2018) Characterizing adversarial subspaces using local intrinsic dimensionality. In: ICLR
258. Madry A, Makelov A, Schmidt L, Tsipras D, Vladu A (2018) Towards deep learning models resistant to adversarial attacks. In: ICLR
259. Mahajan D, Tople S, Sharma A (2021) Domain generalization using causal matching. In: ICML
260. McMahan B, Moore E, Ramage D, Hampson S, y Arcas BA (2017) Communication-efficient learning of deep networks from decentralized data. In: AISTATS
261. McPherson M, Smith-Lovin L, Cook JM (2001) Birds of a feather: homophily in social networks. Annu Rev Sociol 27(1):415–444
262. Mei S, Zhu X (2015) Using machine teaching to identify optimal training-set attacks on machine learners. In: AAAI
263. Meilicke C, Fink M, Wang Y, Ruffinelli D, Gemulla R, Stuckenschmidt H (2018) Fine-grained evaluation of rule-and embedding-based systems for knowledge graph completion. In: ISWC
264. Miao S, Liu M, Li P (2022) Interpretable and generalizable graph learning via stochastic attention mechanism. In: ICML
265. Miao S, Luo Y, Liu M, Li P (2023) Interpretable geometric deep learning via learnable randomness injection. In: ICLR
266. Milgram S (1963) Behavioral study of obedience. J Abnormal Soc Psychol 67(4):371
267. Ming Y, Cai Z, Gu J, Sun Y, Li W, Li Y (2022) Delving into out-of-distribution detection with vision-language representations. In: NeurIPS
268. Ming Y, Fan Y, Li Y (2022) Poem: out-of-distribution detection with posterior sampling. In: ICML
269. Mirzasoleiman B, Cao K, Leskovec J (2020) Coresets for robust training of deep neural networks against noisy labels. In: NeurIPS
270. Misra I, Maaten Lvd (2020) Self-supervised learning of pretext-invariant representations. In: CVPR
271. Miyato T, Maeda Si, Koyama M, Ishii S (2018) Virtual adversarial training: a regularization method for supervised and semi-supervised learning. IEEE Trans Pattern Anal Mach Intell 41(8):1979–1993
272. Mohri M, Rostamizadeh A, Talwalkar A (2018) Foundations of machine learning. The MIT Press, Cambridge
273. Murray CW, Rees DC (2009) The rise of fragment-based drug discovery. Nat Chem 1 3:187–92
274. Nagarajan N, Inderjit SD, Pradeep R (2013) Learning with noisy labels. In: NeurIPS
275. Nagarajan V, Andreassen A, Neyshabur B (2021) Understanding the failure modes of out-of-distribution generalization. In: ICLR
276. Nakkiran P, Kaplun G, Bansal Y, Yang T, Barak B, Sutskever I (2021) Deep double descent: where bigger models and more data hurt. J Stat Mech Theory Exp 2021(12):124003
277. Namkoong H, Duchi JC (2016) Stochastic gradient methods for distributionally robust optimization with f-divergences. In: NeurIPS

278. Netzer Y, Wang T, Coates A, Bissacco A, Wu B, Ng AY (2011) Reading digits in natural images with unsupervised feature learning. In: NeurIPS workshop on deep learning and unsupervised feature learning 2011
279. Nie W, Guo B, Huang Y, Xiao C, Vahdat A, Anandkumar A (2022) Diffusion models for adversarial purification. In: ICML
280. Northcutt C, Jiang L, Chuang I (2021) Confident learning: estimating uncertainty in dataset labels. J Artif Intell Res 70:1373–1411
281. OpenAI R (2023) Gpt-4 technical report. arXiv preprint arXiv:230308774
282. Page L, Brin S, Motwani R, Winograd T (1999) The pagerank citation ranking: bringing order to the web. Tech. rep., Stanford infolab
283. Pal S, Bhattacharya M, Islam MA, Chakraborty C (2024) Ai-enabled chatgpt or llm: a new algorithm is required for plagiarism-free scientific writing. LWW Int J Surg 110(2):1329–1330.
284. Panagopoulos G, Nikolentzos G, Vazirgiannis M (2021) Transfer graph neural networks for pandemic forecasting. In: AAAI
285. Pang T, Lin M, Yang X, Zhu J, Yan S (2022) Robustness and accuracy could be reconcilable by (proper) definition. In: ICML
286. Parascandolo G, Neitz A, Orvieto A, Gresele L, Schölkopf B (2021) Learning explanations that are hard to vary. In: ICLR
287. Park J, Kim MH, Choi S, Kweon IS, Choi DG (2019) Fraud detection with multi-modal attention and correspondence learning. In: ICEIC
288. Parkhi OM, Vedaldi A, Zisserman A, Jawahar C (2012) Cats and dogs. In: CVPR
289. Patrini G, Rozza A, Krishna Menon A, Nock R, Qu L (2017) Making deep neural networks robust to label noise: a loss correction approach. In: CVPR
290. Paudice A, Muñoz-González L, Lupu EC (2019) Label sanitization against label flipping poisoning attacks. In: ECML PKDD
291. Pearl J (2009) Causality. Cambridge University Press, Cambridge
292. Peng X, Wang K, Zhu Z, You Y (2022) Crafting better contrastive views for siamese representation learning. In: CVPR
293. Peters J, Bühlmann P, Meinshausen N (2016) Causal inference by using invariant prediction: identification and confidence intervals. J Roy Stat Soc Ser B (Statistical Methodology) 78(5):947–1012
294. Peters J, Janzing D, Schlkopf B (2017) Elements of Causal inference: foundations and learning algorithms. The MIT Press, Cambridge
295. Pezeshki M, Kaba S, Bengio Y, Courville AC, Precup D, Lajoie G (2021) Gradient starvation: a learning proclivity in neural networks. In: NeurIPS
296. Philip ML, Rocco AS (2008) Random classification noise defeats all convex potential boosters. In: ICML
297. Platonov O, Kuznedelev D, Babenko A, Prokhorenkova L (2023) Characterizing graph datasets for node classification: homophily-heterophily dichotomy and beyond. In: NeurIPS
298. Pokhrel SR, Choi J (2020) Federated learning with blockchain for autonomous vehicles: analysis and design challenges. IEEE Trans Commun 68(8):4734–4746
299. Ponsuresh M, Rajkumar TD (2024) Time and accuracy based impact analysis of poisoning attack on support vector machine algorithm. In: IITCEE
300. Qi X, Huang K, Panda A, Wang M, Mittal P (2023) Visual adversarial examples jailbreak large language models. arXiv preprint arXiv:230613213
301. Qi Z, Tan X, Qu C, Xu Y, Qi Y (2023) Safer: a robust and efficient framework for fine-tuning bert-based classifier with noisy labels. In: ACL
302. Qin T, Wang S, Li H (2022) Generalizing to evolving domains with latent structure-aware sequential autoencoder. In: ICML
303. Qiu W (2022) A survey on poisoning attacks against supervised machine learning. arXiv preprint arXiv:220202510
304. Qu M, Chen J, Xhonneux L, Bengio Y, Tang J (2021) Rnnlogic: learning logic rules for reasoning on knowledge graphs. In: ICLR

305. Qu Z, Li X, Duan R, Liu Y, Tang B, Lu Z (2022) Generalized federated learning via sharpness aware minimization. arXiv preprint arXiv:220602618
306. Radford A, Kim JW, Hallacy C, Ramesh A, Goh G, Agarwal S, Sastry G, Askell A, Mishkin P, Clark J et al (2021) Learning transferable visual models from natural language supervision. In: ICML
307. Rame A, Dancette C, Cord M (2021) Fishr: invariant gradient variances for out-of-distribution generalization. arXiv preprint arXiv:210902934
308. Rebuffi SA, Gowal S, Calian DA, Stimberg F, Wiles O, Mann T (2021) Fixing data augmentation to improve adversarial robustness. arXiv preprint arXiv:210301946
309. Recht B, Roelofs R, Schmidt L, Shankar V (2019) Do imagenet classifiers generalize to imagenet? In: ICML
310. Reddi SJ, Charles Z, Zaheer M, Garrett Z, Rush K, Konečný J, Kumar S, McMahan HB (2021) Adaptive federated optimization. In: ICLR
311. Rice L, Wong E, Kolter JZ (2020) Overfitting in adversarially robust deep learning. In: ICML
312. Rojas-Carulla M, Schölkopf B, Turner R, Peters J (2018) Invariant models for causal transfer learning. J Mach Learn Res 19(36):1–34
313. Rolnick D, Veit A, Belongie S, Shavit N (2017) Deep learning is robust to massive label noise. arXiv preprint arXiv:170510694
314. Ronneberger O, Fischer P, Brox T (2015) U-net: convolutional networks for biomedical image segmentation. In: MICCAI
315. Rosenblatt F (1957) The perceptron - a perceiving and recognizing automaton. Tech. Rep. 85-460-1, Cornell Aeronautical Laboratory
316. Rosenfeld E, Ravikumar P, Risteski A (2022) Domain-adjusted regression or: ERM may already learn features sufficient for out-of-distribution generalization. arXiv preprint arXiv:220206856
317. Rubinstein R (1999) The cross-entropy method for combinatorial and continuous optimization. Methodol Comput Appl Probab 1(2):127–190
318. Sadeghian A, Armandpour M, Ding P, Wang DZ (2019) Drum: end-to-end differentiable rule mining on knowledge graphs. In: NeurIPS
319. Sagawa S, Koh PW, Hashimoto TB, Liang P (2020) Distributionally robust neural networks. In: ICLR
320. Sagui C, Darden T (1999) Molecular dynamics simulations of biomolecules: long-range electrostatic effects. Annu Rev Biophys Biomol Struct 28(1):155–179
321. Sagun L, Evci U, Güney VU, Dauphin YN, Bottou L (2018) Empirical analysis of the hessian of over-parametrized neural networks. In: ICLR, workshop track proceedings
322. Saito K, Yamamoto S, Ushiku Y, Harada T (2018) Open set domain adaptation by backpropagation. In: ECCV
323. Saito K, Kim D, Saenko K (2021) Openmatch: open-set semi-supervised learning with open-set consistency regularization. In: NeurIPS
324. Schlichtkrull M, Kipf TN, Bloem P, Van Den Berg R, Titov I, Welling M (2018) Modeling relational data with graph convolutional networks. In: ESWC
325. Schmidt L, Santurkar S, Tsipras D, Talwar K, Madry A (2018) Adversarially robust generalization requires more data. NeurIPS 31
326. Schubert E, Sander J, Ester M, Kriegel HP, Xu X (2017) Dbscan revisited, revisited: why and how you should (still) use dbscan. ACM Transactions on Database Systems
327. Schütt K, Kindermans P, Sauceda F HE, Chmiela S, Tkatchenko A, Müller K (2017) Schnet: a continuous-filter convolutional neural network for modeling quantum interactions. NeurIPS
328. Schütt K, Unke O, Gastegger M (2021) Equivariant message passing for the prediction of tensorial properties and molecular spectra. In: ICML
329. Sehwag V, Mahloujifar S, Handina T, Dai S, Xiang C, Chiang M, Mittal P (2022) Robust learning meets generative models: can proxy distributions improve adversarial robustness? In: ICLR
330. Serfling RJ (2009) Approximation theorems of mathematical statistics. John Wiley & Sons, London

331. Shah H, Tamuly K, Raghunathan A, Jain P, Netrapalli P (2020) The pitfalls of simplicity bias in neural networks. In: NeurIPS
332. Sharif M, Bhagavatula S, Bauer L, Reiter MK (2016) Accessorize to a crime: real and stealthy attacks on state-of-the-art face recognition. In: CCS
333. Shi C, Holtz C, Mishne G (2021) Online adversarial purification based on self-supervised learning. In: ICLR
334. Shi Y, Seely J, Torr P, N S, Hannun A, Usunier N, Synnaeve G (2022) Gradient matching for domain generalization. In: ICLR
335. Shi F, Chen X, Misra K, Scales N, Dohan D, Chi EH, Schärli N, Zhou D (2023) Large language models can be easily distracted by irrelevant context. In: ICML
336. Shwartz-Ziv R, Tishby N (2017) Opening the black box of deep neural networks via information. arXiv preprint arXiv:170300810
337. Sinha K, Sodhani S, Dong J, Pineau J, Hamilton WL (2019) Clutrr: a diagnostic benchmark for inductive reasoning from text. arXiv preprint arXiv:190806177
338. Smith LN (2017) Cyclical learning rates for training neural networks. In: WACV
339. Sohn K, Berthelot D, Carlini N, Zhang Z, Zhang H, Raffel CA, Cubuk ED, Kurakin A, Li CL (2020) Fixmatch: simplifying semi-supervised learning with consistency and confidence. In: NeurIPS
340. Song Y, Sohl-Dickstein J, Kingma DP, Kumar A, Ermon S, Poole B (2021) Score-based generative modeling through stochastic differential equations. In: ICLR
341. Song Y, Dhariwal P, Chen M, Sutskever I (2023) Consistency models. In: ICML
342. Stacey J, Cheng J, Torr J, Guigue T, Driesen J, Coca A, Gaynor M, Johannsen A (2024) Lucid: LLM-generated utterances for complex and interesting dialogues. arXiv preprint arXiv:240300462
343. Steinhardt J, Koh PWW, Liang PS (2017) Certified defenses for data poisoning attacks. In: NeurIPS
344. Stolfo A, Jin Z, Shridhar K, Schölkopf B, Sachan M (2023) A causal framework to quantify the robustness of mathematical reasoning with language models. In: ACL
345. Struppek L, Hintersdorf D, Correia ADA, Adler A, Kersting K (2022) Plug and play attacks: towards robust and flexible model inversion attacks. In: ICML
346. Stuke A, Kunkel C, Golze D, Todorović M, Margraf J, Reuter K, Rinke P, Oberhofer H (2020) Atomic structures and orbital energies of 61,489 crystal-forming organic molecules. Sci Data 7(1):58
347. Suchanek F, Kasneci G, Weikum G (2007) Yago: a core of semantic knowledge. In: WWW
348. Sun Y, Li Y (2022) Dice: leveraging sparsification for out-of-distribution detection. In: ECCV
349. Sun B, Saenko K (2016) Deep CORAL: correlation alignment for deep domain adaptation. In: ECCV
350. Sun L, Dou Y, Yang C, Wang J, Yu PS, Li B (2018) Adversarial attack and defense on graph data: a survey. arXiv preprint arXiv:181210528
351. Sun Y, Wang S, Tang X, Hsieh T, Honavar VG (2020) Adversarial attacks on graph neural networks via node injections: a hierarchical reinforcement learning approach. In: WWW
352. Sun Y, Guo C, Li Y (2021) React: out-of-distribution detection with rectified activations. In: NeurIPS
353. Sun Y, Ming Y, Zhu X, Li Y (2022) Out-of-distribution detection with deep nearest neighbors. In: ICML
354. Sutton RS, Barto AG et al (1998) Reinforcement learning: an introduction. MIT Press, Cambridge
355. Tanaka D, Ikami D, Yamasaki T, Aizawa K (2018) Joint optimization framework for learning with noisy labels. In: CVPR
356. Tang Z, Wang Y, He X, Zhang L, Pan X, Wang Q, Zeng R, Zhao K, Shi S, He B et al (2023) Fusionai: decentralized training and deploying LLMs with massive consumer-level GPUs. arXiv preprint arXiv:230901172
357. Tang Z, Shi S, Li B, Chu X (2022) Gossipfl: a decentralized federated learning framework with sparsified and adaptive communication. IEEE Trans Parallel Distrib Syst 34(3):909–922.

358. Tang Z, Zhang Y, Shi S, He X, Han B, Chu X (2022) Virtual homogeneity learning: defending against data heterogeneity in federated learning. In: ICML
359. Tang Z, Huang J, Yan R, Wang Y, Tang Z, Shi S, Zhou AC, Chu X (2024) Bandwidth-aware and overlap-weighted compression for communication-efficient federated learning. In: ICPP
360. Tang Z, Kang X, Yin Y, Pan X, Wang Y, He X, Wang Q, Zeng R, Zhao K, Shi S, Zhou AC, Li B, He B, Chu X (2024) Fusionllm: a decentralized LLM training system on geo-distributed GPUs with adaptive compression
361. Tang Z, Zhang Y, Dong P, ming Cheung Y, Zhou AC, Han B, Chu X (2024) Fusefl: one-shot federated learning through the lens of causality with progressive model fusion. In: NeurIPS
362. Tang Z, Zhang Y, Shi S, Tian X, Liu T, Han B, Chu X (2024) Fedimpro: measuring and improving client update in federated learning. In: ICLR
363. Tao L, Du X, Zhu J, Li Y (2023) Non-parametric outlier synthesis. In: ICLR
364. Teney D, Abbasnejad E, Lucey S, van den Hengel A (2021) Evading the simplicity bias: training a diverse set of models discovers solutions with superior OOD generalization. arXiv preprint arXiv:210505612
365. Teru KK, Denis E, Hamilton WL (2020) Inductive relation prediction by subgraph reasoning. In: ICML
366. Tian Y, Sun C, Poole B, Krishnan D, Schmid C, Isola P (2020) What makes for good views for contrastive learning? In: NeurIPS
367. Tian Q, Zhu H, Wang L, Li Y, Lan Y (2023) R3 prompting: review, rephrase and resolve for chain-of-thought reasoning in large language models under noisy context prompting: review, rephrase and resolve for chain-of-thought reasoning in large language models under noisy context. In: EMNLP
368. Timmermann KE, Nowak RD (1999) Multiscale modeling and estimation of poisson processes with application to photon-limited imaging. IEEE Trans Inform Theory 45(3):846–862
369. Tishby N, Zaslavsky N (2015) Deep learning and the information bottleneck principle. In: ITW
370. Tishby N, Pereira FC, Bialek W (2000) The information bottleneck method. arXiv preprint arXiv:physics/0004057
371. Toutanova K, Chen D (2015) Observed versus latent features for knowledge base and text inference. In: PWCVSMC
372. Tsilivis N, Kempe J (2022) What can the neural tangent kernel tell us about adversarial robustness? In: NeurIPS
373. Tsipras D, Santurkar S, Engstrom L, Turner A, Madry A (2019) Robustness may be at odds with accuracy. In: ICLR
374. Tzeng E, Hoffman J, Saenko K, Darrell T (2017) Adversarial discriminative domain adaptation. In: CVPR
375. Vapnik V (1991) Principles of risk minimization for learning theory. In: NeurIPS
376. Vashishth S, Sanyal S, Nitin V, Talukdar P (2019) Composition-based multi-relational graph convolutional networks. In: ICLR
377. Vaswani A, Shazeer N, Parmar N, Uszkoreit J, Jones L, Gomez AN, Kaiser L, Polosukhin I (2017) Attention is all you need. In: NeurIPS
378. Veličković P, Cucurull G, Casanova A, Romero A, Liò P, Bengio Y (2018) Graph attention networks. In: ICLR
379. Volpi R, Namkoong H, Sener O, Duchi JC, Murino V, Savarese S (2018) Generalizing to unseen domains via adversarial data augmentation. In: NeurIPS
380. Vu MN, Thai MT (2020) Pgm-explainer: probabilistic graphical model explanations for graph neural networks. In: NeurIPS
381. Wah C, Branson S, Welinder P, Perona P, Belongie S (2011) The Caltech-UCSD Birds-200-2011 dataset. California Institute of Technology, Pasadena
382. Wald Y, Feder A, Greenfeld D, Shalit U (2021) On calibration and out-of-domain generalization. In: NeurIPS
383. Wang Q, Mao Z, Wang B, Guo L (2017) Knowledge graph embedding: a survey of approaches and applications. IEEE Trans Knowl Data Eng 29(12):2724–2743

384. Wang X, Cheng M, Eaton J, Hsieh C, Wu SF (2018) Attack graph convolutional networks by adding fake nodes. arXiv preprint arXiv:181010751
385. Wang J, Liu Q, Liang H, Joshi G, Poor HV (2020) Tackling the objective inconsistency problem in heterogeneous federated optimization. In: NeurIPS
386. Wang J, Luo M, Suya F, Li J, Yang Z, Zheng Q (2020) Scalable attack on graph data by injecting vicious nodes. In: ECML PKDD
387. Wang Y, Zou D, Yi J, Bailey J, Ma X, Gu Q (2020) Improving adversarial robustness requires revisiting misclassified examples. In: ICLR
388. Wang H, Li Z, Feng L, Zhang W (2022) Vim: out-of-distribution with virtual-logit matching. In: CVPR
389. Wang Q, Liu F, Zhang Y, Zhang J, Gong C, Liu T, Han B (2022) Watermarking for out-of-distribution detection. In: NeurIPS
390. Wang WW, Xu G, Pu R, Li J, Zhou F, Shui C, Ling C, Gagné C, Wang B (2022) Evolving domain generalization. arXiv preprint arXiv:220600047
391. Wang Z, Ma J, Wang X, Hu J, Qin Z, Ren K (2022) Threats to training: a survey of poisoning attacks and defenses on machine learning systems. ACM Comput Surv 55(7):1–36
392. Wang H, Li Y, Yao H, Li X (2023) CLIPN for zero-shot OOD detection: teaching clip to say no. arXiv preprint arXiv:230812213
393. Wang S, Tan Z, Guo R, Li J (2023) Noise-robust fine-tuning of pretrained language models via external guidance. In: EMNLP
394. Wang Y, Chen H, Heng Q, Hou W, Fan Y, Wu Z, Wang J, Savvides M, Shinozaki T, Raj B et al (2023) Freematch: self-adaptive thresholding for semi-supervised learning. In: ICLR
395. Wang H, Huang Z, Lin Z, Liu T (2024) Noisegpt: label noise detection and rectification through probability curvature. In: NeurIPS
396. Wang Q, Lin Y, Chen Y, Schmidt L, Han B, Zhang T (2024) A sober look at the robustness of clips to spurious features. In: NeurIPS
397. Weerasinghe S, Erfani SM, Alpcan T, Leckie C (2019) Support vector machines resilient against training data integrity attacks. Pattern Recognit 96:106985
398. Wei J, Tay Y, Bommasani R, Raffel C, Zoph B, Borgeaud S, Yogatama D, Bosma M, Zhou D, Metzler D et al (2022) Emergent abilities of large language models. arXiv preprint arXiv:220607682
399. Wencel-Delord J, Glorius F (2013) C–H bond activation enables the rapid construction and late-stage diversification of functional molecules. Nat Chem 5(5):369–375
400. Winterbach W, Mieghem P, Reinders M, Wang H, Ridder D (2013) Topology of molecular interaction networks. BMC Syst Biol 7(1):90
401. Wong E, Rice L, Kolter JZ (2020) Fast is better than free: revisiting adversarial training. arXiv preprint arXiv:200103994
402. Woodworth BE, Patel KK, Srebro N (2020) Minibatch vs local SGD for heterogeneous distributed learning. In: NeurIPS
403. Wu Z, Xiong Y, Yu SX, Lin D (2018) Unsupervised feature learning via non-parametric instance discrimination. In: CVPR
404. Wu D, Xia ST, Wang Y (2020) Adversarial weight perturbation helps robust generalization. In: NeurIPS
405. Wu T, Ren H, Li P, Leskovec J (2020) Graph information bottleneck. In: NeurIPS
406. Wu B, Li J, Hou C, Fu G, Bian Y, Chen L, Huang J (2022) Recent advances in reliable deep graph learning: adversarial attack, inherent noise, and distribution shift. arXiv preprint arXiv:220207114
407. Wu Q, Zhang H, Yan J, Wipf D (2022) Handling distribution shifts on graphs: an invariance perspective. arXiv preprint arXiv:220202466
408. Wu Y, Wang X, Zhang A, He X, Chua TS (2022) Discovering invariant rationales for graph neural networks. In: ICLR
409. Wu Y, Xia X, Yu J, Han B, Niu G, Sugiyama M, Liu T (2023) Making binary classification from multiple unlabeled datasets almost free of supervision. arXiv preprint arXiv:230607036

410. Wu Z, Qiu L, Ross A, Akyürek E, Chen B, Wang B, Kim N, Andreas J, Kim Y (2023) Reasoning or reciting? Exploring the capabilities and limitations of language models through counterfactual tasks. arXiv preprint arXiv:230702477
411. Wu Y, Yao J, Han B, Yao L, Liu T (2024) Unraveling the impact of heterophilic structures on graph positive-unlabeled learning. In: ICML
412. Wu Y, Yao J, Xia X, Yu J, Wang R, Han B, Liu T (2024) Mitigating label noise on graphs via topological sample selection. In: ICML
413. Xia X, Liu T, Han B, Wang N, Gong M, Liu H, Niu G, Tao D, Sugiyama M (2020) Part-dependent label noise: towards instance-dependent label noise. In: NeurIPS
414. Xia X, Liu T, Han B, Gong M, Yu J, Niu G, Sugiyama M (2022) Sample selection with uncertainty of losses for learning with noisy labels. In: ICLR
415. Xiao H, Xiao H, Eckert C (2012) Adversarial label flips attack on support vector machines. In: ECAI
416. Xiao C, Zhu JY et al (2018) Spatially transformed adversarial examples. In: ICLR
417. Xiao C, Chen Z, Jin K, Wang J, Nie W, Liu M, Anandkumar A, Li B, Song D (2023) Densepure: understanding diffusion models for adversarial robustness. In: ICLR
418. Xie T, Grossman JC (2017) Crystal graph convolutional neural networks for an accurate and interpretable prediction of material properties. Phys Rev Lett 120(14):145301
419. Xie B, Bian Y, Zhou K, Chen Y, Zhao P, Han B, Meng W, Cheng J (2024) Enhancing neural subset selection: integrating background information into set representations. In: ICLR
420. Xie B, Chen Y, Wang J, Zhou K, Han B, Meng W, Cheng J (2024) Enhancing evolving domain generalization through dynamic latent representations. In: AAAI
421. Xie B, Wang Y, Chen Y, Zhou K, Li Y, Meng W, Cheng J (2024) HORSE: hierarchical representation for large-scale neural subset selection. In: NeurIPS
422. Xiong W, Hoang T, Wang W (2017) Deeppath: a reinforcement learning method for knowledge graph reasoning. In: EMNLP
423. Xu H, Caramanis C, Mannor S (2009) Robustness and regularization of support vector machines. J Mach Learn Res 10:1485–1510
424. Xu W, Evans D, Qi Y (2017) Feature squeezing: detecting adversarial examples in deep neural networks. arXiv preprint arXiv:170401155
425. Xu K, Li C, Tian Y, Sonobe T, Kawarabayashi K, Jegelka S (2018) Representation learning on graphs with jumping knowledge networks. In: ICML
426. Xu Y, Shang L, Ye J, Qian Q, Li YF, Sun B, Li H, Jin R (2021) Dash: semi-supervised learning with dynamic thresholding. In: ICML
427. Xu K, Chen H, Liu S, Chen P, Weng T, Hong M, Lin X (2019) Topology attack and defense for graph neural networks: an optimization perspective. In: IJCAI
428. Xu K, Hu W, Leskovec J, Jegelka S (2019) How powerful are graph neural networks? In: ICLR
429. Yang Z, Zhang Y, Zheng Y, Tian X, Peng H, Liu T, Han B (2023) Fedfed: feature distillation against data heterogeneity in federated learning. In: NeurIPS
430. Yang H, Yan X, Dai X, Chen Y, Cheng J (2021) Self-enhanced GNN: improving graph neural networks using model outputs. In: IJCNN
431. Yang J, Zhou K, Li Y, Liu Z (2021) Generalized out-of-distribution detection: a survey. Int J Comput Vision 132(12):5635-5662
432. Yao Y, Yu B, Gong C, Liu T (2021) Understanding how pretraining regularizes deep learning algorithms. IEEE Trans Neural Netw Learn 34(9):5828–5840 Systems
433. Yao H, Choi C, Cao B, Lee Y, Koh PW, Finn C (2022) Wild-time: a benchmark of in-the-wild distribution shift over time. arXiv preprint arXiv:221114238
434. Yehudai G, Fetaya E, Meirom E, Chechik G, Maron H (2021) From local structures to size generalization in graph neural networks. In: ICML
435. Yerlikaya FA, Bahtiyar Ş (2022) Data poisoning attacks against machine learning algorithms. Expert Syst Appl 208:118101
436. Ying Z, Bourgeois D, You J, Zitnik M, Leskovec J (2019) Gnnexplainer: generating explanations for graph neural networks. In: NeurIPS

437. Yoo J, Kim J, Yoon H, Kim G, Jang C, Kang U (2021) Accurate graph-based PU learning without class prior. In: ICDM
438. Yoon J, Hwang SJ, Lee J (2021) Adversarial purification with score-based generative models. In: ICML
439. Yu X, Liu T, Gong M, Tao D (2018) Learning with biased complementary labels. In: ECCV
440. Yu J, Xu T, Rong Y, Bian Y, Huang J, He R (2021) Graph information bottleneck for subgraph recognition. In: ICLR
441. Yu Q, Ikami D, Irie G, Aizawa K (2020) Multi-task curriculum framework for open-set semi-supervised learning. In: ECCV
442. Yu T, Kumar S, Gupta A, Levine S, Hausman K, Finn C (2020) Gradient surgery for multi-task learning. In: NeurIPS
443. Yuan H, Tang J, Hu X, Ji S (2020) XGNN: towards model-level explanations of graph neural networks. In: KDD
444. Yuan H, Yu H, Wang J, Li K, Ji S (2021) On explainability of graph neural networks via subgraph explorations. In: ICML
445. Yuan S, Feng L, Liu T (2024) Early stopping against label noise without validation data. In: ICLR
446. Yuan S, Feng L, Liu T (2023) Late stopping: avoiding confidently learning from mislabeled examples. In: ICCV
447. Yuan Y, Jiao W, Wang W, Huang Jt, He P, Shi S, Tu Z (2023) Gpt-4 is too smart to be safe: stealthy chat with LLMs via cipher. arXiv preprint 230806463
448. Yuan S, Feng L, Han B, Liu T (2025) Enhancing sample selection by cutting mislabeled easy examples. arXiv
449. Yuan S, Lin R, Feng L, Han B, Liu T (2025) Instance-dependent early stopping. In: ICLR
450. Zügner D, Günnemann S (2019) Adversarial attacks on graph neural networks via meta learning. In: ICLR
451. Zagoruyko S, Komodakis N (2016) Wide residual networks. arXiv preprint arXiv:160507146
452. Zeng Y, Lin H, Zhang J, Yang D, Jia R, Shi W (2024) How Johnny can persuade LLMs to jailbreak them: rethinking persuasion to challenge ai safety by humanizing LLMs. arXiv preprint arXiv:240106373
453. Zhai R, Cai T, He D, Dan C, He K, Hopcroft J, Wang L (2019) Adversarially robust generalization just requires more unlabeled data. arXiv preprint arXiv:190600555
454. Zhai R, Dan C, Kolter JZ, Ravikumar P (2022) Understanding why generalized reweighting does not improve over ERM. arXiv preprint arXiv:220112293
455. Zhang M, Chen Y (2018) Link prediction based on graph neural networks. In: NeurIPS
456. Zhang Y, Yang Q (2021) A survey on multi-task learning. IEEE Trans Knowl Data Eng 34(12):5586–5609
457. Zhang Y, Yao Q (2022) Knowledge graph reasoning with relational directed graph. In: WWW
458. Zhang R, Zhu Q (2017) A game-theoretic analysis of label flipping attacks on distributed support vector machines. In: CISS
459. Zhang X, Zitnik M (2020) Gnnguard: defending graph neural networks against adversarial attacks. In: NeurIPS
460. Zhang H, Cisse M, Pradeep R, Dauphin NY, Lopez-Paz D (2018) mixup: beyond empirical risk minimization. In: ICLR
461. Zhang C, Bengio S, Hardt M, Recht B, Vinyals O (2017) Understanding deep learning requires rethinking generalization. In: ICLR
462. Zhang H, Yu Y, Jiao J, Xing EP, Ghaoui LE, Jordan MI (2019) Theoretically principled trade-off between robustness and accuracy. In: ICML
463. Zhang S, Tay Y, Yao L, Liu Q (2019) Quaternion knowledge graph embeddings. In: NeurIPS
464. Zhang Y, Jia R, Pei H, Wang W, Li B, Song D (2020) The secret revealer: generative model-inversion attacks against deep neural networks. In: CVPR
465. Zhang Y, Yao Q, Chen L (2020) Interstellar: searching recurrent architecture for knowledge graph embedding. In: NeurIPS

466. Zhang Y, Yao Q, Dai W, Chen L (2020) Autosf: searching scoring functions for knowledge graph embedding. In: ICDE
467. Zhang B, Wang Y, Hou W, Wu H, Wang J, Okumura M, Shinozaki T (2021) Flexmatch: boosting semi-supervised learning with curriculum pseudo labeling. In: NeurIPS
468. Zhang C, Bengio S, Hardt M, Recht B, Vinyals O (2021) Understanding deep learning (still) requires rethinking generalization. Commun ACM 64(3):107–115
469. Zhang J, Zhu J, Niu G, Han B, Sugiyama M, Kankanhalli M (2021) Geometry-aware instance-reweighted adversarial training. In: ICLR
470. Zhang Z, Liu Q, Huang Z, Wang H, Lu C, Liu C, Chen E (2021) Graphmi: extracting private graph data from graph neural networks. In: IJCAI
471. Zhang J, Lopez-Paz D, Bottou L (2022) Rich feature construction for the optimization-generalization dilemma. In: ICML
472. Zhang L, Wang H, Muniz M, Panagiotopoulos A, Car R et al (2022) A deep potential model with long-range electrostatic interactions. J Chem Phys 156:124107
473. Zhang M, Sohoni NS, Zhang HR, Finn C, Ré C (2022) Correct-n-contrast: a contrastive approach for improving robustness to spurious correlations. In: ICML
474. Zhang Y, Gong M, Liu T, Niu G, Tian X, Han B, Schölkopf B, Zhang K (2022) Adversarial robustness through the lens of causality. In: ICLR
475. Zhang C, Bauer S, Bennett P, Gao J, Gong W, Hilmkil A, Jennings J, Ma C, Minka T, Pawlowski N, Vaughan J (2023) Understanding causality with large language models: feasibility and opportunities. arXiv preprint arXiv:230405524
476. Zhang X, Wang L, Helwig J, Luo Y, Fu C, Xie Y, Liu M, Lin Y, Xu Z, Yan K, Adams K, Weiler M, Li X, Fu T, Wang Y, Yu H, Xie Y, Fu X, Strasser A, Xu S, Liu Y, Du Y, Saxton A, Ling H, Lawrence H, Stärk H, Gui S, Edwards C, Gao N, Ladera A, Wu T, Hofgard EF, Tehrani AM, Wang R, Daigavane A, Bohde M, Kurtin J, Huang Q, Phung T, Xu M, Joshi CK, Mathis SV, Azizzadenesheli K, Fang A, Aspuru-Guzik A, Bekkers E, Bronstein MM, Zitnik M, Anandkumar A, Ermon S, Liò P, Yu R, Günnemann S, Leskovec J, Ji H, Sun J, Barzilay R, Jaakkola TS, Coley CW, Qian X, Qian X, Smidt TE, Ji S (2023) Artificial intelligence for science in quantum, atomistic, and continuum systems. arXiv preprint arXiv:230708423
477. Zhang Y, Li Y, Cui L, Cai D, Liu L, Fu T, Huang X, Zhao E, Zhang Y, Chen Y, et al. (2023) Siren's song in the ai ocean: a survey on hallucination in large language models. arXiv preprint arXiv:230901219
478. Zhang Y, Zhou Z, Yao Q, Chu X, Han B (2023) Adaprop: learning adaptive propagation for graph neural network based knowledge graph reasoning. In: KDD
479. Zhao Y, Li M, Lai L, Suda N, Civin D, Chandra V (2018) Federated learning with non-iid data. arXiv preprint arXiv:180600582
480. Zhao X, Chen F, Hu S, Cho JH (2020) Uncertainty aware semi-supervised learning on graph data. In: NeurIPS
481. Zhao WX, Zhou K, Li J, Tang T, Wang X, Hou Y, Min Y, Zhang B, Zhang J, Dong Z, et al. (2023) A survey of large language models. arXiv preprint arXiv:230318223
482. Zheltonozhskii E, Baskin C, Mendelson A, Bronstein AM, Litany O (2022) Contrast to divide: self-supervised pre-training for learning with noisy labels. In: WACV
483. Zheng H, Saparov A (2023) Noisy exemplars make large language models more robust: a domain-agnostic behavioral analysis. In: EMNLP
484. Zheng S, Wu P, Goswami A, Goswami M, Metaxas D, Chen C (2020) Error-bounded correction of noisy labels. In: ICML
485. Zhou Z, Zhou C, Li X, Yao J, Yao Q, Han B (2023) On strengthening and defending graph reconstruction attack with Markov chain approximation. In: ICML
486. Zhou Z, Yao J, Liu J, Guo X, Yao Q, He L, Wang L, Zheng B, Han B (2024) Combating bilateral edge noise for robust link prediction. In: NeurIPS
487. Zhou J, Zhu J, Zhang J, Liu T, Niu G, Han B, Sugiyama M (2022) Adversarial training with complementary labels: on the benefit of gradually informative attacks. In: NeurIPS
488. Zhou X, Lin Y, Pi R, Zhang W, Xu R, Cui P, Zhang T (2022) Model agnostic sample reweighting for out-of-distribution learning. In: ICML

489. Zhu J, Rosset S, Tibshirani R, Hastie T (2003) 1-norm support vector machines. In: NeurIPS
490. Zhu Z, Zhang Z, Xhonneux L, Tang J (2021) Neural Bellman-Ford networks: a general graph neural network framework for link prediction. In: NeurIPS
491. Zhu J, Yao J, Han B, Zhang J, Liu T, Niu G, Zhou J, Xu J, Yang H (2022) Reliable adversarial distillation with unreliable teachers. In: ICLR
492. Zhu J, Li H, Yao J, Liu T, Xu J, Han B (2023) Unleashing mask: explore the intrinsic out-of-distribution detection capability. In: ICML
493. Zhu J, Yao J, Liu T, Xu J, Han B et al (2023) Combating exacerbated heterogeneity for robust models in federated learning. In: ICLR
494. Zhu J, Yu G, Yao J, Liu T, Niu G, Sugiyama M, Han B (2023) Diversified outlier exposure for out-of-distribution detection via informative extrapolation. In: NeurIPS
495. Zou X, Zheng Q, Dong Y, Guan X, Kharlamov E, Lu J, Tang J (2021) Tdgia: effective injection attacks on graph neural networks. In: KDD
496. Zou A, Wang Z, Kolter JZ, Fredrikson M (2023) Universal and transferable adversarial attacks on aligned language models. arXiv preprint arXiv:230715043
497. Zügner D, Akbarnejad A, Günnemann S (2018) Adversarial attacks on neural networks for graph data. In: KDD
498. Ji Y, Zhang L, Wu J, Wu B, Huang LK, Xu T, Rong Y, Li L, Ren J, Xue D, Lai H, Xu S, Feng J, Liu W, Luo P, Zhou S, Huang J, Zhao P, Bian Y (2022) Drugood: out-of-distribution (OOD) dataset curator and benchmark for ai-aided drug discovery—a focus on affinity prediction problems with noise annotations. arXiv preprint arXiv:220109637
499. Zhang Y, Yang Z, Tian X, Wang N, Liu T, Han B (2024) Robust training of federated models with extremely label deficiency. In: ICLR